Mix Smart

Mix Smart

Pro Audio Tips for Your Multitrack Mix
Alexander U. Case

AMSTERDAM • BOSTON • HEIDELBERG • LONDON
NEW YORK • OXFORD • PARIS • SAN DIEGO
SAN FRANCISCO • SINGAPORE • SYDNEY • TOKYO

Focal press is an imprint of Elsevier

ELSEVIER

Focal Press

Focal Press is an imprint of Elsevier
225 Wyman Street, Waltham, MA 02451, USA
The Boulevard, Langford Lane, Kidlington, Oxford, OX5 1GB, UK

Notices
Knowledge and best practice in this field are constantly changing. As new research and
experience broaden our understanding, changes in research methods, professional practices, or
medical treatment may become necessary.

Practitioners and researchers must always rely on their own experience and knowledge in
evaluating and using any information, methods, compounds, or experiments described herein.
In using such information or methods they should be mindful of their own safety and the safety
of others, including parties for whom they have a professional responsibility.

To the fullest extent of the law, neither the Publisher nor the authors, contributors, or editors,
assume any liability for any injury and/or damage to persons or property as a matter of products
liability, negligence or otherwise, or from any use or operation of any methods, products,
instructions, or ideas contained in the material herein.

Library of Congress Cataloging-in-Publication Data
Application submitted

British Library Cataloguing-in-Publication Data
A catalogue record for this book is available from the British Library.

ISBN: 978-0-240-81485-8

For information on all Focal Press publications
visit our website at www.elsevierdirect.com

11 12 13 14 15   5 4 3 2 1

Printed in the United States

Dedication

For my brother Charles, the avid concertgoer, omnivorous record listener, passionate guitar player, and resourceful recordist whose rock-star success in his day job has never interfered with instilling a love of music in the next generation. How cool is that?

Contents

Acknowledgments

The reader should know that if—and let's just leave it at *if*, shall we?—if your author were a disorganized, long-winded, distracted, typagrafickle error, um, typographical error–generating machine, this book hits the shelves beautiful and organized only because of the essential skill, patience, care, commitment, and professionalism of Catharine Steers, Kate Iannotti, and—especially—Carlin Reagan, at Focal Press. Deadlines for this title grew increasingly specific, so that by the end, they were telling me the month, date, time, and time zone the next deliverable was due. I have no idea why … thanks, Catharine, Kate, and Carlin.

Nicholas LaPenn's beautiful and fun, yet always accurate and informative drawings are enduring and were so well-loved when published in *Sound FX—Unlocking the Creative Potential of Recording Studio Effects* (Focal Press, 2007) that we reuse the relevant ones here, about 75 of them! He makes the figures a welcome study break from the text, but we still learn while daydreaming and enjoying his work. I admire that he—an audio engineer and musician—has such a good eye.

Jim Anderson brought to this book his decades of experience working at the top of this field, plus his wisdom as an audio educator, in many ways. He performed a detailed technical review of the manuscript, offering many valuable suggestions. I fear that when he agreed to it, he thought *Mix Smart* was some sort of bartender's guide. When he discovered its true content, he still read on. He's also been a generous colleague, always willing to talk shop at a level of detail that reveals both his talent as an engineer and his commitment as an educator.

My cohorts at the University of Massachusetts Lowell—Will Moylan, John Shirley, Bill Carman, Paul Angelli, and Alan Williams—steadily advance my knowledge of sound recording through the work they do, the art they create, and the collaborations we have in the halls, classrooms, and studios at school. They've no doubt long suspected this, but let's admit it officially: I am always skimming their expertise.

My graduate and undergraduate students at the University of Massachusetts Lowell forever increase the pace of my lifelong learning—restoring Edison wax, decoding the soul of Stax, recording stunning tracks, and pushing every mix to the max. It's my good fortune to be included in their culture of intellectual curiosity, constant musicality, and a contagious commitment to hard work and high quality.

I am indebted to my many colleagues at the Audio Engineering Society (AES) who provide a home for the essential research, dialog, and healthy debate about audio. The countless interactions with the luminaries in our field through committees, tutorials, workshops, conferences, and the casual conversations in between has been essential to my professional development. I am particularly grateful to AES movers-and-shakers Jim Anderson (again), Roger Furness, Bill McQuaide, Steve Johnson, John Krivit, and Dave Greenspan, whose hard work make the AES invaluable to those who actively participate.

My parents, Dolores and Joe, bought me a fixer-upper car before I could drive. We worked on it so hard that by the time I had a license, I also had great respect for the machine. This is typical of how they raised me. What a lucky kid, to get a car from the git-go! But they provided it with an eye on mastery. How does a son thank someone for that? I am spoiled by them to this day, but—through their care and wisdom—I am not spoiled rotten. Right, Mom?

My best beloved bride Amy connects me with two of my favorite, most missed students—Patricia and Stan.

You wanna mix?

You gotta get smart!

Engineers take the achievements of scientists and find ways to *use* them, to *do* things with them, and to *make* things from them. Wielding science to engineer new works of art is one of our species' highest achievements. Glad to have you joining in on the fun.

Mix engineers must create music based on a deep knowledge of the informing disciplines: music, acoustics, electrical engineering, computer science, and digital signal processing, and—to a lesser extent—psychology, philosophy, business, game theory, and a little dream interpretation and pain management consulting helps. This call to action—this need to act, commit to, and create something in a world governed by a rich set of rules, theories, properties, laws, and equations from many related fields—separates the engineer from the scientist and the mix engineer from the mere recordist. It isn't easy.

In fact, it is such a difficult endeavor that many who attempt it fail. Only the most talented, most visionary, most committed, most inspired, most driven, best educated, hardest working, most creative people succeed. If you want to mix at a professional level and to make the sort of music that holds its own with your favorite records, films, games, and concerts of all time, you must not take the task lightly. There is much to learn. There are many disciplines to master. Doing so enables your mix finally to be an expression of you—a creative expression that builds on a collaboration between and among those who compose, those who perform, and those who know how to play the studio as a musical instrument.

You've probably heard the uninspired monotony that comes from some kid assaulting a piano or guitar with uninspired, untrained, two-finger noodling. You've also heard what happens when a hard-working prodigy spends time with an instrument—practicing, studying, exploring, and jamming. We hear the difference.

Same thing goes for mixing.

It is such a complicated process that it overwhelms many. It can't be learned in a day. It can't be learned in a week. Come to think of it, it's been more than two decades and I'm still learning. The temptation to cut corners seduces many. We live in a time in which the tools for sound recording are so capable that many can achieve mediocre results. Just owning the gear, reading the manuals, and browsing the Web makes average audio available to all. But excellence belongs only to the most passionate—those willing to practice, study, explore, and jam. People will hear the difference.

Mixing isn't just having the tools and toys; it's really knowing how to use them in service of the art. We have to get beyond the manual and beyond the web chatter. When a musician takes a solo, they aren't repeating what they read about and talked about, they are *creating*. When a musician mixes, they aren't imitating setting by setting what they heard others did; they are too are *creating*.

In addition to the musicality of the effort, there is a necessary, very practical side to mixing. It's not all grand gestures, special effects, and ear candy. Mix engineers probably spend most of their time solving problems. Our mixes sound great when we learn to spot and know how to solve the myriad little rubs, clunky bits of conflict, and those occasional, fundamental failings that will dog the tracks we are asked to mix. We don't blindly/deafly repeat settings from past mixes in today's mix. But we do solve some of the same problems we've seen/heard in recent projects, tailoring the mix gestures as needed to address this problem in this project.

Our final mix is achieved incrementally, based on thousands of individual decisions we'll make along the way. Some are creative. Most are corrective. Each one represents a slight change in direction for the final mix. Each one contributes to the final sound of the mix. To be a great mixer, you have to get most of these decisions right, most of the time.

This book tries to up your odds. We study the core tools of the mix studio in enough detail that we can truly master them beyond anything the user manual tells us and get so comfortable with them that we can work fast without mental distraction, and then mix! No top-secret settings are revealed here, because none exist. Be suspicious of anyone who tries to tell you otherwise. Offered here are organizing principles and highly focused strategies for navigating the infinite possibilities presented to us with each mix.

When you get to work on Chapters 3 through 9, studying key effects one at a time, you'll see a common way of organizing the mix. Every signal processor has multiple applications. Any given device might have been invented for one purpose, but the day after they made the first one, somebody on some session found other interesting possibilities. We categorize and prioritize all the options by recognizing three alliterative mix motivations.

THE FIX

Problems arise. Some might have been anticipated. Some are surprises. Some could have been fixed by the performer; others by the tracking engineer. Some are big; some are small. Some are frequent; some are rare. But as mix engineer, the buck stops here. Whatever went wrong, it stops with us. We seek out problems in need of fixing, and we fix them.

THE FIT

Multitrack mixing offers something of a jigsaw puzzle. Some number of audio components must be made to fit together so that they make sense to a listener wearing ear buds on the train, streaming through traffic in the car, holding down the couch watching TV, clicking about on their laptop, or relaxing with popcorn in a surround-sound cinema. Squeezing more than 100 tracks of audio into a few loudspeakers, frankly, is doomed from the start. We dedicate careful thought, critical listening, and countless signal processes to getting the tracks to fit in and around each other, falling into place as a greater-than-the-sum-of-its-parts whole.

THE FEATURE

With the tactics of fixing and fitting implemented, we then get to the more expressive part of the mixing craft: we massage mangle, decorate, deteriorate, emphasize, synthesize, and in all ways refine the sounds for artistic impact. We use our mixing studio to feature any aspect of any signal that we think benefits the production.

Along the way, we flag four notable moments:

📢	The loudspeaker icon flags moments in the text when having an actual listen would be helpful. It indicates that you might want to take the book with you to the studio and find similar examples in your projects.
▶	The play button encourages you to do exactly that. Go to the studio and try out these specific session suggestions.
💡	Of course the whole book is a tip. Your author has made sure of it. But small moments of insight in which a memorable take-away thought has been offered are flagged by this lightbulb.
!	History repeats itself—in world events, and in our own mixes. This exclamation point tries to warn us so that we learn from past mistakes (hopefully the mistakes of others).

Alex U. Case
Portsmouth, New Hampshire, 2011

CHAPTER 1

The Mix Mindset

Be of sound mind, heart, and soul.

Mixing is one of the most fulfilling activities we experience when creating recorded works of art, but it isn't easy. This essential production task requires broad technical mastery guided by inspired musical creativity. To be successful, we must mix smart.

1.1 MUSICAL BALANCE

When listening to a completed mix, engineers and music fans alike may be drawn to its unique features and exaggerated sonic details—the surreal aura around the vocal, the urgent push of the electric guitar, the adrenaline rush of the snare, the thunderous crush of the bass, and all things lush on the synths and samples. That's fine when listening to the final product—it's one way to enjoy the music. But when we are in the studio—when we are creating our own mix—we don't start there. None of those bits of ear candy have any value if the mix isn't balanced.

1.1.1 The Mix Arrangement

Before we get to the fun stuff, we must tend to the fundamental stuff. A mix is balanced when each and every element of the tune sonically serves its musical purpose. First and foremost, that means that each and every track that *should* be heard *can* be heard, without effort from the listener. It's simple in concept, but keeping all tracks audible turns out to be a serious challenge (see also Chapter 2).

Loud tracks make it harder to hear soft tracks. The low-frequency richness in the bass guitar diminishes the listener's enjoyment of similar low frequencies in the kick drum. Each blast of energy from every powerful snare hit briefly obscures other details in the mix. An out-of-tune piano can sour our sense of what was sung. An out-of-time hand percussion performance will wreck the rhythm of what the drummer played. Similar sounds coming from similar locations, front to back, left to right, are hard to segregate.

This constant interaction among the multitrack components, in which some tracks might obscure others and where some tracks clash with others, is a pervasive challenge and must be kept under watchful ear by the mix engineer and sorted out through three fundamental balancing tools: faders, pan pots, and—let us not forget—mute switches. The fader adjusts the overall amplitude of the signal. The pan pot adjusts the relative level among mix busses, left versus right and front versus back. The mute switch removes amplitude altogether, deleting the signal from your mix entirely.

Any fader pushed too high can lead to a clumsily loud track that overpowers those tracks sitting at lower levels, possibly robbing the tune of some other key elements of the mix. Any fader left too low can relegate a track to obscurity and near-inaudibility. Spectrally similar tracks will likely compete and might need panning to different locations left to right and—if the project is in surround sound—front to back. Ultimately, if there is no fader setting and pan pot position that works, or if the track remains technically or musically distracting, don't be afraid to hit that mute button, silencing the problem track and freeing the rest of the tracks to play their necessary roles.

The first step in building a mix is to push up the faders and begin listening to the song. Notice that the goal is to hear the song as a whole, not the tracks individually. The band, the composer, the producer, or some combination thereof, has a vision for the tune. In the lyrics and the instrumentation, there is a message and a range of intended human emotions. In the groove and the beat, there is a pace and range of intended dance floor gyrations. Through a music recording, the artist is trying to communicate sophisticated and sometimes subtle thoughts to the music fan. It doesn't work if the guitars are drowning out the vocals, or if the bass is obscuring the beat. There is within the tracks an arrangement that supports what the artist is trying to say. That 48-track project represents 48 related musical ideas that need to come together in symphony to realize the artist's vision. It's a puzzle at first. So pushing up the faders and listening for the overall song is a challenging step not to be skipped or taken lightly.

If you are hearing the piece for the first time, this is an exciting—but high-pressure—moment. You must find an effective multitrack arrangement through terrific concentration, governed by respect for the producers, engineers, and musicians who have put their hearts into the tracks you hear, sustaining a tireless curiosity to learn what their musical vision might be and motivated by a creative drive to enhance, embellish, refine, or redirect the project as appropriate with your own mix ideas. There is nothing in the tracks that tells you what the right mix arrangement will be. The humble first step of auditioning the tracks and assembling them into a balanced whole is in fact a challenging and creative process. The immature mix engineer is eager—too eager—to start playing with reverb and compression. The experienced mix engineer—the musical mix engineer—recognizes that balancing the mix is the essential first step that sets the creative context for all that is to follow.

Upon hearing the discrete tracks that will make up the mix, we are expected to have a point of view on what might sound best. We form an internal aural image of what the song could sound like, backed up with the technical know-how to realize that goal.

The process is wonderfully nonlinear. The mix engineer must iteratively adjust and readjust the volume of each and every track until the combination starts to make musical sense. When you have fine-tuned the level and panning of the core elements of the multitrack arrangement, thoughtfully muting those tracks that undermine the quality of the production, the mix is said to be "balanced."

In the course of mixing a tune, a single song may be played several hundred times. The mixdown session—in which a final stereo or surround recording is created by processing and combining the individual multitrack components into an artistically meaningful whole—is such a complicated process that in most pop productions, it takes from several hours to several days to mix a three- to four-minute song.

In the course of just that first playback of the piece, however, you must begin to find the fader levels and pan pot positions that enable the song to stand on its own. The goal is to empower each and every track to make its contribution to the overall music without undermining or obliterating other parts of the multitrack production. Balancing a mix is a fundamental skill that all engineers must develop.

1.1.2 Level

It is essential to get the fundamental elements of the mix under control through careful setting of levels across all of the tracks. Although there is no single right answer, it would be fair to say that in a typical pop mix, the vocal and the snare sit pretty high in the mix and are often the loudest two tracks in the whole arrangement. The kick drum and the bass guitar come close behind, just a tick lower in level.

These four elements—vocal, snare, kick, and bass—are essential ingredients for almost every style of music. As such, they are usually among the loudest tracks, no matter how many tracks are in the production. Place them at a level you like—one that feels correct to you. No single track should drown out the other.

! Mix engineers focus on these tracks constantly, even as they work on other elements of the mix arrangement. As each additional track is introduced to the mix—a sax solo here, some hand percussion there—you always listen back with focus on these core components of the mix. When a sweet reverb is instantiated here, and an ambitious echo effect dialed in there, listen not only to the effects themselves, but also to the impact those effects have on these all-important tracks. No new track, no added effect is allowed to rob any other track of its significance in the mix. The mix is kept balanced.

Of course, no two songs are alike, and the range of musical styles around the world is wonderfully vast. The emphasis on these four tracks, then, is just a rule of thumb—but it is a big, strong, wise, time-proven, bossy thumb that is telling us what to do. It applies to almost all songs we mix. Know the exceptions, and deviate from these guidelines deliberately, not accidentally.

▶ Dance music understandably has a louder kick drum than folk music. Country music typically has louder vocals relative to the rhythm section than death metal. In some tunes, the basic tempo is hi-hat-driven, not snare-driven. In others, it might be led by a percussive acoustic guitar part. In jazz, the soloing instrument often sits top of the mix, playing the role of the vocal where none is present. Find the important tracks for the genre of music you are mixing and keep an ear on them.

The core elements should always be independently audible, never interfering with each other. The rest of the arrangement—the synths, the guitars, the background vocals—live just beneath, around, or behind, never interfering with these four critical tracks. Introduce harmony instruments, countermelodies, hand percussion, sweeteners, and other details without diminishing the overall impact of the core tracks.

It is common practice when mixing, that when you stop for the night—when you think you have finished—you print the mix, leave everything set up, and plan to check it in the morning before moving on to the next tune. That first playback the next day too often reveals mistakes: perhaps it is hard to hear the bass, or the vocal is too loud, or both. If this has happened to you, you are not alone. You've experienced what we've all endured at some point in our mix careers: an unbalanced mix.

Balancing a mix is far more difficult than expected. Sorting through the infinite decisions you have to make in creating a killer mix of the 96 tracks the band provided, it is frustratingly easy to lose sight of the bass. You had it sounding great at the beginning of the mix session, but 4 to 8 hours later, as you tweaked the reverb on the tambourine, you may have stopped listening as carefully to the bass. You know what it sounds like. You know it sounds amazing. Unlike the listener who downloads the mix, you had the pleasure of hearing the bass in isolation, uncorrupted by other distractions and interferences. As we migrate our attention through the many elements that make up the mix, shifting our focus to one, then another, then the next detail in the tracks, it understandable that we might lose site of the first, fundamental elements in the mix—the vocal, the snare, the kick drum, and the bass. Although it might be understandable, it remains unacceptable. We must pay attention to these tracks continuously, no matter what else we might do while mixing.

💡 The vocal must reign supreme. Beware of other tracks with similarly strong spectral presence in the middle frequencies. Two tracks occupying similar frequency ranges will compete for our attention and one track may partially obscure another because of this. If any track is allowed to compete too much with the vocals, the words may become difficult to understand. That result rarely supports the artist's goals for communicating with their audience. Keep an ear out for other instruments playing in the same range as the vocal. A piano, guitar, sax, or cello can make a vocal difficult to hear if the parts are too similar. Distorted electric guitars and reedy saxophones are also common vocal-busters.

Solo sections and instrumental tunes often have a featured solo instrument in lieu of a lead vocal. Give that solo instrument the same priority a lead vocal would get. Find fader positions for the tracks that keep midrange harmony instruments out of the way enough so that the listeners can understand and enjoy the phrasing of the singer or the soloist.

1.1.3 Panning

The location in the stereo or surround field is controlled by the pan pot, which, like the fader, is a simple level control device. We localize discrete mix elements of the mix toward the pan-pot-specified louder location. Hard-panned tracks, for which the pan pot is turned fully to one side or the other, come entirely from one loudspeaker and listeners will reliably localize that sound at that location. As we back off the hard-panned setting to one that it is, for example, slightly louder in the left speaker than the right, a perceptual illusion of location between speakers is created. Listeners sitting on the median plane—those locations available in the room that are the same distance from the left and right loudspeakers, both the front pair and the surround pair—will hear the sound coming from an intermediate position between the left and right speakers. When we pan a track so that it is a bit louder on the left, we are trying to create an image of a track is located slightly off to the left. When the level is the same, left versus right, listeners sitting on the median plane, will localize at a phantom center image.

STABILITY

Typically, the most important tracks—vocal, snare, kick drum, and bass—are panned front and center.

Again, the style of music may suggest exceptions. It is not unusual in jazz recordings to coax the kick drum slightly off to one side, perhaps the bass slightly to the other. The Beatles' early stereo experiments of the 1960s regularly had the drums and/or vocals off to the side. But the vast majority of mixes pan these key tracks dead center, so that the core of your production is anchored and centered, providing stability for the rest of your tracks.

AUDIBILITY

Pan the various nonvocal tracks competing in the middle frequencies off to opposite sides, out of the way of the lead vocal and each other. Instruments with similar pitch and spectral content panned to the same location will likely blur together and cloud the enjoyment of each individual instrument. Pan them to different locations for an immediate improvement in clarity and multitrack independence.

Panning leads to a somewhat fragile illusion. Although listeners sitting on the median plane in rooms that are well-behaved using playback systems that are hooked up properly will hear a range of left to right panning as the mix engineer intended, the left to right sonic horizon can be easily corrupted: someone sits off-center; the room is far from symmetric, full of noise, reverberation, or echoes; the sound system isn't connected properly, the left/right levels aren't matched, and the loudspeakers are placed at less-than-ideal locations.

We pan core instruments such as the vocal, the snare, the kick drum, and the bass to the center, not just for reasons of aesthetics—many songs make sense when they are in the center—but also for reasons of reliability and audibility: panning these tracks to the center means that they are coming out of both the left and right loudspeakers at the same level. The hope is that no matter where anyone sits, in any room, on any system, these key tracks will still be relatively audible, keeping the overall mix therefore relatively reliable across systems. For vinyl releases, center panning has the added benefit that low frequencies are less likely to toss the needle out of the groove than a hard-panned low-frequency signal.

! When you dare to pan the vocal off to the side, it makes a dramatic statement for the well-tuned listening experience. But it risks leaving some listeners unable to hear the vocal based on where they sit in their room, the quality of that room, and the correctness of their hi-fi hookup. Mix engineers are willing to take that risk for special elements of a mix—a rhythm guitar, a synth pad, a conga pattern—but the foundation of the song needs to be stable across as broad a range of end-user sound systems as possible. As a result, the vocal, the snare, the kick drum, and the bass are panned center—or very close to center—for the vast majority of the time in the vast majority of musical styles.

🔊 When spreading other instruments left to right and front to back, we want to maintain spatial interest without letting the mix get lopsided. Similar parts fighting for attention can be easier to hear, and more fun to listen to, when panned to different locations. The hi-hat performance slightly to the left can be counterweighted by the rhythmic acoustic guitar performance slightly to the right. The double-tracked rhythm electric guitar part grows more interesting when the first guitar track is panned to one side and the second guitar track is panned to the other. Layered harmony vocals sometimes sound best when each part is panned to a different location. You create a spatially engaging illusion through the careful panning of the various elements of the multitrack arrangement to worthy locations. There is no right answer. Try different panning locations and listen for exciting new levels of horizontal interest in the production.

But beware of conflict when busy but similar musical elements are colocated and can't be easily enjoyed. Panning them apart from each other is an effective solution. Beware also of distraction, when unusual tracks or unexpected performance elements sit too far to the side or too far to the rear, drawing too much attention to themselves and undermining the rest of the musical composition. Panning them to safer locations closer to the center and more toward the front attenuates the distraction.

SOUND STAGE

Some recordings are made with panning in mind, seeking to make specific use of the horizontal dimension of the mix, the sound stage. Left/right pairs of stereo tracks are recorded with a stereo mix in mind. Surround sets of four, five, or more tracks are recorded with the surround release format in mind. Typically, these stereo and surround recording approaches are trying to capture a section of performers (an entire orchestra, a string section, a brass section, gang vocals) and/or an interesting recording space (a live room, a concert hall, an empty swimming pool). At mixdown, we evaluate these stereo and surround tracks carefully. It helps to know something about how they were recorded—the microphones used, the type and size of room, and the placement of the musicians and the microphones within the space. The mix engineer's ability to fully realize the intended spatial qualities of any stereo and surround elements of the multitrack arrangement depends on the original tracking engineer's ability to record them properly and to communicate their intent through useful documentation.

When mixing stereo tracks, there is no obligation to pan them all the way out to hard-panned positions (Figure 1.1). A drum kit may be recorded for maximal theatrical width, but the mix engineer is free to toe-in those pan pots a little if the drum kit is too wide for the mix or if it distracts from or conflicts with other elements of the mix arrangement that need to live outside of the drums. The tracking engineer keeps all options open by recording a kit whose image fills the entire space between the loudspeakers when hard-panned. This approach provides maximum flexibility later because the mix engineer can use that full left/right width or a narrower width, coax it off to the side, or even collapse it to a mono image of the kit. Analogous decisions are made for surround sound sets

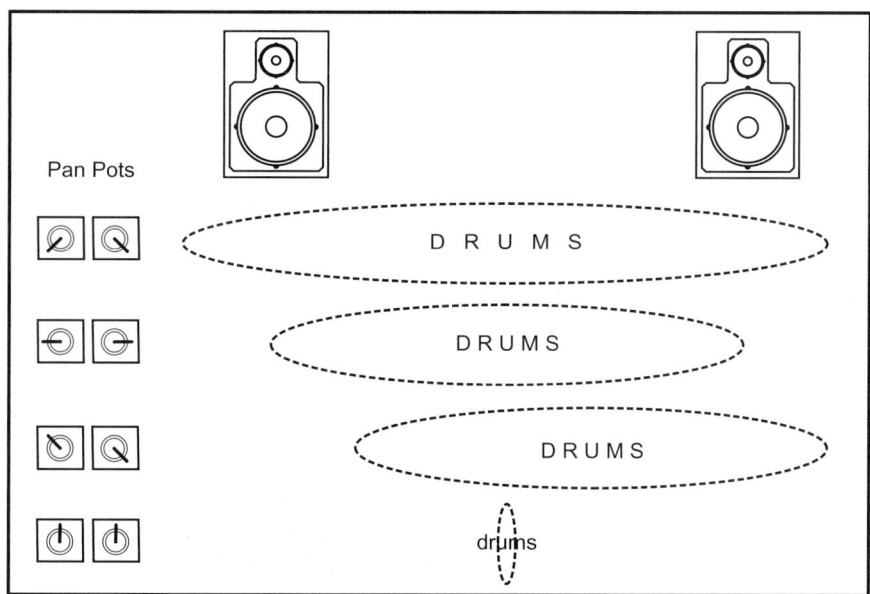

FIGURE 1.1
Sound stage: pan pots for stereo tracks aren't always hard-panned.

of tracks. The mix engineer must place stereo and surround track pan pot positions based on their needs and wishes for these and the many dozens of other tracks. A surround sound string section must still be made to fit in a mix with vocals, drum loops, keyboards, and banjos. A stereo room track for drums lives in context with guitars, bass, scratched samples, and ukulele. The pan pot positions are a free choice for the engineer: wide, narrow, mono, or reversed.

EQUAL OPPORTUNITY PANNING

The net sum of all your panning decisions across all the tracks in the multitrack project must lead to a fairly equal use of the left and right sides, and it typically sits more in the front than in the rear.

! As you devise a strategy for panning keyboard parts and guitar parts out to different locations, you usually shouldn't pull the whole mix off to one side. A left-heavy mix, in which the bulk of the energy of the mix seems to come from slightly to the left of center, can be distracting for the listener. It also suggests a missed opportunity technically, as it means the right loudspeakers were under-utilized. If the music fan listening at home or in his or her car has gone to the trouble and expense of setting up left and right playback channels, we should take advantage of them—both of them.

We make panning decisions that, track by track, make sense for that instrument yet also steadily counterbalance the panned locations of the other tracks. We want the mix to have a center of gravity that stays fairly stable, centered left to right, and at least slightly forward front to back.

The standard disclaimer appears here, of course: the only rule in music is that there are no rules in music. Tell an artist what they shouldn't do, and that is what they will most want to do. So if you are feeling a passionate desire to make a rear-heavy surround mix that pulls off to the right, feel free to do so *if* you have a compelling reason to do so. It will sound like a mistake, it will distract listeners, and it fails to fully exploit the full potential of the end-user's playback system, but it may make a bold statement. If you wish to get the attention of the listener, unnerve them, and set them up for the next verse, the next movement, or the next song that contrasts with this and blows them away, then allow yourself an off-center moment. But recognize that we have a steady set of guiding principles: key tracks dead center, with supporting tracks preserving a balanced left-to-right feeling on average, and for surround projects, we usually define a net focus of attention that is more in the front than in the rear.

1.1.4 Mute

There was a time, more than a century ago, when recordings were made live to mono. An entire band gathered around a wax cylinder or disk cutter and played. One take to one track. Done. Decades later, analog tape enabled 2-, 3-, 4-, 8-, 16-, and eventually 24-track complexity. The greed for tracks was not yet satisfied. Two 24-track machines were synchronized and 48-track analog productions became possible, if clumsy, inconvenient, and expensive.

You and I are lucky to be a part of audio at this moment in history. Our multitrack recorder is often the computer, a quickly evolving device that has many technical leaps and profound innovations ahead of it. All such advances will directly benefit our studios. One form of progress: more and more tracks. Where a track count of 48 was a realistic limit in the 1990s, we see 96 and more tracks on a casual basis today. In the hands of a great artist, new arrangements become possible and new forms of art are invented. More tracks, in the hands of a mediocre musician, make mixing much more monotonous.

We fight back with the mute button. Just because the track was recorded doesn't mean it is a contributing part of the production. Mix engineers need to know not only how to place tracks in the mix—level and panning—but also when to get rid of them by muting. When the mix is sounding cluttered and you are unable to sort out the conflicts, try removing one of the busier tracks. If you experience an involuntary, stress-reducing sigh of relief when you press the mute button, it's probably the right mix move.

! Mature artists and talented producers know what they need in the arrangement. They plan the parts ahead of time and allow for in-the-moment improvising, but stop when the arrangement is complete. Musicians new to the multitrack experience often record every idea they can think of, without regard to how it might fit in the mix. A guitarist sets up, and guitar parts are recorded across all the songs on the album. A trombonist shows up, and trombone ideas are plopped onto every song on the album. The piano gets tuned, and … you get the idea. This issue can be problematic, as it is possible that guitar, trombone, and piano don't belong on every song on the album. It is even possible that trombone doesn't belong on *any* song on *any* album.

Each and every track should be recorded with a musical purpose in mind—a vision for the overall mix arrangement. In reality, the multitrack becomes bloated with good, bad, redundant, distracting, stale, and incomplete ideas. The mix engineer must sort this out, making strategic, musical use of the all-important mute button.

1.2 MENTAL BALANCE

Mixing snaps a tightrope of tug-of-war between your creative side and your technical side. A successful mix results when you deliberately, consciously, and actively balance all the conflicts that ensue (Figure 1.2). Remember, tug-of-war can be fun.

To realize the full artistic potential of the project that appears on the mix engineer's doorstep on the first day of a project, one must see and understand all of it—the big picture, the proverbial forest. That is, one must be able to translate the many individual tracks of audio, melodic fragments, and musical details into a fully formed overall vision for the mix without distraction. That big picture point of view is in direct conflict with the obsessive-compulsive perspective needed to make the mix a technical success—the zoomed-in point of view, the individual trees in that forest. Monitoring CPU loads, linking surround

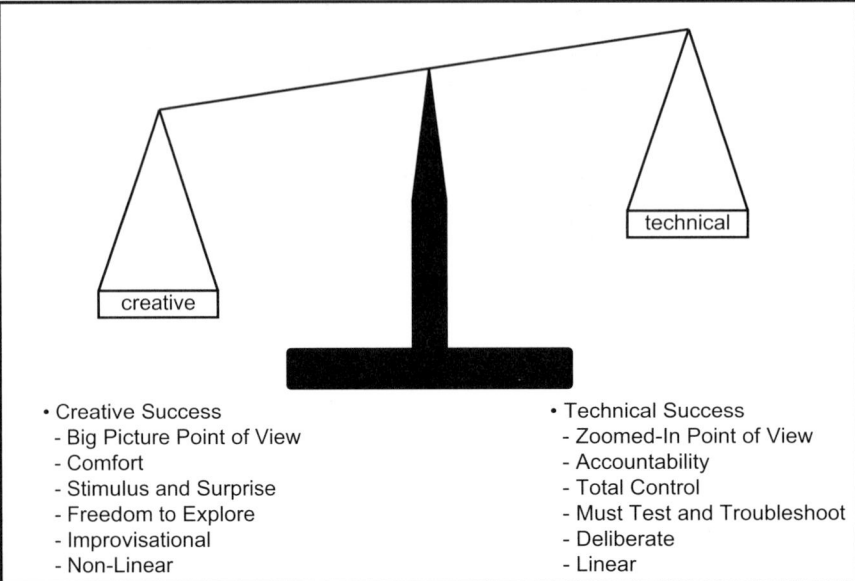

- Creative Success
  - Big Picture Point of View
  - Comfort
  - Stimulus and Surprise
  - Freedom to Explore
  - Improvisational
  - Non-Linear

- Technical Success
  - Zoomed-In Point of View
  - Accountability
  - Total Control
  - Must Test and Troubleshoot
  - Deliberate
  - Linear

FIGURE 1.2
Mixing requires balance between the creative and the technical.

gates, balancing stereo reverb returns, gain staging each device perfectly, and fine-tuning the thousands of settings on the hundreds of signal processors used across many dozens of tracks is done with an eye on the details. Mix engineers must see the trees *and* the forest. And we fix each and every tree we don't like in an effort to make an even better forest.

Creativity comes, in part, from total comfort. You'll do your best mixes bolstered by the confidence that comes from understanding—and hopefully really liking—the style of the music. You'll discover wholly new signal processing techniques only after you have total mastery of all the gear in the studio. It really does help to situate your workspace so that you like being there—supportive chairs, good lighting, a big screen, fresh air, and whatever color lava lamp suits your taste.

We can't live buffered by total comfort for very long in our business—we instead live buffeted by constant change. With the never-ending evolution of musical styles and genres, and an ever-growing list of studio plug-ins and equipment, we must work hard on the details, too. The engineer is the one responsible for knowing how to use all of the equipment in the room right now, in support of today's particular project. The mix engineer is the one held accountable when mistakes are made, from accidental distortion to over-compression, from computers crashing to out-of-tune tracks clashing. Nothing comforting about it; we mix engineers must stay on our toes.

The creative side of you needs to react instinctively to stimulus and surprise, happily getting pushed into new sounds and alternative approaches that yield unexpected results and take the music to a new and better place. Your technical side, on the other hand, needs total control, actively preventing those unexpected things from cluttering the difficult-to-organize sonic jigsaw puzzle before you.

To create, you must occasionally give yourself the freedom to explore. When you are happy with the mix, when you think it is okay enough to satisfy the band and their fans, you might take it to the next level by unleashing your creative drive—that part of you full of mix ideas, sonic aspirations, and musical innovations that led you to start mixing in the first place. You dial-in one of those crazy patches you've always wanted to try for the reverb on the vocal. You experiment with audio blog–inspired new approaches with the snare. You have a vision for a different overall texture in the chorus. You know that with a few edits, you could create a stronger introduction for the song. You poke and prod it in new directions and see whether you can redefine some of the mix's qualities, even in small ways, and thereby transform the entire piece of music.

That is how we get from a mix that is just okay to one that is stunning. The band and their fans stop comparing it to the last album and see it as the next profound step in the progression of the artist.

Such explorations are in direct conflict with your technical requirements as an engineer. Each new plug-in you instantiate and each new compressor you patch in must be tested and validated. Any problems found must trigger a careful ritual of troubleshooting and repairing.

The creative mix engineer improvises, but technical success is built on a foundation of deliberate and organized process. Nonlinear thinking leads to creative outcomes, but linear rigor is required to keep it all in line. Every desire to explore a "What if …" is met by an internal voice saying "But don't forget to … ."

💡 As mix engineers, we must stretch to develop a mental agility that enables us to leap back and forth between the big picture, comfortable, chance-taking, exploratory, improvisational, nonlinear thinking that lets us get creative and the zoomed-in, on-edge, under control, check and double-check, deliberate, linear thinking that keeps things working. Such is the bipolar life of a mix engineer.

1.3 MIX SMART

If you are looking for the right way to mix—*the* right way to mix—I have good news and bad news.

💡 Right, the bad news first: there is no right way to mix.

Ah, but here's the good news: there is no right way to mix.

Mixing is wonderfully open-ended and creative. There isn't one way to mix. There are as many ways to mix a song as there are mix engineers eager to mix that song. One of the challenges you have ahead of you is the development of *your* way to mix. Some engineers start with the drums. Some start with the vocal. Among those who start with the drums, some start with the kick, others start with the snare, still others start with the overheads.

Experienced engineers have their proven strategies and routines, but great engineers don't have just one; they use a variety of approaches, flexing their system in reaction to the music. They can approach any given multitrack project from many different starting points and still build a successful mix.

Learning to mix isn't learning a step-by-step process. If you need a linear, step-by-step, rigorous process, become an accountant. Mixing is for those willing to tackle a big, broad, seemingly intractable problem and find the art within. It isn't easy, but it is a very rewarding pleasure.

1.3.1 Creative Vision with Technical Expertise

Learning to mix requires mastery of two forms of achievement:

- Creative vision: see the artistic goals for the whole work.
- Technical expertise: derive the myriad technical steps that might be used to achieve that vision.

These two challenging activities are constantly *on* in the mix engineer's head. Listening to the raw tracks of the music, you formulate a creative goal. You immediately begin to put into place the necessary signal processing. Though tackling the technicalities of hooking up the gear and refining the settings, you still allow yourself the freedom to listen as a passionate music fan and concoct still more ideas and new directions for the tune. The subsequent and steady need for modifying settings and adding plug-ins must never be allowed to slow you down or prevent you from experimenting.

Mixing requires constant progress down these two paths in parallel: refine and reinvent the creative goals while designing and implementing the technologies necessary to make them so. Neither the creative goal formulation nor the technical signal processing implementation has a single, correct answer. You were hired to have an artistic opinion and the technical chops to deliver that vision. You aren't implementing the sole approach for that band's multitrack production; you are creating your own personal interpretation.

To succeed, you need two things: mix experience and a mix strategy. Experience comes over time, naturally. Don't wait for it; advance it actively by assisting other engineers—the best and busiest engineers you know, working at the best studios in your area. While being helpful to the session, absorb their whole process. Pay attention to the musical decisions they make as well as the tweaky techniques that ensue. You need to advance your experience as a musician, and as an engineer; as a creator, and as a technician.

Many enter the field of sound engineering with strong musical opinions. You know the bands you love, and your favorite songs among them. An engineer needs to have similarly advanced, passionate, and well-articulated opinions about sonics as well. You need to know the mixes you love and have a sense for how that mix was achieved. You learn about the compressors that sound best on a reggae snare drum and the reverbs that can decorate a pop ballad.

It is a different way to enjoy music, but it is no more difficult. If you are a music fan, you can adapt your passions to be a mix fan too (see the "Mix Icons" section in Appendix D). You will eventually be hired because others will value your accomplishments in marrying the musical with the technical:

- Your piano sound is the most lush, most ear-tingling piano sound they know, yet it always complements and never overshadows the female vocalist.
- You understand and respect traditional jazz and know how to get a contemporary sound without clashing with expectations or interfering with the interaction among the musicians.
- You know how to recreate the larger-than-life thrill of a live, power trio through the humble playback system of mere average home loudspeakers.
- You can motivate a thrilling amount of low end to come out of vintage drum machines and keyboards without ever letting it get too muddy or overwhelmingly bass-heavy.

Develop genre-specific and gear-specific expertise in one area and work your way out from there. Get more experience by mixing, and remixing, and re-remixing your own music. Volunteer for community organizations and mix all of their events—from single-microphone lectures to full-on live music concerts. Most importantly, mix as many multitrack projects for others as you can. Over time—with more experience—as the quality of your mixing improves, the tier of artist calling on you improves. You get better, and the bands get better. It is a virtuous circle.

! Don't take shortcuts as you develop your knowledge and experience. Doing so will rob you of your chance to succeed. Skip steps as needed to get to work sooner, but loop back and fill in the blanks when you have time. Anyone can get decent-sounding mixes today, without a strong understanding of what they are doing. But they'll hit a technical and expressive limit. Don't tolerate blind spots in your understanding of how things work. Never stop studying the advancing state of recorded art. It takes awhile, but develop your skills and expertise thoroughly and continuously. Then there is no limit to what you might achieve.

Your mix strategy comes from you, informed by all of your experiences and hard work, including what you are doing right now: reading this book. These chapters lay out strategies for bringing order and clarity to things as open-ended and seemingly vague as equalizing an acoustic guitar, or compressing a kick drum. You, the mix engineer, have to decide each and every step to take, and the order in which to take them. But when you make those decisions informed by an overarching strategy, you'll find you make good decisions. You don't follow checklists—you solve problems. You don't imitate settings by others on other projects—you implement settings by *you* for *this* project. You know how to mix—every step of the way—because you can see the goal and you know which actions need to be taken to support that goal. You have a smart mix strategy.

1.3.2 Production Motivations

Still, we need to bring some order to the broad challenge of mixing. Throughout this book, we break mixing down into three production motivations, that is, three reasons to do anything in a mix:

- The fix
- The fit
- The feature

First we scan the tracks, the performances, the arrangement, the composition for things with errors, musical and technical, and we *fix* them. Then we make sure the pieces *fit* together, reshaping the sounds as needed. Finally, we *feature* any elements that need to be highlighted to help our mix tell the musical story. Fixing, fitting, and featuring are the jobs of the mix engineer, and the studio is the tool for getting these things done.

These three motivations help us know when and how to use equalization (Chapter 3), distortion (Chapter 4), compression and limiting (Chapter 5), expansion and gating (Chapter 6), delay (Chapter 7), pitch shifting (Chapter 8), and reverb (Chapter 9).

Mixing will forever remain wonderfully vast and open-ended—the ultimate creative challenge. Through the approaches articulated here, the limitlessness is reigned in, and the infinite is neatly organized so that rather than being overwhelmed by it, we might instead make sense of it and form art from it.

CHAPTER 2
Approaching Your First Mix

There's no obvious place to start.
Let's get started.

Mixing is perhaps the single greatest challenge we face in our career. How can we possibly succeed at something so large, variable, open-ended, and creative while owning the necessary and vast technical details of hardware, software, room acoustics, and critical listening?

Doing your first mix is a lot like getting that first job. Nobody will hire you without experience, but you can't get experience until someone hires you. You won't know how to mix until you mix, but you can't mix until you know how to mix. We've got to start somewhere, so here we go.

There is no single right way to do it. There are no hard and fast rules to follow. There is, however, something of a standard approach to mixing a contemporary recording that is worth reviewing. It isn't the only way. In some cases, it isn't even the correct way. But it is a framework for study, a template that any engineer can modify freely to suit his or her needs for the current project. This chapter offers a starting point in the highly nonlinear, unpredictable creation of a successful multitrack mix.

2.1 MIX ROOM SETUP

Although it is sometimes useful to view the world through rose-colored glasses, we can't mix that way. We need to see—well, we actually need to *hear*—our mixes exactly as they are, uncolored by a flattering monitoring environment. We don't want rose-colored loudspeakers and rooms, unless they go nicely with our shoes. We definitely don't want rose-sounding mix rooms.

We can't hear what our mix really sounds like unless we work in a room that is predictably accurate, honest, and revealing. We need to hear the recording we are creating, independent of the equipment and room acoustics of the mix space. We can't create a great-sounding mix until we create a work environment capable of sounding great.

If you've seen carefully designed recording studios, you know that the architecture and construction techniques get pretty complicated and pretty funky. Books have been written and teams of specialists stand by ready to help anyone properly address the many details associated with building a mix room. Acoustics is a deep but fun field. It is also a sometimes counterintuitive branch of physics and a discipline rich with poorly informed people peddling products and myths that don't help the room. Study room acoustics. Hire qualified specialists. Your mixes will sound better when your room sounds better. If you want to get good at mixing, get a good room built. Such a task is vast and can best be done as a collaboration among the requisite experts. Key things to keep an eye on include:

- Vibration control
    - Locate your studio as far away from noisy machines as possible (air conditioners, heaters, refrigerators, helicopters, fire stations, shooting ranges, trombone factories, etc.). Choose the location for your room that best situates yourself versus all things noisy. Specialists can install the machines on necessary isolators to minimize the propagation of their earth-shaking vibrations, but it's cheaper to start with a vibration-free space.
- Noise control
    - You need a quiet space to work so that you hear even the quietest, most subtle details within the mix. Build appropriately designed sound-isolating walls, floors, ceilings, windows, and doors to keep unwanted noises out of your room, and—conveniently—to keep your sometimes-loud music out of neighboring spaces. Specialists can install proper heating, venting, and air conditioning equipment that provide comfort without distracting noise. Computer fans sonically drape a blanket of obscurity over your entire mix, unless you locate them in a well-isolated and well-ventilated closet or other space.
- Room shape
    - We like to have as much left-to-right symmetry as possible, as viewed from the mix position. You sit on the center line of the room. When you compare the left side of the room to the right side—the loudspeakers, the walls, the windows—you want to see (and hear) a left side that is a mirror image of the right side—or as close to that mirror image as reasonably achievable.
- Reflection control
    - We typically design control rooms so that we hear the sound from the speakers without corruption from the room. One common defense is to prevent the first (and second) reflections of the sound radiating from the loudspeakers toward the walls, floor, ceiling, and furniture from bouncing toward the mix position. Carefully designed angles and bumps redirect the sound elsewhere. Strategically placed absorption attenuates the reflections as well. Consultants, architects, and good contractors can come up with clever solutions in this area.

- Reverb time
  - Reverberation is typically very short (less than $1/3$ of a second or so, depending on the size of your room) so that we hear the reverb in the recording, not the reverb in the mix room. Specialists can help you hit the right target across all frequencies of interest. Beware of fuzz- and fabric-only solutions, as they may absorb high frequencies without taming the low frequencies appropriately, leading to the common problem of a short middle- to high-frequency reverb time, undermined by a boomy, lingering low-frequency reverb time.
- Modal resonances
  - Rooms resonate at some frequencies better than others, and in the smallish rooms typical for a mix room, those resonances are well within the audible frequency range. Acousticians can calculate, estimate, and measure these room modes and design ways to minimize their effect on what you hear in the room.
- Loudspeakers
  - Loudspeakers present a great challenge, as they must be consistent in behavior from the lowest to the highest frequencies. A flat frequency response indicates that the output amplitude from the speaker is fairly consistent, neither overly emphasizing nor attenuating certain spectral regions.
  - The frequency response isn't the whole story, however, as it says nothing about time. Transient detail and image accuracy come only from loudspeakers that do not blur things in time. It is not enough that the loudspeaker have good output down to 40 Hz. We also require that 40 Hz energy to be created with the same agility as the middle and high frequencies.
  - The laws of physics can never be broken, no matter how rebellious you otherwise are. Beware of speakers that look too small to sound so good. The large wavelengths of low frequencies need large drivers, significant power, and/or extra design emphasis to have not only amplitude output, but also time integrity.
  - The placement of the loudspeakers within your room is a decision that has a profound effect on sound quality. Should they be against the front wall, or away from it? How far from the side walls is best? There is no single correct answer. These important decisions are made with the help of specialists and are tailored to your specific room size, geometry, treatment, and the type of loudspeakers you've acquired. It's never simple.
  - There's still more: power response, directivity, power handling, efficiency, and the like. As with room acoustics, don't hesitate to read up on loudspeakers and talk to experts.
- What about headphones?
  - ! Headphone listening is quite different, physically and physiologically, from loudspeaker listening. And although current trends point towards increased use of ear buds as a mode of listening to music, the vast majority of listeners will hear your mix over loudspeakers. It is very unwise to

mix in headphones hoping to predict your mix's efficacy through the completely different user interface that loudspeakers offer. It is fine to occasionally listen to your mix on headphones to hear how your mix decisions affect the headphone listener's experience. So mix through loudspeakers, and occasionally put on some headphones to hear this other listening modality.

■ Use headphones as a way to check for hard-to-hear details that might be missed when listening to loudspeakers. Distortion, channel-specific problems, brief errors, instantaneous clicks, brief pops, short-lived drop-outs, and other details are quite effectively revealed in headphones. For quality control, headphones are a great way to zoom in, magnify, and scan for problems. But allow your overall mix aesthetic to developed on loudspeakers.

Working in an inferior room with mediocre loudspeakers holds back the quality of your mixes, the pace of your learning, and the success of your business. Room design and loudspeaker selection are critical first steps, worth a thoughtful investment of time and resources and likely relying on specialists to get right. Don't skip this step.

2.2 MIXER SETUP

Next, lay out the console or preset the *digital audio workstation (DAW)*. Whether using a mixing console or a DAW, it helps to do some of our hooking up and plugging in ahead of time. We predefine as much of the signal flow as can be anticipated. You'll need to be creative later, when you dive into the tracks, but you've got to hook everything up correctly, too. The technical steps tend to interfere with the creative process, so it helps to do a chunk of the tedious and technical thinking ahead of time so that it doesn't interfere with the flow of inspiration while mixing.

2.2.1 Global Effects

▶ All of the various signal-processing effects that you need in a mix can't be fully predicted, but some basic, staple effects are so likely to be used that they can be set up ahead of time. Many mixes make use of a long reverb (hall-type program with a reverb time around two seconds, simulating a symphony hall); a short reverb (plate or small to medium room with a reverb time around one second); a spreader (explained shortly); some echo (dotted eighth-note, quarter-note, or quarter-note triplet in time); and any special effects you suspect this tune will need—I'm thinking we may reach for a guitar amp simulator, so I'll get one ready.

Before digging into the details of the mix, you can launch the appropriate plug-ins and patch up the appropriate hardware to get these anticipated effects going (see Appendix A). By setting them up ahead of time, these effects are available with minimal additional effort. Want some long reverb on the strings? It's already there; just raise the appropriate effects send. Want to put a little echo on the guitar? It's as simple as turning a knob, clicking, and dragging. If the mix calls for it, you can effortlessly send a bit of vocal, snare, and lead guitar to the same effect.

The way to have all these effects handy is to use aux sends. For example, you might lay out the mixer so that aux 1 goes to the long reverb, aux 2 feeds to the short one, and aux 3 sends a signal to the spreader—arrange them in whatever order is most comfortable for you. The effects are returned to empty, available monitor paths, labeled, panned as desired, and assigned to the mix bus on the mixer or DAW.

🔊 An essential part of this prehookup phase is to then confirm that you've done it correctly. Make sure that the right signal is going to the right place. Check stereo outputs to make sure that they both work, are returning at the same level, and are sound-balanced left versus right. Send pink noise through the reverb, watch meters, and listen for left/right balance. Send a snare hit panned dead center and listen to the left/right balance of the device. A reverb that is right-heavy will likely undermine your intent when mixing. In the heat of mixing later, when the vocals, drums, guitars, and bass are cranking along and sounding great, it is easy to add a touch of 'verb to the tambourine and fail to notice that it is lopsided. Check all the effects now in this clinical early phase of mixing, so that your mind is free to stay creative later, unburdened by the details of chasing a faulty left output.

2.2.2 Special Effects

▶ What's the spreader? It is often desirable to take a mono signal and make it a little more stereo-sounding by sending a track through two unique, short, pitch-shifted, hard-panned delays. Each delay time is set to a different value somewhere between about 10–50 milliseconds (ms); if too short, it starts to flange/comb filter (see "Short Delay" in Chapter 7), and if too long, it pokes out as an audible echo. One delay output is panned left and the other panned right. The idea is that these quick delays add a kick of supportive energy to the mono track being processed, sort of like the early sound reflections that would be heard if the instrument were played in a real room. When musicians perform acoustically, without sound reinforcement systems, they tend to do so in live, sound-reflective spaces. It's more fun to sing in a church or in the shower than it is to sing in the sound-reflection-less outdoors, or in a sponge factory. Sound-reflective walls can acoustically amplify and perceptually enlarge a sound. The strong sound reflections can make the performance sound better, stronger, more exciting. Outside, there are no walls and no ceiling to reflect the sound back. The sponges in a sponge factory absorb the sound.

Enclosed, sound-reflective spaces are very good at augmenting a sound in this flattering way. The *spreader* simulates two reinforcing reflections coming from the left and right by taking a single mono sound and sending it to two slightly different, short delays. Any two delay lines can be used as a basis for the spreader effect. You probably have a delay plug-in that is itself capable of two or more unique delay times. Use it as the basis for your spreader.

The effect is further refined by pitch-shifting the delayed signals ever so slightly. That is, detune each delay by an all but imperceptible amount, maybe 5–15 cents. There are 100 cents in a half step, the pitch difference between two adjacent keys on a piano or two adjacent frets on a guitar. So 5 to 15 cents is a tiny fraction of a half step, far from being a musically relevant new pitch. Again, a stereo effect is the goal, so the spreader requires slightly different processing on the left and right sides. Just as a slightly different delay time was specified for each output, dial in a slightly different pitch shift as well (maybe the left side goes up 9 cents and the right side goes down 9 cents). It is customary to keep things symmetric—pitch the left side up by the same amount the right was shifted down—but not strictly required.

Introducing these slight pitch shifts to the short delays abandons any basis in acoustic reality that the spreader might have had. The effect isn't just simulating early reflections anymore. With this spreader, you are taking advantage of signal-processing equipment to create a widened stereo sound that only exists in music reproduced through loudspeakers. This sort of sound doesn't happen in the physical world. This way of thinking is a source of great creative power in mixing: consider a physical effect and then manipulate it into something that is removed from and hopefully better than reality (good luck, and listen carefully).

As long as we've given ourselves permission to abandon reality, let's introduce a slow and small amount of delay time modulation to each side of this spreader. Most delay lines have the ability to sweep their delay time, so let's do so slowly (less than 1 Hz), with medium to small depth. Again, we'll set them to slightly different values, left versus right.

A side effect of delay time modulation is a bit of pitch shift. As the delay time is slowly increased, the pitch droops slightly down. As the delay time is slowly decreased, the pitch wobbles back up.

Now we have a unique concoction of left/right differences to intrigue the ears of our listeners: relative to the signal you send to the spreader, you get two unique outputs that are slightly delayed, and that delay time constantly shifts, and is slightly detuned, and that pitch constantly changes. The result: a perceptual spreading out, left to right, of any signal sent to it.

This effect might be added, at least in a small amount, to the lead vocal track, among others. You are likely to pan lead vocal track straight up the middle. With a touch of the spreader effect, the vocal should stay centered but perceptually widen from an imagined point in between the loudspeaker to a larger, apparently wider sound source.

🔊   In order for the spreading effect to keep the vocal centered, it helps to be strategic with the delay and pitch settings. Consider the delay portion of the spreader only. If you listen to the two panned short delays, the resulting stereo image pulls toward the shorter delay. For example, with a 30-ms delay panned left and a 20-ms delay panned right, a listener sitting on the median plane can hear the sound coming from the right, even as the unprocessed track sits panned dead center.

Now consider just the pitch-shifted part of the spreading equation. The higher pitch tends to dominate the image. With a 9-cent pitch up to the left and 9-cent pitch down to the right, the image shifts ever so slightly left, toward the higher, brighter sound.

The full spreader calls for both delay and pitch shift. Arrange it so that the two components balance each other out (e.g., delay pulls right while pitch pulls left). This way, on net, the main track stays centered. Experiment with different amounts of delay and pitch change. Each offers a unique signature to your mix. When overused, the vocal will sound too mechanized and too processed. When conservatively applied, the voice becomes wider and more compelling, filling up more space between the loudspeakers; singers like this, and so does the music-downloading public.

2.2.3 Specific Effects

As you become a more experienced mix engineer, and when you also have some experience with the tracks being mixed—perhaps you were the tracking engineer, or you've already mixed a couple of similar songs by the same artist—you may also anticipate some track-specific effects: equalizers and compressors, for example.

💡 *Insert* (see Appendix A) the type, make, and model of effects devices that you think you might use: compressor X on the vocal, equalizer Z on the snare, and so on. Through experience, you will begin to have a sense of which one will sound best on these tracks for this sort of tune. It's a time-saver and creativity catalyst to place them in their intended places so that compressing the vocal is as simple as pressing or clicking the ON button. Your head and heart stay focused on the music generally and the sound of the vocal specifically, undistracted by thinking through which compressor, which track, and how do I hook it up? Again, check each instantiation and patch now to be sure the software will launch, the hardware doesn't buzz, and so on. Set the device to neutral parameter settings so that when you turn it on, it doesn't suddenly wreck the sound of the track.

2.2.4 Hearing Effects

Experienced mix engineers work fast, hear deep into the mix, and ... well, they make it look easy. You'll get there, eventually. But mastering the subtle audible nuances of the various mix strategies described in this book is tough. So although our goal is to create a mixed recording that we love, we're going to spend some time listening to something else. We are going to solo pieces of the mix and loop parts of the tune.

SOLO

To explore an effect without distraction, we solo the track, with the effect. It is true: end users will never hear the tracks soloed, only the final, full mix. But if we are to tweak the parameters of the various signal processors and have a fighting chance of hearing it, we need to temporarily persuade all the other tracks to shut up for a minute, thank you.

Make ample use of the solo feature when you are wrestling with a tricky problem, exploring a new piece of gear, or treading new territory in your professional development. It's allowed.

🔊   When you think you've mastered the immediate challenge, be sure to then check your work without the solo. The ultimate decision for what is good and what is bad for your mix must be made in the context of the full mix. Working in solo mode is a step—a necessary early step—in the mixing process.

LOOPING

Your hearing system has a short memory. No offense. So does mine.

It can be difficult to evaluate the impact of our mix decisions—an equalization (EQ) change here (see Chapter 3 for an in-depth look at EQ), and compressor tweak there—when the tracks themselves are constantly changing. Loop a four- or eight-bar segment, however, and you'll memorize every nook and cranny, every wart and wonder in the track. You can then begin to hear with extraordinary confidence and precision, figuring out exactly what it takes to convert the existing sound into your imagined, target sound.

As with the use of solo, view this as an acceptable step, but never the final step. Looping four bars of a verse might make it possible for you to do wonders for the verse. But you've got to check out what all those tweaks do to the chorus, the bridge, and the other verses. The ultimate decision for what is good and what is bad for your mix must be made across the full timeline of the tune.

2.3 A BASIC MIX APPROACH

With your room setup, the mixing environment prepared, and the wacky spreader explained, let's tackle a basic mix. Consider a 48-track production that includes the following representative set of mix challenges: drums, bass, doubled rhythm guitars, lead guitar, assorted samples and loops, lead vocal, and background vocals. Some nagging questions immediately begin to, um, nag: Where do I start? When do I work on vocals? When do I do drums? Which effects should I use?

2.3.1 Balance

The essential first step is to get the mix balanced. Push up the faders, find the musical makeup of the song, and adjust the level, panning, and necessary muting until you can easily make sense of what the artist is trying to do with these tracks. We made mention of this in Chapter 1. We dig into specifics here.

Vocal, snare, kick drum, and bass will be—as always—critical tracks to keep an ear on. Although the vocal is typically the most important single element in the entire production and therefore often the loudest single track in the entire mix, it is not unusual for the snare drum to rival it in loudness. In fact, the snare might even be a bit louder than the lead vocal in some songs—but please don't tell the singer that. A loud snare hit can make it all but impossible to enjoy the rest of the tracks in the mix, during the snare hit. Because the snare is so short in duration, its total takeover of the mix is quite temporary. So it is sometimes (but certainly not always) acceptable to target a rather ambitious level for this important instrument, if you dare.

Vocal and snare live above the fray of your mix, but—no matter which one is louder technically—the vocal ultimately must reign supreme, musically. It should always be easy to understand the words. No matter how energetic the band is underneath, that mere mortal, humbly human singer must somehow never be outshone. Throughout the rest of the session, as you bring in the various other elements of the mix arrangement, mentally loop back to the vocal and revise its fader setting (and associated effects) to maintain the singer's place at the top. You can't just keep turning the vocal up. You'll have to restrain all the tracks as you go, introducing each track to its own audible, appropriate-sounding level and pan pot position, without overpowering the vocal and other tracks.

It cannot be overstated: keeping a mix balanced is not easy, and it's not limited to the vocal.

There is a natural tendency in the course of mixing to push the snare fader up, and up, and up. A snare that sounds great while soloed may sound disappointing when placed back into the full context of the mix. Keys and guitars will fight the snare in the midrange. Cymbals and distortion (see Chapter 4) can create unwanted spectral competition in the higher frequencies. Bass guitars and keyboard parts can seem to rob the snare of low-frequency punch and power. But do not turn the snare up too much; if too loud, the snare loses musical value, becoming a distraction rather than a source of rhythm and energy. Like the vocal, the snare drum gets constant attention. Finesse the snare fader so that the snare sound is always an enjoyable part of the multitrack production. We discuss in later chapters all the other ways to keep the snare exciting and audible, especially through EQ, compression, gating, and reverb (Chapters 3, 5, 6, and 9). Keeping a mix balanced means returning often to the snare fader to keep it carefully placed relative to the vocal and the rest of the band.

! The kick drum and bass guitar may enter a competitive low-frequency contest. Set their levels very carefully. A song with a lot of bass can be thrilling on first listen, but if the bass guitar or low synth line is too loud, the kick drum loses impact. The tracking engineer, the producer, the drummer, and the bassist should choose instruments, tunings, playing styles, musical parts, and recording techniques that aim for at least slightly different frequency ranges, kick versus bass. We will use signal processing at mixdown to ensure the result.

Typically, the deep tone of the kick drum is lower in frequency than any low emphasis of the electric bass, so we track and mix toward that goal. But there are exceptions; there are always exceptions. A small kick drum, tuned to an upper low-frequency heartbeat in a production using a five-string bass with the low string tuned down to the B slightly more than three octaves below middle C (and therefore having a fundamental frequency of about 31 Hz) can turn things around, spectrally. In cases such as these, it might make sense to place the bass guitar below the kick drum, spectrally. In either case, these two instruments must be shown how to get along together. There is a natural tendency for them to fight. Target slightly different characters and spectral footprints so that they fit together more naturally.

The fader positions for these two instruments, kick and bass, are given deliberate attention. Raise their levels to points that are clearly too loud and then pull them back. Lower their level until they are too soft, and then push them back up. Somewhere between these two positions lives the most effective musical level. Engineers constantly massage the levels, making fine adjustments up—then down—throughout the early part of a mix session both to teach themselves the dangers of losing control of these tracks and to make sure their levels make the most musical sense. The entire mix achieves new levels of refined clarity when the kick drum and bass cooperate with each other and maintain sufficient modesty not to overpower any other track.

The other parts of the multitrack arrangement (both tracks and effects) fill in around and underneath these most important tracks. If the guitar is louder than the vocals, the band is probably going to sell fewer disks and downloads. If the music fans cannot hear the piano when the sax plays, the song loses musical impact (and the engineer probably loses the chance to work with that piano player on the next album). Work the faders hard to find a balance that is fun to listen to, supports the music, and reveals all the complexity and subtlety of the song.

Balancing a mix sounds so straightforward in concept. Engineers who are early in their career will soon discover that it is not easy. Keeping a mix balanced requires experience, excellent listening skills, patience, focus, and a strong musical opinion. Throughout the mix session, the balance directly drives everyone's opinion of the quality of the project. The engineer is very much on the spot, expected to balance the tune and keep it balanced at all times.

Yet in the course of the session, there are other points of focus. The drum tracks need attention at one phase, the bass at another. The vocal track always demands careful attention. Compressors, reverbs, and all of our other plug-ins and processors are complex entities that can distract us when dialing in new settings. No matter what other session priorities are present, no matter what other difficulties and headaches arise, we are always required to keep the production balanced. When making tough decisions on a vocal equalizer/compressor/reverb signal processing chain, we might temporarily turn up the vocal, turn down some other tracks, or even solo the vocal with effects. As soon as the vocal sound is processed and tweaked into shape, however, that vocal must be turned down and tucked back into the mix so that the vocal, snare, kick drum, bass, and all other supporting elements are again balanced and the overall mix can be enjoyed.

Never underestimate the production power of the mute button. The mix engineer is not obligated to push up every fader available. In fact, our job as mix engineer really begins with a close study of the music. What is the song about? How does each sound support the feelings and the goals of the artist? Scrutinize all the parts being played by all the musicians to understand their role within the arrangement. Playing the track through a few times, you must develop complete mastery of what each player is doing and why. You must have deep knowledge of all the parts, as if you were a member of the band, as if you'd been at every rehearsal. Great artists have a technical or creative motivation behind every note. Drummer and bass player interact closely. A soloist weaves in and out of a tune, staying clear of the vocal, embellishing countermelodies. The rhythm guitar might slip into a groove highly motivated by the hi-hat part. The keyboards and horns might riff off each other. There is no limit to the musical interactions that might be designed into the production. We must notice and support these ideas.

Study each track and evaluate its interaction with every other track. Musical clashes are common, and muting one or both of the offending tracks is often the only solution. Delete the tracks that distract. Just because it's there doesn't mean you have to use it. Musical judgment informed by technical mastery guides you through the entire set of tracks so that you can figure out which ones get muted for some or all of the tune. A strategic mute here or there can make balancing a mix much, much easier.

It's time to move on to the rest of the tracks. The following discussion shows a way of thinking about a mix. It is, in fact, foolish to talk specifics (frequencies, ratios, reverb times) without hearing the specific tracks and knowing the artistic intent of the specific tune. Follow along, though, because a valid problem-solving, action-taking way of working is revealed. The specifics vary from project to project, but the level of thought and the formulation of strategies are always present.

2.3.2 Drums

With the mix balanced, it's time to advance the mix beyond faders, pan pots, and cut buttons. Even though the vocal is almost always the most important element of every pop song, no matter how many tracks there are, we typically start with the drums. Starting with the vocal makes good sense, because every track should support it. But easily 99 percent of all pop mixes start with the drums.

The drums are often the most difficult instrument to get under control in the recording studio. It's an instrument with at least eight (generally more) separate instruments playing all at once in close proximity to each other (kick drum, snare drum, hi-hat, two or three rack toms, a floor tom, a crash cymbal, a ride cymbal, and all the other various add-ons the drummer has managed to collect over the years). It's hard to hear all the problems and thoughtfully refine the sounds of the drums without listening to them in isolation. When the faders for the vocals and the rest of the rhythm section are up, it's difficult to dial in just the right amount of compression on the rack toms. So after balancing the overall mix for a good starting point, we almost always dig in to a mix by tackling the drums in isolation. We need this complicated instrument, which is spread across so many tracks, to be tamed before we can even think about moving on to other instruments.

What does an engineer do with the drums? Here's one approach.

▶ The kick drum and snare are generally the source of punch, power, and tempo for the entire tune—they've simply got to inspire awe. So it is natural to start with these tracks. Step one: keep them dead center in the mix, or very close to it. The kick and snare, like the bass and vocal, are so important to the mix that they almost always live center stage. The kick needs both a clear, crisp attack and a solid low-frequency punch. EQ and compression are the best tools for making the most of what was recorded. The obvious: wide EQ boost of 3 to 6 dB somewhere around 3 to 5 kHz for more attack, plus another EQ boost at about 40 to 80 Hz for more punch. Not so obvious: EQ cut with a narrow bandwidth around 200 Hz to get rid of some muddiness and reveal the low frequencies beneath (see Chapter 3).

Compression does two things for the kick. The first goal of compression on kick might be to manipulate the attack of the waveform so that it sounds punchy and cuts through the rest of the mix. Chapter 5 describes the sort of low-threshold, medium-attack, high-ratio compression that sharpens the amplitude envelope of the sound. Second, compression controls the relative loudness of the individual kicks, making the slightly weaker kicks sound almost as strong as the more powerful ones. Drumming is physical work. Drummers understandably get tired in the fifth chorus of take 17. Compression can help them out.

Placing the compressor after the equalizer lets the engineer reshape the sound in some clever ways. The notch around 200 Hz keeps the compressor from reacting to that unwanted murkiness. As the engineer pushes up that 40 to 80 Hz low-frequency boost on the EQ, the compressor reacts. With an aggressive low-frequency boost, the compressor is forced to yank down the signal hard. In this way, heightened low-frequency punchiness makes the drums sound larger than life.

▶ The snare is next. As a starting point, it likely gets a similar set of effects: EQ and compression. The metallic clatter of the snare is broadband, especially rich with energy from 2 kHz on up. Pick a range that sounds good to you on this particular drum for this particular tune: 8 kHz might sound too harsh or trashy, but much above 10 kHz and it starts to sound too fine, too sweet. Try 5 kHz. It is a subjective, musical decision very much influenced by the flavors of middle and upper frequencies offered by the other instruments: cymbals, acoustic guitars, saxophones, ukuleles, and so on. With the exciting buzz and rattle tastefully highlighted, a low-frequency boost for punchiness is also common for snare. Typically the engineer looks higher in frequency than was done on the kick, maybe 80–160 Hz or so—sometimes as high as 200 Hz. It is wise to also look for some unpleasant frequency ranges to cut. Somewhere between 600 or 1,200 Hz lives a biting zing that often doesn't help the snare sound and is only going to fight with the vocal and guitars anyway. One simply finds a narrow band to cut, the snare tone improves, and the rest of the mix goes more smoothly.

The snare definitely benefits from the addition of a little ambience. In many mixes, it is not unusual to send it to the short reverb you already set up and/or hope to find some natural ambience in the other drum tracks. A plate reverb patch has such a dense dose of upper middle frequency content that it also reshapes the timbre of the snare in often flattering ways (see "Timbre Through Reverb" and "Texture Through Reverb" in Chapter 9).

The drum overhead microphones are a good source of supportive, natural snare sound. Any recorded ambience on room tracks should be considered now. A gate across the room tracks keyed open by the close microphone on the snare can create a subtle touch of ambience on each snare hit (see "Keyed Gating" in Chapter 6).

With the kick and snare made powerful and punchy and equalized for any careful timbral touchups, it's time to raise the faders for the overheads and hear the kit fall into a single, powerful whole. The overheads often have the best "view" of the entire drum kit and the snare regularly sounds phenomenal within these tracks. Carefully blend the overhead tracks with the kick and snare tracks to make the overall drum performance congeal into a single, powerful event. It is tempting to add a gentle high-frequency boost across the overheads to keep the kit crisp—a modest shelf boosting 3 or 4 dB at 8 to 10 kHz and above is a natural beginning (see "Shelving EQ" in Chapter 3). Listen carefully first. If the tracks were recorded with plenty of high-frequency content already, additional high-frequency emphasis at mixdown will lead to a harsh, unpleasant drum sound. In fact a gentle (less than 3 dB) and wide presence boost somewhere a bit lower, between 1 and 5 kHz, might be the magic dust that makes the drummer happy.

When the toms are on separate tracks, the engineer will again reach for those tried-and-true equalizers and compressors. Creativity is required, but nominally the engineer might EQ in a little bottom and maybe some crisp attack around 6 kHz. One should consider EQing out some muddiness with a narrow cut somewhere in the 200- to 400-Hz range, analogous to what was done with the kick earlier. Compress the toms for attack and punch (see "Reshaping Amplitude Envelope" in Chapter 5) and the drum mix is all set for now.

2.3.3 Bass

With a first draft of the drum mix tentatively set, we direct our attention to the bass guitar. The bass line often needs compression to make it a more steady, foundational low-frequency mix element. Except in the hands of that rare, great bass player, some notes are accidentally louder than others. On all but the best instruments, some strings on the bass are quieter than others.

Gentle compression (4:1 ratio or less) can even out these problems. A medium to slow attack time adds punch to the bass in exactly the same way it did on the drums (see Chapter 5).

Release is tricky on bass guitar. Many compressors can release so fast that they follow the sound as it cycles through its low-frequency oscillations. That is, a low note at, say, 40 Hz, cycles so slowly (once every 25 ms) that the compressor can actually release during each individual cycle. This cycle-by-cycle reshaping amounts to a kind of harmonic distortion (see Chapter 4). Engineers typically slow the release down so that it doesn't distort the waveform in this way. The typical goal is for the compressor to ride the sound from note to note, not cycle to cycle.

On bass, the obvious EQ move is to aim for more low end. But we must be careful, as the track might already possess more than enough low-frequency content. The bass player likely seeks it out when they choose and set up their instrument for the session. The tracking engineer likely emphasized it further through microphone selection and placement when it was recorded. The trick at mixdown is to get a good balance of low frequencies from 30 Hz through 300 Hz. It is our responsibility as mix engineers to evaluate all of the decisions made before us. In particular, we listen for an outlying hump in the frequency response—either too much or too little in a single low-frequency pocket. We find it and simply equalize in the appropriate correction, creating low-frequency bliss without overdoing it.

💡 At this point, it makes sense to glance back—with our ears—at the kick drum. If the kick sound is defined in the low end by a pleasing emphasis around 65 Hz, then we might need to make room for it in the bass guitar with a complementary but gentle cut. The trick is to find EQ settings on both the kick and the bass so that the punch and power of the kick doesn't disappear when the bass fader is brought up.

It is not unusual in some styles of music to add a touch of chorus to the bass (see "Chorus" in Chapter 7). This method is most effective if the chorus effect doesn't touch the lowest frequencies and instead just provides a bit of spectral motion in the middle frequencies. The bass provides important sonic and harmonic stability in the low frequencies. A chorus, with its associated pitch bending, would undermine this. The solution: place a filter on the send to the chorus and remove everything below about 250 Hz. The chorus effect works on the overtones of the bass sound, adding that desirable richness without weakening the song's foundation at the low end.

2.3.4 Rhythm Guitars

For this basic mix, the rhythm guitars are doubled. It's a rock-and-roll cliché to track the same rhythm guitar twice. The two tracks might be identical in every way except that the performance is ever so slightly, humanly different. Panned apart, the result is a rich, wide, ear-tingling wall of sound. The effect is better still as the subtle differences between the two tracks are stretched slightly. Perhaps the second track is recorded with a different pickup setting, different tuning, different guitar, different amp, different microphones, different microphone placement, or some other slightly different sonic approach.

In mixdown, we make the most of this doubling by panning them to opposite extremes: one goes hard left, the other hard right. It is essential to balance their levels so that the net result stays centered between the two speakers. A touch of compression might be necessary to control the loudness of the performance, but often electric guitars are recorded with the amp cranked to its physical limits, giving it amplitude compression effects already. Complementary, complicated EQ contours (boost one where the other is cut and vice versa) can add to the effect of the doubled, spread sound.

2.3.5 Samples and Loops

The sweetening and groove enrichment that comes from various samples and loops completes the rhythm section in this basic mix. These sounds are often already aggressively processed and might sound fine without further signal processing. Just tuck them into their place with fader finesse and pan pot patience. If anything, you may want to give them a unique sound—low fidelity or highly filtered—through EQ and other effects to ensure they get noticed, but without outshining the rest of the band. Effects to consider: add some flange using a short delay (see "Flanger" in Chapter 7) and/or introduce a tasty bit of distortion by sending the sample or loop through a guitar amp with a distortion stomp box or, more conveniently, using an amp simulation plug-in (luckily, we already set this one up). In this way, we upgrade sterile samples and loops into an ear-tingling, buzzing source of musical caffeine.

Pan the samples for spatial interest. They might be composed to interact with and provide counterpoint to the guitars. Or are they call-and-response sounds with the lead vocal? Assess their interaction with what the rest of the band is doing and pan them to a logical location so that they complement rather than conflict with the companion tracks.

Loops generally interact with the rhythmic drive of the drum and bass performance. Tuck them all in together for a bigger, better groove machine. Or consider a hard-panned gesture to give the groove emphasis, letting it stand on its own. Depending on how the performances interact, a good approach is to pan the loop opposite the hi-hat or the most active toms or the solo guitar—whatever it takes to keep the spatial counterpoint most exciting. Add a short delay (maybe tuned to a sixteenth-note, or eighth-note, or even a quarter-note duration) low pass filtered and panned to the opposite side of the loop if it enhances the groove.

With drums, bass, guitars, samples, and loops carefully placed in the mix, a first draft of the rhythm section is complete. It is time at last to address the fun and important parts of vocals and lead guitar.

2.3.6 Lead Vocal

The all-important vocal gets our attention, at last. The voice must be present, intelligible, strong, and exciting. Presence and intelligibility come from a healthy amount of spectral energy in the upper-middle frequencies. If the singer/vocal microphone pairing didn't do the job, we reach for EQ to ensure the consonants of every word cut through our stunning—but spectrally competitive—wall of rhythm guitars. A careful search from 1 kHz to maybe 6 kHz should reveal a region suitable for boosting that raises the vocal up and out of the chaotic chatter of the guitars and cymbals. We'll almost certainly have to go back and modify the drum and guitar equalizer settings to tailor the critical upper midrange.

Mixing requires this sort of iterative approach. The vocal highlights a problem in the guitars, so we loop back and fix it. Trading off effects among the competing tracks, we constantly seek out a balance between crystal-clear lyrics and perfectly crunchy guitars, visceral beats, and world-dominating bass.

Mix supremacy for the vocal will come from panning it to the center, adding compression, and maybe boosting the upper lows (around 250 Hz) and some breathy highs (above 10 kHz). Compression controls the dynamics and raises the loudness of the vocal performance so that it fits in the crowded, hyped-up mix that's screaming out of the loudspeakers. All of this compression and EQ track by track has so maximized the energy of the song that it won't forgive a weak vocal. Natural singing dynamics and expression are often too extreme to work because either the quiet bits are too quiet or the loud screams are too loud, or both. Compressing the dynamic range of the vocal track makes it possible to turn the overall vocal level up. The soft words become more audible, but the loud words are pulled back by the compressor so that they don't overdo it.

The vocal, a tiny point in the center, risks seeming a little small relative to the drums and guitars. The spreader effect is designed to combat this problem. Sending some vocal to the spreader helps it take on that much desired larger-than-life sound. As with a lot of mix moves, you may find it helpful to turn the effect up until it's clearly too loud and then back off until it's just audible. Too much spreader is a common mistake, weakening the vocal with a chorus-like sound. The goal is to make the vocal more convincing, adding a bit of width and support in a way that the untrained listener wouldn't notice as an effect.

Additional polish and excitement might come from a very-high-frequency EQ boost (10 kHz or 12 kHz or higher) and some slick reverb. The high-frequency emphasis will highlight the breaths the singer takes, revealing more of the emotion in the performance. It is not unusual to add short reverb to the vocal—try the plate we already set up. It adds midrange complexity and enhances the stereo width of the voice further still. Depending on the style of tune, we might try to add a touch of long reverb to give the vocal added depth and richness. Sending the vocal to an additional echo (or two) is another common mix move. Tune the delay to the song by setting it to a musically relevant delay time—maybe a quarter note in this case. It is mixed in so as to be subtly supportive, but not exactly audible. If there is room for more mix complexity, the next step might be to add some regeneration to the delay so that it gracefully repeats and fades. Taking it further still, you can send the output of the delay to the long reverb too. Now the singer's every word is followed by a wash of sweet reverberant energy, pulsing in time with the music. Too much, and it is Velveeta-cheesy. Get it just right, though, and you've created a loudspeaker illusion that is larger than life.

EQ, compression, delays, pitch shifting, and two kinds of reverb represent, believe it or not, a normal amount of vocal processing. It's going to require some experimentation to get it all under control. By turning up the various pieces of processing until they are clearly too loud and then backing off, we learn the role each effect plays. It certainly requires us to go back and forth among every piece of the long processing chain. Change the compression, turn up the delay, turn down the reverb, and back to the compressor again. With patience and practice, we find even ornate combinations of effects easy to control.

And that's just a basic patch. Why not add a bit of distortion to the vocal? We've already got that amp simulator running. Or flange the reverb? We've already got a flanger setup for the samples and loops. Or distort the flanged reverb? Anything goes.

2.3.7 Background Vocals

The background vocals might get a similar treatment, but we get to push them a little more. The various background vocal parts are typically panned out away from center, and the whole set of effects can be made more prevalent in the mix. One reliable approach is to hit the spreader and the long reverb a little harder with background vocals to help give them more of that magic pop sound. Intelligibility might be less important for background vocals. Repeated words or call and response lyrics can often be understood through context, freeing you to pull out some presence in the already crowded middle frequencies and emphasize other timbral details in less-competitive spectral spaces, if you wish. With background vocals, we get more ambitious and more creative than we allowed ourselves to be with the lead vocal. These tracks can be mixed as hybrid instruments: part vocals, part string section; half human, half synth.

2.3.8 Lead Guitar

We find—as required by the Rock and Roll Standards Bureau (don't worry, it doesn't really exist)—that the song has an electric guitar solo. We also find that, when the solo is shredding, that the singer sits out. In fact, if the singer is singing while the guitarist is taking a solo, it is a pretty good bet that the band will break up soon. The typical mix serves up singer *or* solo—rarely both.

The lead guitar can be thought of as replacing the lead vocal during the solo. Because they do not occur simultaneously, the solo guitar is allowed to rival the lead vocal for attention. The mix challenge is to get the lead guitar to sit clearly above the din of the band, very much analogous to our already solved challenge with the lead vocal. An EQ contour like that of the lead vocal is a good strategy: presence and low-end strength. Unlike the vocal, however, many styles of guitar require little to no compression. Electric guitars are naturally compressed at the amp when tracked at maximum volume. Additional reverb is also optional for guitars. The overall tone of the guitar might be fully set at the time of tracking by the guitarist when the amp was set up and the settings were dialed in—that includes the reverb built into the amp.

Solo guitar might get sent to the spreader, and it might feed a short slapback delay (see "Slap" in Chapter 7). The slap delay might be somewhere between about 100–200-ms long. It adds excitement to the sound via a just-perceptible echo reminiscent of live concerts, big spaces, and large crowds. It can be effective to pan the solo about halfway off to one side and the slap a little to the other. If the singer is also the guitarist playing the solo, it might make more sense to keep the solo panned to center—they solo where they sing. Of course, you can always add a touch of phaser, flanger, wah-wah, or something from the long list of effects in the digital multieffects unit. You can even add additional distortion. Guitar welcomes an aggressive application of effects, but significant tonal changes should probably be made with the guitarist present to get the blessing of an expert in the field.

2.3.9 The Whole Mix

The entire mix might get a touch of global EQ and compression. As this can be done in mastering, it is wise to resist the temptation at first. With experience, you should feel free to put a restrained amount of stereo or surround effects across the entire mix. As mix engineers, we are always trying to make the mix sound the best it possibly can, after all. But the best-sounding mix comes from a talented mix engineer who also knows how to best take advantage of the mastering stage of record production. Leave some room for the mastering engineer to lend wisdom, talent, and some excellent signal processing to refine and enhance the overall sound of your mix.

For EQ, a little push at the lows around 80 Hz or a tick below and the highs around or above 10 kHz can be the right sort of polish. If you've got this sort of global EQ applied to the whole mix, you'll need less of this applied individually across all the other tracks. Soft compression with a ratio of 2:1 or less, slow attack, and slow release can ride the gain almost invisibly while helping the mix coalesce into a more professional sound that is well-suited to ear buds on subways and car stereos on highways.

As the entire mix is going through this equipment, it is essential to use very good-sounding, low-noise, low-distortion effects, devices, and applications. If the studio doesn't yet have some very high-end effects, save these processes for a mastering studio that does.

Don't do any loudness maximizing at this phase. You should focus on getting the most pleasing mix out of these tracks, fully independent of loudness. If you want it louder, turn up your control room level. Those signal processors that raise loudness belong in a different phase of production, well clear of the mix session. Loudness is such a crude and irresistible effect that it sonically blinds us from doing other necessary mix work. It is hard to affect the necessary and often subtle refinements that every mix needs when the whole mix is subjected to aggressive loudness maximization.

On the whole mix, try only the conservative, restrained applications of EQ and compression that you think would benefit the tune while preserving opportunities for the whole set of mastering decisions at the next session.

2.4 SUMMARY

This chapter wraps up our representative approach to one mix. It is meant to demonstrate a way of thinking about a mix, revealing one way through the open-ended maze that is mixing, not the step-by-step rules for mixing. Maybe you noticed how easy it is? Or is it?

When you start mixing, expect to be stumped occasionally by some decisions—even too many decisions: you push up a vocal fader and have no idea what to do. Sounds fine. I guess I'll EQ it, compress it, and add some reverb. That's what most people do, right? Wrong. That is not how successful mixers think through their mix decisions. In the chapters that follow, you'll see very specific reasons for mix moves—EQ, distortion, compression, expansion, delay, pitch shift, reverb, and so forth. When you have a specific strategy in mind—when your mind's ear knows what needs to be achieved sonically and your mind's mind knows how to do it technically—you'll mix with speed and purpose.

! Warning: even when you figure out how to mix one tune and mix it well, it doesn't give you the keys to the mix-everything machine. Different songs demand different approaches. The frustrating fact is that every multitrack production triggers so many ideas that we may not know where to start. Don't sweat it. You've got a lot of learning ahead, but a lot of fun awaits you as well.

Use this chapter as a map through the process so that you are never overwhelmed by the long list of creative and technical options that pop up in the course of any mix. All of the remaining pages in this book are meant to give you the technical depth, problem-solving prowess, and aesthetic context needed to form your own crisp mix strategy with the workflow momentum and artistic confidence needed to make orchestrated use of the broad array of tools that constitute your mix studio.

CHAPTER 3
Equalization

The most intuitive effect —
OK EQ is EZ
U need a Hi EQ IQ

MIX SMART QUICK START: Equalization
GOALS

- Fix spectral problems such as rumble, hum and buzz, pops and wind, proximity, and hiss.
- Fit things into the mix by leveraging of any spectral openings available and through complementary cuts and boosts on spectrally competitive tracks.
- Feature those aspects of each instrument that players and music fans like most.

GEAR

- Master the user controls for parametric, semiparametric, program, graphic, shelving, high-pass and low-pass filters.
- Choose the equalizer with the capabilities you need, focusing particularly on the parameters, slopes, and number of bands available.

Home stereos have tone controls. We in the studio get equalizers. Perhaps better described as a "spectral modifier" or "frequency-specific amplitude adjuster," the equalizer allows the mix engineer to increase or decrease the level of specific frequency ranges within a signal. Having an equalizer is like having multiple volume knobs for a single track. Unlike the volume knob that attenuates or boosts an entire signal, *the equalizer is the tool used to turn down or turn up specific frequency portions of any audio track.*

Out of all signal-processing tools that we use while mixing and tracking, EQ is probably the easiest, most intuitive to use—at first. Advanced applications of equalization, however, demand a deep understanding of the effect and all its possibilities. Don't underestimate the intellectual challenge and creative potential of this essential mix processor.

When you want more of something spectrally, EQ offers the solution. The successful engineer must also be able to make a related, more difficult judgment: how to identify when a mix element needs *less* of something. So don't be fooled. EQ is straightforward at first, but with its ability to both boost and cut, it requires careful thought, practice, and excellent critical listening skills while also demanding an accurate and informative studio-monitoring environment.

This chapter defines the functions of the user-adjustable controls found on most equalizers in the recording studio and then focuses on how the mix engineer uses EQ to support the art by various means of fixing, fitting, and featuring elements of the mix as desired.

3.1 TOOLS FOR SPECTRAL MODIFICATION

The equalizer is easy and intuitive to use, and when pronounced settings are applied it is easy to hear. A collection of equalizers—whether plug-ins or outboard gear—is likely to include a few variations on the user interface: parametric, semiparametric, program, shelf, graphic, as well as simple—but no less important—high-pass and low-pass filters.

3.1.1 Parametric EQ

The most flexible type of equalizer is the parametric equalizer, so called because it provides three adjustable parameters for altering the spectral shape of any audio signal. The other types of equalizers have just one or two of these three parameters, so mastering these three parameters makes understanding all other types of equalizers much easier. When you learn how to use a parametric equalizer, you are pretty much learning how to use all types of equalizers.

FREQUENCY SELECT

Perhaps the most obvious parameter needed is the one that determines the frequency range to be altered. The center frequency of the spectral region being affected is dialed up on a knob, switch, or slider labeled *frequency*. In search of bass, you might decide that the signal needs additional low-frequency content in the area around 80 Hz. Or is it closer to 40 Hz? These decisions are made at the frequency select control.

CUT/BOOST

▶ Having aimed for the desired frequency range, you determine how much to alter its amplitude by adjusting a second parameter called *cut/boost* or *gain*. This control determines the amount of decrease or increase in amplitude at the center frequency chosen via the frequency selection parameter mentioned previously. To reduce the shrillness of a brassy horn track, select a high frequency (perhaps around 8 kHz) and *cut* it by a small amount (maybe about 3 dB). To add a lot of bass, *boost* 9 to 12 decibels at the low frequency that sounds best, somewhere between 40 and 120 Hz perhaps. These two parameters alone—frequency select and cut/boost—give the engineer a terrific amount of spectral flexibility.

Q

! The parametric equalizer's third and final parameter is just a little trickier to understand. Consider a boost of 6 dB at 3,000 Hz. This step could be the EQ move needed to add presence to a ukulele track so that it becomes more distinct from the similar, competitive performance on the guitar. Choosing the amount of gain and the center frequency isn't enough to completely define this modification to the ukulele track's frequency content. Figure 3.1 demonstrates two possible results for the same frequency select and cut/boost settings. When a center frequency for this increase in level is selected, it affects not just that single frequency but also the neighboring frequencies as well. The degree to which it also boosts other frequencies nearby is defined by the third parameter, *Q*. The *Q* describes the frequency width of the cut or boost region. It is perhaps counterintuitive, but the industry standard is that a small value for *Q* indicates a wide spectral area being boosted. A large value for *Q* specifies a narrower, more specific area of influence. Engineers frequently say "low Q" and "high Q" to describe wide and narrow equalization settings, respectively.

3.1.2 Multiband EQ

Frequency select, cut/boost, and Q are the three basic, adjustable parameters needed to achieve almost any kind of alteration to the spectral content of a signal, from broad and subtle enhancements to pronounced and aggressive

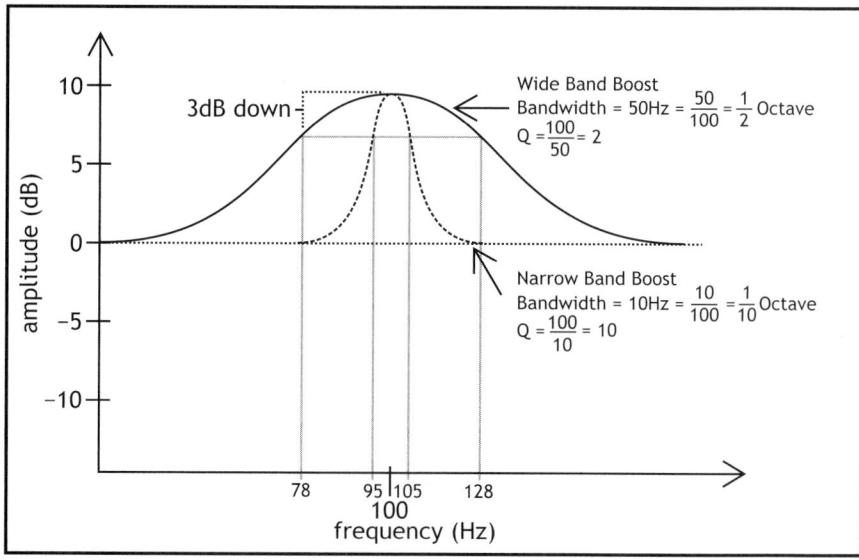

FIGURE 3.1
Any cut or boost at any frequency may be wide or narrow.

notches. This trio of controls makes up a single band of parametric equalization. A parametric equalizer may then offer several (usually three or four) of these bands, with overlapping frequency ranges. You coordinate the action of these multiple bands of EQ to achieve a single, beneficial result for the signal.

! A four-band parametric EQ has *12* controls on it: the 3 parameters × 4 bands = 12 controls in all! It offers the three parameters four different times so that you can select four different spectral targets and shape each of them with their own amount of cut or boost, and each with whatever unique Q is needed. In the frequency ranges where any two adjacent bands overlap, the equalization changes accumulate. Boosts overlapping with boosts lead to still more of a boost at those frequencies. Cuts overlapping with cuts behave similarly. When the boost of one parametric band overlaps with the cut of an adjacent band, they work against each other, and the net effect is simply the algebraic sum of the boost minus the cut. The larger magnitude wins.

The result, if our ears can follow it all, is the ability to drive a tremendous amount of change in the spectral content of a signal. Figure 3.2 shows one possible result of four-band parametric equalization. Clearly, there is no limit to the spectral reshaping an engineer can explore. The terrific amount of sonic shaping power that four bands of a parametric equalizer offer makes it a popular piece of gear in any studio. However, other useful EQ control sets exist as well.

3.1.3 Semiparametric EQ
Some equalizers set the Q internally, providing the mix engineer access to only the frequency select and cut/boost parameters. Because of this reduction from three user-adjustable parameters to two, this type of EQ is sometimes called a

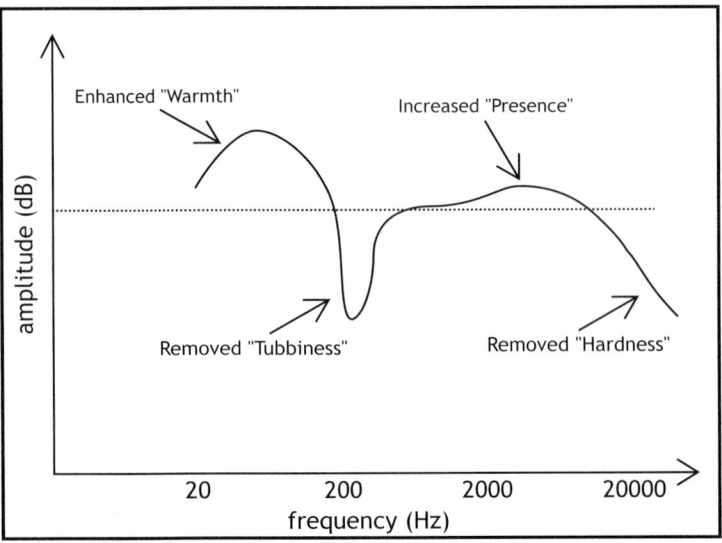

FIGURE 3.2
A four-band parametric EQ has great production potential.

semiparametric equalizer. You can't directly adjust the bandwidth—just the center frequency and the amount of cut or boost. Emphasizing the adjustability in the frequency domain, this type of equalizer is also sometimes called a "sweepable" EQ.

🔆 This configuration in which each band offers only two adjustable parameters is still quite desirable. With the removal of controls that might be distracting, the engineer can work faster and thereby stay creative without as much searching and fumbling for the right parameter. The simpler semiparametric design also represents an opportunity for the manufacturer to offer an equalizer at a somewhat lower price or to sell a higher-quality design at the same price. Make no mistake: the semiparametric is not simply a compromise, rather it is still a very useful tool in the sound recording.

🔊 A semiparametric equalizer's lack of a Q control doesn't mean the concept of a broad or narrow cut or boost is gone. There is still a region of neighboring frequencies affected by the selection of frequency and gain on a semiparametric equalizer. However, the spectral width of any EQ adjustment is set by the design of the equalizer, not by an adjustable control available to us as we mix. Through careful listening, we must develop a sense of the width of each make and model of semiparametric EQ we use. This knowledge influences our thinking when we reach for EQ in a future recording session. Over time, you must collect knowledge of the approximate bandwidth of all semiparametrics available to you. You'll choose Brand X, Model A over Brand Y, Model B because you know it to have the Q best-suited to your production needs at that instant—and these various production needs are discussed in detail later in this chapter.

🔊 Much fuss is made about the sound quality of certain equalizers. Some sound better than others, it is said. To be sure, the quality of the design, components, and manufacturing processes can have a noticeable effect on the sound quality of the device. But in the case of semiparametric equalizers, a critical distinguishing characteristic is the design-determined Q. Some music production situations leave us in need of a wide cut or boost—time to reach for a low-Q equalizer. Other production challenges are solved with a sharp notch to more surgically remove a troublesome resonant frequency. In this case, a high-Q processor is needed. The better-sounding EQ is the one that is right for the task at hand. Some semiparametric EQs have low Q; others have higher Q. Know your gear, and listen carefully. An equalizer ill-suited to one task might sound great on another. Accidentally using too narrow a bandwidth when a wide one was needed will sound disappointing. But it isn't fair to blame the EQ, categorically condemning it as a bad-sounding equalizer. When a production situation requiring a narrow Q arrives, this "bad-sounding" equalizer may sound perfect. We must defend ourselves from rumors and misinformation and must not be too quick to judge the quality of audio devices until we have had the chance to study, use, and hear them across a range of actual mix sessions.

❗ A semiparametric equalizer's lack of a Q control doesn't mean that this parameter is fixed. Some equalizer designs deliberately allow the Q to increase (narrower bandwidth) as the cut or boost becomes more extreme. The idea— and it's a good one—is that a small gain setting, say +/−3 dB or less, reflects an engineer's need for subtle reshaping of spectrum. A low Q accompanies such a gain setting. A larger gain setting, perhaps +/–12 dB, implies that more specific frequency ranges must be highlighted or removed. A high Q is offered. This useful design cleverly allows Q to increase as the absolute value of gain increases to extreme cut/boost settings. This feature—limited to only a few semiparametric designs—can come in handy.

In the hands of a thoughtful engineer, semiparametric equalizers offer great spectral shaping flexibility.

3.1.4 Program EQ

There is room for further simplification of the user interface on an equalizer. Sometimes we are offered control only over the amount of cut/boost and can adjust neither the frequency nor the Q of the equalization shape. Generally called *program EQ*, this is the sort of equalizer found on, for instance, your parents' stereo (labeled "treble" and "bass"). This type of EQ is also found on many consoles, vintage and new. It appears most often in a two- or three-band form: three knobs labeled high, mid, and low that are fixed in frequency and Q and offer the recording engineer only the choice of how much cutting or boosting they are going to apply. In the case of mixing consoles, remember that there may be the same equalizer repeated over and over on every channel of the console. If it costs an extra 20 dollars to make the equalizer sweepable, that translates into a bump in price of more than $600 on a 32-channel mixer. If it costs $100 to make them fully parametric, and it's a 64-channel console… now we are talking about a price increase equivalent to that of a really good used car. The good news is that well-designed program equalization can sound absolutely gorgeous. It often offers frequencies that are close enough to the ideal spectral location to get the job done on many tracks. Sometimes the engineer doesn't even miss the frequency select parameter; program EQ does the job just fine.

3.1.5 Graphic EQ

▶ A variation on the program EQ concept is the *graphic equalizer*. Like program EQ, this device offers the engineer only the cut/boost decision, with no user-adjustable controls for Q or frequency. On a graphic EQ, the several frequency bands are presented not as knobs, but as sliders, like faders on a console. The graphic EQ presents sliders from low to high frequency, left to right. The visual result of such a design is that the fader positions provide a good visual description of the frequency response modification that is being applied—hence the name "graphic." Handy also is the fact that the faders can be made quite compact. It is not unusual to have from 10- to 30-band graphic equalizers that fit into one or two rack spaces: compact, visually informative, and easy to use.

When the graphic equalizer lives in software, adjustments can be as simple as clicking on a frequency response plot and dragging the line into the shape that is desired.

Graphic EQ is an extremely intuitive and comfortable way to work. Being able to see an outline of the current frequency response modification makes it easier and quicker to achieve the spectral goal at hand. Turning knobs on a four-band parametric equalizer is more of an acquired skill.

There are moments in the course of a project when one must reshape the harmonic content with great care, using a parametric EQ. In other instances, there is no time for such careful tweaking, and a graphic EQ is the perfect, efficient solution. All engineers should plan to master both.

3.1.6 Which Is Best?

When building a studio and investing over time in the hardware and software needed to create ever better recordings, one naturally wonders: what type of EQ is best? Are parametric EQs better than semiparametric? Are graphic EQs better than program EQs? Life isn't that simple. All design types have the potential to be very-high-quality equalizers—or not.

It's a bit complicated. In all types of hardware EQ, we find models on which the knobs move smoothly and continuously across their available range of settings and other models, which have knobs that "click" to discrete values. Frequency select may be continuously sweepable from, say, 125 Hz to 250 Hz, or the knob may snap directly from 125 Hz all the way over to 250 Hz, offering nothing in between. If, in the latter case, you wanted the equalization contour to be centered on exactly 180 Hz, you are out of luck.

What seems to be a reduction in engineering control and flexibility, however, may offer an improvement in equalizer sound quality. Clicking a knob on the faceplate may physically select different electronic components inside the device. The equalizer is literally swapping components in the circuit path for different frequency selections. It isn't just adjusting some variable piece of the circuit; it is physically changing the circuit. In choosing which type of equalizer to acquire and use, engineers have to trade off sound quality versus price and processing flexibility versus ease of use. Some companies have such high standards for sound quality that they take away a little bit of user flexibility (continuous controls) to get a better sound. Conversely, if one finds an equalizer that is fully parametric and sweepable across four bands yet costs less than a large pizza, it would be wise to investigate how the manufacturer made the EQ so infinitely adjustable and how much sound quality was sacrificed in the name of this flexibility. Don't value an equalizer based on the number of controls it has. A simple program EQ with cut and boost controls might contain only extremely high-quality algorithms or components inside. The best studios acquire the gear that sounds best and keeps the engineer most comfortable.

Don't fall for the sexiest graphics or coolest user interface; what you see isn't necessarily what you hear. Don't be seduced by the power a multiband, fully parametric equalizer offers when it might only lead to confusion in the heat of a session and might represent some sacrifice in sound quality necessary to offer all of those knobs and switches. Listen carefully and consult experienced experts.

3.1.7 Shelving EQ

The most elaborate, feature-rich equalizer envisioned so far—the one with the most knobs on the faceplate—is the multiband parametric equalizer. With four bands of parametric EQ, the device provides us with 12 controls. There is room for still more processing flexibility. Here's how.

Each parametric band offers a region for spectral emphasis when boosting or deemphasis when attenuating. This shape is called a *peak/dip contour* because of the visual change it makes in the frequency response. Roughly shaped like a bell curve, it offers a bump up or down in the frequency response. Alternative EQ contours exist.

▶ The *shelving equalizer* offers half of a peak/dip response on one side of the selected center frequency, and a flat cut or boost region on the other (Figure 3.3). You choose a frequency to boost or cut, and the shelving equalizer achieves that

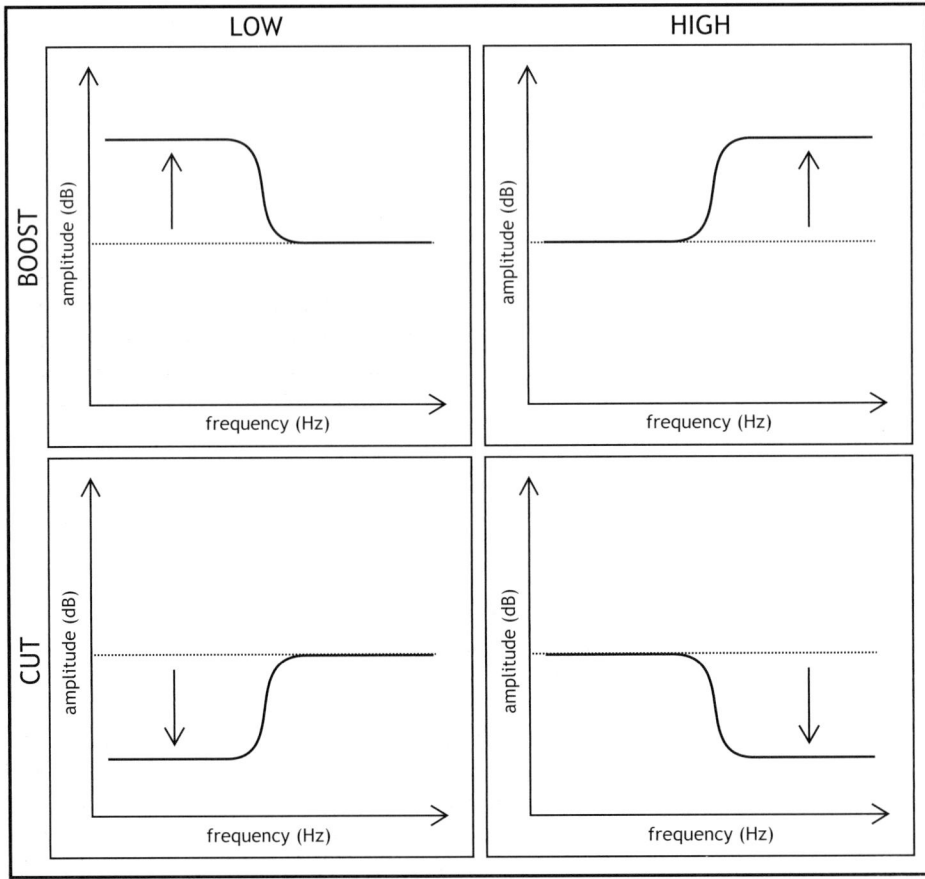

FIGURE 3.3
Shelving equalization.

and then holds the boost or cut beyond that frequency. A broad equalization desire might be to add brightness to the sound in general. A high-frequency shelving EQ bumped up 4 to 6 dB at 8 kHz will raise the output from about 8 kHz and above. It isn't limited to a center frequency and its associated bandwidth; rather, it raises 8 kHz and the frequencies above. The resulting alteration in the frequency response is flat (like a bookshelf) beyond the selected frequency.

As Figure 3.3 shows, the concept of shelving EQ applies to low frequencies as well as high, and to cuts as well as boosts. In all cases, there is a flat region beyond the selected center frequency—above the selected high frequency or below the selected low frequency—that may be boosted or attenuated.

3.1.8 Filters

Beyond the peak/dip and shelf, an important third option exists for reshaping the frequency content of a signal: the *filter*. Engineers speak generally about filtering a signal whenever they change its frequency response in any way. Under this loose definition, all of the equalizers discussed so far are audio filters. But to be more precise, a true filter must have one of the two shapes shown in Figure 3.4. A high-pass filter (Figure 3.4a) allows high frequencies through with no change in amplitude but attenuates lows. A low-pass filter does the opposite. A low-pass filter (Figure 3.4b) allows low frequencies to pass through the device without a change in amplitude, but high frequencies are attenuated.

▶ Because the sonic result can be rather similar to a shelf EQ cutting out extreme high or low frequencies, there is the potential for some confusion between them. Filters distinguish themselves from shelf EQ in two key ways. First, filters are attenuation-only devices. They do not deliberately raise the amplitude at any frequency. In contrast, shelf EQ can cut or boost. Second—and this is more important—filters offer a never-ending amount of attenuation beyond the selected frequency. They do not flatten out like the shelf. They just keep cutting and cutting. If there is some unwanted low-frequency air conditioner rumble on a track that you never, ever want to hear, a filter can offer significant attenuation. A shelf equalizer has a finite limit to the amount of attenuation it can achieve—perhaps only 12 or 16 dB down. The weakness of using a shelving equalizer in this case is easily revealed on every quiet passage whenever that track is being played, as it might still be possible to hear the air conditioner rumbling merrily along faintly in the background.

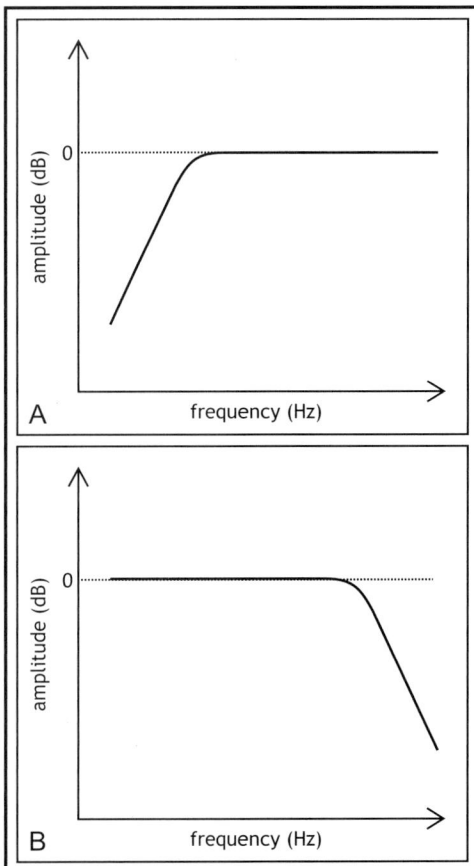

FIGURE 3.4
(a) High-pass filter;
(b) low-pass filter.

! Filters and shelving equalizers introduce some complexity to the faceplate of an equalizer. The four-band parametric (12 controls) gets upgraded with a low-pass and a high-pass filter at each end of the spectrum, as well as a switch that toggles the lowest parametric band from a peak/dip bell curve to a low-frequency shelf and the top parametric band from a peak/dip bell curve to a high-frequency shelf. Such an equalizer contains a rich amount of capability with which the engineer can freely alter the spectral content of any signal in the studio. These knobs and switches enable you to bend and shape the frequency response of the equalizer into almost any contour imaginable.

3.2 TECHNIQUE

With knowledge of what EQ can do for us, what are we to do with EQ?

3.2.1 Non-EQ Equalization

Before the details of applying EQ are discussed, it should be noted that many of the goals associated with equalization are realized though other means. If we are mixing somebody else's tracks, the spectral content of the audio provided to us is very much influenced by a range of factors beyond our control. The frequency makeup of each track was largely determined by the performer who played it and the engineer who recorded it.

When you are the tracking engineer *and* the mix engineer, things go more smoothly. You have a vision for the final mix and can track accordingly so that every incremental decision at the tracking session coaxes the tone towards that vision. Ideally, when you are at last ready to mix the project, no EQ will be needed. Realistically, we've done our job well if we can at least track things so that *less* EQ will be needed.

Less is better because the electrical circuit or signal-processing code associated with the equalizer isn't without sonic drawbacks. Phase distortion, noise, harmonic distortion, and worse sneak into the signal when we insert an equalizer. As our listening skills develop, our understanding of EQ matures, and our experience with recording and mixing deepens, we can loop back around through these decisions so that—from the beginning of the project—all relevant actions support the desired spectral outcome:

Equalizing without EQ

- Instrument selection
    - Instrument
        - Make
        - Model
        - Vintage
        - Size
        - Maintenance
        - Tone settings
        - Effects

- ■ Picks
- ■ Bows
- ■ Strings
- ■ Pickups
  - Type
  - Selection
- ■ Sticks
- ■ Mallets
- ■ Heads
- ■ Mouthpiece
- ■ Mute
- ■ Tuning/alternative tuning
- ■ Composition
- ■ Arrangement
- ■ Musicianship
  - ■ Chops
  - ■ Technique
  - ■ Intonation
  - ■ Embouchure
  - ■ Confidence
  - ■ Influences
  - ■ Strength
  - ■ Control
  - ■ Maturity
- ■ Performance decisions
  - ■ Left-hand and right-hand techniques
  - ■ Articulation
  - ■ Voicings
  - ■ Extended techniques
- ■ Room acoustics
- ■ Microphone selection/placement
- ■ Microphone preamplifier
- ■ Other effects, especially
  - ■ Distortion (see Chapter 4)
  - ■ Compression (see Chapter 5)
  - ■ Delay (see Chapter 7)
  - ■ Reverb (see Chapter 9)

💡 Navigating through all these spectrum-driving decisions toward the goal of getting the right tone for every track in the mix—perhaps even obviating any need for EQ—seems to require time travel. How can we make the right decisions during a guitar overdub regarding its ultimate place in the mix when we have yet to record the piano and the accordion and the mix session might be months away? Simple: we guess. We make an informed guess based on our experience in past projects. We try very hard to have a strong vision for the final mix from the very first recording session. And we chase that vision diligently, refining it as necessary as we learn more about the piece and as we welcome the surprise moments of inspiration and exploration that come from creative recording sessions along the way. To the best of our ability, we make all the decisions that influence tone in ways that lead us toward an appropriate goal, and we rely on things other than EQ to achieve it. You know you are getting good at EQ when you hardly use it. Leverage every line in the list above from the beginning of the project to make the final tone you desire at mixdown an inevitability from the first tracking session.

3.2.2 Patching and Plugging In

Equalizers are often found built into the console or DAW, available on each and every strip. Additional outboard and plug-in equalizers are typically incorporated into the audio signal flow using the *insert* of the mixer or digital audio workstation. In most applications, we use EQ to alter the signal from its current form into a better-sounding tone. We no longer want to hear the old sound. We don't want to hear both the improved spectral content and the old, non-EQed signal—just the equalized track please.

The insert is the patch that effectively forces the track through the EQ, altering its tone and giving us no chance to accidentally mix together both the equalized and the nonequalized versions of the sound. Alternatively, EQ can be patched between the multitrack return and the mixer line input, which effectively inserts the EQ into the signal flow. The use of auxiliary sends or buses is unusual for this type of effect, as we rarely need to send multiple different tracks to the same equalizer, and we don't care to return the equalized signal to the mix in parallel with the nonequalized signal (see Appendix A).

3.2.3 Boost, Search, and Set

The most common approach for finding a useful EQ setting is to *boost*, *search*, and then *set* the equalizer. *Boost* by a clearly audible amount, maybe 9 dB to 12 dB or more. *Search* by sweeping the frequency select control until the desired spectral region is audibly highlighted. Finally, *set* the EQ to the level and bandwidth needed. Either attenuate the gain and cut the frequency if it is unwanted or back off to a less radical amount of boost and fine-tune the bandwidth if it is a positive contributor to the mix. It's that simple.

Over time, through experience and ear training, you'll find that you can still be productive in this process while boosting by less than 12 dB, eventually skipping the boost and search steps entirely and instead reaching directly for the frequency range you wish to manipulate.

3.2.4 Anticipate

Early in your career, you should occasionally challenge yourself to first listen to the problematic tracks. Then, before touching any equipment, imagine—in your mind's ear—the frequency ranges in need of alteration. You reach for the frequency select first, not the cut/boost control, and aim immediately for those frequencies. With the frequency set, you raise/lower the gain and tailor the bandwidth and create—on your first try—the right alteration to tone. The goal is to skip the boost and search parts.

This premeditated approach to equalization is the method of the experienced engineer. The less-experienced engineer should try this from time to time during low-pressure recording sessions. In fact, audio listening skills develop more quickly when we discipline ourselves to make an informed guess of the target frequency without boosting and searching first. Learning from mistakes made this way will help you develop a deep feeling for the qualities of different spectral regions.

However, until you've had several years in the mix position, know that there is no harm in taking advantage of those first two steps. Doing so is fast, relatively easy, and has the extra benefit that the producer and any band members who might also be in the control room will be able to follow along sonically and be supportive of your EQ decisions. Even the famous, expensive engineers resort to the boost, search, and set approach on occasion. Remember, the music-buying public doesn't care how it is done—just that it sounds great.

3.2.5 Improve

! Three things slow our progress as we become proficient with equalization. The first challenge is the sheer range of possibilities that EQ offers. Using boosts and cuts, high Q to low Q, peak/dip, shelf, or filter, the humble effect known as EQ represents a nearly limitless range of options from 20 Hz to 20,000 Hz. Such wide-open possibilities can paralyze. Don't get lost wandering among the spectral possibilities with EQ, waiting to see/hear if anything interesting happens; seek out specific motivations to equalize a signal, as we discuss in the following section ("Mix Strategies—EQ").

🔊 Second, learning how to hear with high resolution isn't easy. Critical listening skills are developed over a lifetime and require careful concentration, good equipment, and a good monitoring environment. It is straightforward to distinguish the difference between 1 kHz and 10 kHz and even between 1 kHz and 4 kHz. But can you easily hear the difference between a 6 dB boost at 1 kHz versus a 6 dB boost at 2 kHz? What if it's just a 2dB boost? How 'bout 1 kHz versus 1.2 kHz? No one learned the difference between 1 kHz and 1.2 kHz overnight. It's a journey of practice and increasing resolution. Take ear training seriously.

Third, interfering with this already challenging learning process is the temptation to imitate others or repeat equalization moves that worked on the last song. "Magic" settings that make every mix sound great simply don't exist. If a rookie got the chance to write down the equalizer settings used on, say, Stevie Ray Vaughan's guitar track on *Pride and Joy*, it might be tempting to apply it to some other guitar track, thinking that this unique equalizer setting goes a long way toward improving the tone. But the fact is, the tone of Stevie's guitar is a result of countless factors: the playing, the tuning, the type of strings, the kind of guitar, the amp, the amp settings, the stomp boxes, the placement of the amp within the room, the room, the microphones used, the microphone placement, the tape machine, the tape formulation, the tape calibration, and so on. The equalizer alone doesn't create the tone. In fact, in the scheme of things, EQ played a relatively minor role in the development of Stevie Ray Vaughan's tone. Those many other factors have a much bigger influence. So repeating EQ moves that worked in the past simply will not ensure a good sound today—they are pretty much irrelevant.

The way to take out the guesswork and gain fluency with this infinitely variable, difficult-to-hear thing called EQ is to develop a *process* that provides a crisp strategy for when and how to equalize a sound. We do that next.

3.3 MIX STRATEGIES: EQ

When does a recording engineer boost, search, and set? What are they listening for? Why and when do they equalize? EQ is simple in concept but difficult in application. Some important words of comfort: all engineers have a lot to learn about EQ. All engineers—apprentices, hobbyists, veterans, and award winners—are still exploring the sonic variety and musical capability of equalization.

As discussed in Chapter 1, distilling the infinite possibilities available to an engineer into specific actions to take while mixing can be made much simpler by focusing efforts into the three categories of the smart mix mindset: the fix, the fit, and the feature.

It is a mistake to insert an equalizer on the vocal just because the vocal "is supposed to be EQed." The use of EQ on a vocal depends on the qualities of that vocal track that might need spectral adjustment. And there is certainly a chance that no EQ is needed at all. When the vocal could benefit from equalization, what specifically is to be done when the EQ is inserted? A deliberate strategy should motivate the use of EQ on vocals or any other track. Mix engineers don't EQ just for the sake of EQing. They EQ to solve problems and support the music. The challenging task of how best to solve any problems and somehow support the music is made easier when we cycle through three questions:

- What, if anything, needs to be *fixed* using EQ?
- Can EQ be used to help the various elements of the multitrack mix *fit* together better?
- Do any tracks need some aspect or sound quality *featured* through EQ?

3.3.1 Fix

A major motivation for engaging an equalizer is to correct things spectrally, getting rid of problems that lie within specific frequency ranges. For example, outboard equalizers, console channel inputs, microphone preamplifiers, and even microphones themselves often have low-frequency roll-off filters. Their prevalence suggests a common need for a high-pass filter. These devices remove low-frequency energy less for creative "this'll sound awesome" reasons and more to fix the common problems of rumble, hum, buzz, pops, wind, and proximity effect. We'll take these in order.

RUMBLE

In many recording situations, the microphone picks up a very-low-frequency (40 Hz and *below*) rumble. This low-end energy likely comes from the building's heating, ventilation, and air conditioning system, or the vibration of the very structure of the building due to traffic on nearby highways and train tracks. (Note to self: Don't build a studio next door to major freeways, train stations, or helipads.)

! Rumble can be really-low-frequency stuff, so low that singers and most musical instruments are incapable of creating it. Rumble is often so deep that many loudspeakers—even expensive professional-tier models—can barely reproduce it, if at all. Because the loudspeakers won't reproduce it, rumble can be difficult to notice from the comfort of the control room. To keep it out of the audio tracks, rumble must be carefully diagnosed in the recording room by the tracking engineer.

Because it is so difficult to hear and nearly impossible to reproduce, it is appropriate to wonder whether removing rumble is worth the trouble. If the rumble is easily overlooked, can't the problem be ignored? There are two motivations for keeping deep rumble out of our recordings: dynamic range and production standards.

💡 The dynamic range issue is straightforward. Just because music listeners can't easily hear rumble doesn't mean that the audio system doesn't use up precious dynamic range in its effort to process the signal. In fact, an inaudible low-frequency signal might drive an audio device into distortion. The listeners at home may not hear the cause, but they will hear the effect. The distortion creates upper harmonics that are easily within the more audible mid and high frequencies (see Chapter 4). The audio engineer's careful use of the dynamic range of all of their audio devices throughout the whole recording process is corrupted by the presence of unwanted, low-frequency rumble. When we record a vocal, snare drum, guitar, or whatever signal a session presents, we want to raise the level of the signal well above the noise floor of the system but not so much that the device runs out of headroom and begins to clip. Rumble will use up some of that precious headroom, causing us to run into overload at lower signal levels. To prevent rumble from causing overload at any stage—within the microphone, the preamp, the console, any signal processors, the storage device, the storage medium, the power amplifiers, or the monitors—it should be prevented, attenuated, or removed if possible.

Regarding production standards and rumble, is it really acceptable to allow rumble into your recorded production knowing that maybe 80 or 90 percent of listeners won't notice it because their systems and listening environments won't reveal it? Most audio engineers have a natural desire to look out for that 10 or 20 percent of other listeners who will hear it. If you set your production standards high enough to satisfy the pickiest listeners and to far exceed the expectations of the rest, then you share the audio standards of most successful engineers. Rumble is unwelcome because it undermines those high standards.

🔊 To prevent problematic rumble, recording engineers listen carefully to the recording space even as the band is being situated and the microphones are being selected and placed. During the sometimes-chaotic time of getting ready to record, we must assess the noise floor of the recording space. If a deep sound is heard and/or possibly felt (in the chest cavity or through the feet), try to identify the source and, if possible, to stop it. It is not uncommon in some facilities and especially in some on-location recording sessions (e.g., in symphony halls or houses of worship) to turn off the heating or air conditioning and/or plan to use one of the bass roll-off features of your recording chain (on the microphone, the microphone preamp, the console, or an equalizer). Because very little music happens at such low frequencies, it is often appropriate to insert this high-pass filter to remove most of the problematic, very-low-frequency energy. The intent is for this filter to address *mechanical* problems in the room and have little to no effect on the *musical* signal sitting higher up the frequency axis.

! Listen carefully, however. Filters can have a noticeable effect on the signal far above the cutoff frequency. Compare the signal with and without the high-pass filter engaged. Try not to be seduced by the obvious benefit of rumble removal and pay careful attention to the audio signal, even though it may be well above the filter frequency. If the signal isn't diminished in quality, the rumble removal is a success. Unwanted attenuation of nonrumble portions of the signal and phase distortions that blur the transients of the signal, removing realism and detail, are a potential side effect of these filters. Try a small set of different filters that you may have available and choose the one that attenuates the rumble while doing the least damage to the audio that remains in the frequencies above.

That's the rumble fix. A slightly different problem is hum.

HUM AND BUZZ

Hum is typically the interference from our power lines and power supplies that is born from 60-Hz AC power (or 50-Hz power, depending on your locale). The alternating current in the power provided by the utility company often leaks into our audio through damaged, poorly designed or failing power supplies and audio interconnects. It can also be induced into our audio through proximity to electromagnetic radiation of other power lines, transformers, electric motors, light dimmers, frozen margarita machines, and such. As more harmonics above 60 Hz appear—120 Hz, 180 Hz, and 240 Hz—the hum blossoms into a full-grown *buzz*. Buzz finds its way into almost every old guitar amp, helped out a fair amount by fluorescent lighting and single-coil guitar pickups.

💡 The best solution is to stop the hum or buzz by identifying and removing the cause. This removal can be done through steps as simple as turning off all lights on dimmers, turning off all fluorescent lights, or asking the guitarist to move or turn slightly. Sometimes plugging the equipment into a different outlet helps. On the other hand, chasing hum and buzz problems can point to more significant, difficult solutions that can't be implemented during a recording session: rewire the patch bay, redesign the power distribution and grounding for the entire recording studio, modify the equipment, and so forth. Such solutions require deep knowledge of audio circuitry and power and grounding. These solutions can be dangerous and expensive if done incorrectly. Clearly, a specialist is needed. There is no single cure.

▶ When the source of the hum can't be found, and therefore the cause of the hum can't be directly stopped, a high-pass filter helps. To remove hum, engage a roll-off filter—a high-pass filter—starting at a frequency just above 60 Hz (50 Hz) or perhaps an octave above, at 120 Hz (100 Hz). Although rumble is so low that this is not a big problem, filtering out hum requires a filter that is high enough in frequency that it can audibly affect the musical quality of many types of sound. Exercise care and listen carefully when filtering out hum. Many instruments (some vocals, most saxophones, and a lot of hand percussion, to name a few examples) may not be changed much sonically by a high-quality filter. But low-frequency-based instruments (e.g., kick drum, bass guitar) aren't going to tolerate this kind of spectral alteration. The lowest note on a four-string bass has a fundamental frequency of 40 Hz, clearly below the hum frequency. To filter out the hum is to filter out some spectral portion of the bass guitar tone. The lowest note on a six-string guitar has a fundamental of about 80 Hz, frustratingly close to the fundamental hum frequencies. Fortunately, the hum might be less noticeable on a guitar, as the performance can mask a low-level hum.

When hum and buzz can't be prevented and must be removed by EQ, it is worth noting that the spectral components are rather discrete. They are born of pure-tone sine waves. A 60-Hz hum is not a complex sound. It has energy at 60 Hz (or very close to it), and that's it. Buzz built on 60-Hz hum will retain its spectrally discrete character. It will have energy at 60 Hz and multiples thereof. It contributes no noise in between these frequencies, which suggests that very-narrow-band notch filters placed at these frequencies might remove the audible artifacts of hum and buzz yet allow the music through in and around these notched frequencies.

It's never comforting to introduce deep spectral cuts in your favorite tracks. Notching out the problem is our last resort. Clearly, it would be far better to fix the hum and buzz at the source.

POPS AND WIND

🔊 Other low-frequency problems fixed by a high-pass filter are the woofer-straining pops of a breath of air hitting the microphone whenever the singer or voiceover artist articulates a plosive sound. One must pay attention to words containing a "P" or a "B." A similar problem can occur whenever your recording work takes you outdoors (gathering news, doing live sound in outdoor venues, or collecting natural sounds in the field). Any breeze across the microphone isn't just bad for the microphone. It creates unwanted, distracting, dynamic-range-consuming, low-frequency noise. The typical defense is a pop filter and/or windscreen. These devices attempt to keep breath and breeze off of the capsule allowing the desired audio signal through. In fact, pop filters and windscreens are acoustic high-pass filters. As with the insertion of filter circuits, listen carefully to the pass band—the portion of the audio signal that is supposed to be above the filter cutoff frequency. Windscreens and pop filters may have side effects in the mid and high frequencies as they take care of the low-frequency problems caused by pops and wind. Try several different ones and listen for the best low-frequency control with the least mid- and high-frequency compromise. If these acoustic devices can't keep the wind off the microphone sufficiently, then filter the low frequencies out with a low-frequency roll-off switch, preferably at the microphone itself.

PROXIMITY

When the instrument being recorded is positioned very near a directional microphone, an increase in low-frequency content caused by proximity effect appears. Sometimes this is sonically beneficial. Radio DJs love it—it makes them sound larger than life. Sometimes the low-frequency boost is just too much.

▶ Place a directional microphone about 6 inches or less from an acoustic guitar, and the resulting sound will have a distracting, low-frequency thud associated with each full strum of the guitar. This thud generally sounds unpleasant and can mask or interfere with the otherwise great sound of the rest of the captured tone of the instrument. If possible, reduce or remove the unwanted proximity effect by backing the microphone away from the instrument or switching to an omnidirectional microphone (omnidirectional microphones based on pure pressure transduction do not suffer from proximity effect). Alternatively, reach for a filter. Roll off some low end with a gentle high-pass filter and tame the unwanted proximity effect.

HISS

Not all spectral headaches live in the low frequencies. The noise floor of many devices in the studio—most especially from microphones, microphone preamplifiers, and analog magnetic tape—is often random energy, or a signal containing a broad range of frequencies. Such white noise is hopefully so low in level that it is rarely noticed. If a track has a range of dynamic levels from very loud to very soft, with any periods of silence in between phrases, the low-level noise may become audible and distracting.

The spectral character of white noise points to a partial fix. Though it may contain energy across the entire audible spectrum, white noise (the noise of many simple electronic components) is perceived as having a bias toward the higher frequencies. It sounds bright, possessing more energy in the high octaves than the low octaves (versus pink noise, which contains equal energy per octave).

▶ A low-pass filter, or high-frequency shelf set up for gentle attenuation of the highs, might be coaxed into settings (high-frequency, gentle filter slope or low Q) that attenuate enough highs to make the noise less distracting without rendering the audio signal itself too dull. The hiss isn't removed, but it is substantially reduced.

GENERAL FIXES

The list of "fixes" motivating the use of filters and equalizers goes on. Any undesirable quality of any sound that we can attribute to a specific frequency range may be addressed through equalization. The most common problems were discussed earlier, but other challenges will appear. To fix these spectral problems with EQ, you need to learn the spectral qualities of all the signals that you encounter. It's daunting at first, but with focus and practice, you can learn the major spectral features of the instruments you record and mix most.

Contemporary pop production is strongly oriented toward some common instruments: vocals, drums, bass, guitar, and piano, and so on. We discuss the major instruments in detail later in this chapter, but it should be noted that with each new track you encounter in your multitrack mix, you will scan it (through critical listening) for isolatable frequency problems in need of correction. Each problem found may motivate a specific use of EQ to fix it. As you develop experience with these popular instruments, you will encounter some common problems that repeat themselves.

💡 Ever faced a snare drum with an annoying ring? In some tunes, the ring sounds great. In others, such a snare sound is decidedly unpleasant. If the ringing tone is unwanted, it is of course best to fix it at the drum. Before recording the music, dampen or retune the drum. All too often, however, the mixing engineer is not the same person as the tracking engineer. We can inherit problems not fixed at the earlier recording sessions. In this situation, we simply find the frequency range most responsible for the ring (boost, search, etc.) and try attenuating it (as much as 12 dB or more) at a narrow bandwidth.

What qualifies as a narrow Q? Greater than 4. Many equalizers don't offer a Q any sharper. Rare EQs will offer a Q of 10 or greater. These are powerful tools for this kind of EQ approach. A high-Q notch at the troublesome frequency that is wide enough to remove the unwanted ring but narrow enough not to diminish the rest of the instrument's tone is ideal. Often, turning down that ring reveals an exciting snare sound across the rest of the frequency range.

Ever track a guitar with old strings? Dull and lifeless, it is unlikely that this is fixable, but don't rule it out until you've tried a bit of a boost somewhere up between 6 kHz and 12 kHz.

▶ Sometimes a gorgeous spectral element of a sound is hidden by some other, much-less-appealing frequency component. A good example of this can be found in drums. Big-budget commercial releases often have wonderfully powerful, punchy drum sounds. Yet when the less-experienced engineer goes searching for the correct low frequency to boost on their lower-budget projects, it never quite sounds right. The low-frequency stuff that makes a drum sound punchy often lives just a few hertz lower than some rather muddy energy. Boosting the lows invariably boosts some of the mud.

The solution is to keep the low-end power but remove the mud. Search at narrow bandwidth for the ugliest, muddiest component of the drum sound (likely somewhere between about 200 and maybe as high as 400 Hz for kick and toms) and cut it. As this problematic frequency is attenuated, listen to the low end below the mud-range. Often this approach reveals plenty of low-end power and punchiness that just wasn't audible before the well-placed cut was applied. This application highlights low-frequency elements of the sound by applying a strategic EQ cut at a frequency range just above it.

🔊 Removal of a narrow band somewhere within this often undesirable 200- to 400-Hz "mud-range" must be done carefully. Cutting too wide an area or attenuating too deeply can rob the instrument of fullness.

This effect must be used sparingly as well. Applying it to all tracks leaves the entire production thin and powerless. This spectral region must have some energy somewhere in the mix. Save this frequency range for the vocals, guitars, strings, and/or keys but consider pulling it out of some of the drums (e.g., kick drum and tom-toms).

In the instrument-specific discussions that follow, you'll see that we identify some parts of the sound that might—but won't always—be considered a negative attribute. Squeaks, thumps, clunks, clicks, clatter, and other mechanical noises, excessive breathiness, and so on need your thoughtful attention. If they aren't adding to the sound is some way, identify their dominant general spectral locations and try to attenuate them with an ear for the potential impact that this noise has on other desirable sounds in the same spectral area.

3.3.2 Fit

A key reason to equalize tracks in multitrack production is to help fit all these different tracks together. One of the simplest ways to bring clarity to a component of a crowded mix is to get everything else out of the way spectrally. The presence of one signal in some specific range of frequencies can make it difficult to hear any other signals at that same range of frequencies. EQ is one of the key tools that a mix engineer uses to minimize these spectral conflicts.

COMPLEMENTARY CUTS AND BOOSTS

Wanting listeners to be able to hear the acoustic guitar while the synthesized string pad is sustaining, you might find a satisfyingly present midrange boost for the guitar and perform a complementary cut in the mid frequencies of the string pad. This EQ cut on the string pad keeps the sound from competing with or drowning out the acoustic guitar.

In this way, we piece together a spectral jigsaw puzzle. The trick is to find a spectral range that highlights the good qualities of the guitar without doing significant damage to the tone of the synth patch when subjected to the necessary cut. It'll take some trial and error to get it just right, but this approach allows clever recording engineers to layer several details into a mix.

▶   This EQ strategy often motivates us to simply narrow the frequency range of some tracks. Consider a crowded mix, with 48 or more tracks of audio. With so much going on, it is difficult to create a mix in which the various tracks aren't competing with each other and undermining each other, spectrally. There is room for some aggressive equalization. Maybe a piano track is one contributor to a particularly full part of the musical arrangement, when guitars, strings, synths, and background vocals have all joined in. When this sort of everybody-play-at-once crescendo happens in a pop production, the piano might be treated to a rather severe EQ. A low-frequency shelf might be engaged to attenuate everything from about 200 Hz and below by a solid 12 dB or more. So radical an EQ move may strike you as surprising at first. Solo the equalized piano track and it likely has a low-quality, cheap, thin sound. Listen to the piano in the context of the mix, however, when the drums, bass, guitars, strings, synths, vocals, and so on are all playing along, and the piano sounds, well, like a piano. With so much else happening, the listener's experience seems to fill in or assume the low-frequency energy is there.

So fitting tracks together with complementary boosts and cuts can happen at even a crude level. It's not always a complicated jigsaw puzzle of spectral shapes. Sometimes you simply lop off a broad portion of the lows or the highs and let the remaining parts of the production fill those spectral spaces. Shelf EQ and filters are great for getting this job done.

Expect to apply this thinking in a few critical areas of the mix. Around the bass guitar, engineers almost always encounter low-frequency competition that needs addressing. Anyone who plays guitar or piano and does solo gigs as well as band sessions has likely discovered this already. Performing alone, the musician has low-frequency responsibilities as he or she covers the bass line and pins down the harmony. In the band setting, on the other hand, musicians are free to pursue other chord voicings, such as leaving the root of the chord out entirely, or inverting the chord so that the root is an octave or two higher—well out of the way of what the bass player is doing. The guitarist and the pianist don't want to compete with the bass player musically, and the same is true spectrally, which means that the engineer might be able to pull out a fair amount of low end from an acoustic guitar sound to make room for the bass track. Alone, the acoustic guitar might sound too thin, but with the bass guitar playing, all is well. There is spectral room for the low frequencies of the bass because the acoustic guitar no longer competes there. But the acoustic guitar still has the illusion of being a full and rich sound because the bass guitar is playing along, providing uncluttered, full bass for the song and for the mix.

🔊   In the highs, competition appears among the obvious high-frequency culprits like the cymbals, hand percussion, and steel-string acoustic guitar, as well as the not-so-obvious: distorted electric guitars. It is always tempting in rock music to add distortion to guitars, vocals, and anything that moves. Spectrally speaking, this kind of distortion (see Chapter 4) leads to the addition of some upper harmonic energy. This harmonic distortion will overlap with the cymbals and any other distorted tracks. You get a spectral pileup. Make them fit by using the same complementary EQ moves. Maybe the cymbals are informally allocated the highs above 10 kHz, the lead guitar has emphasized distortion around 8 kHz, and the rhythm guitar hangs out at 6 kHz. Mirror-image cuts on the other tracks will help ensure that all these high-frequency instruments are clearly audible in the mix.

💡 Fitting together the spectral pieces this way requires us to focus on what we *can't* hear. When we delete—or at least diminish—the parts that are perceptually less important, our mix gets a lot easier. With clutter and conflict removed, the other tracks become easier to hear. We better reveal additional sophistication and detail in the mix.

We don't always seek out spectrally distinct tracks. We might allow spectral overlap in an effort to create a *metainstrument*. These are built from the deliberately vague blending of complementary, similar-sounding tracks. The most common example is the way we often connect the kick drum attack to the electric bass sustain. Some drummers and bass players are so tight, musically, that we mix them into a single instrument. The performers know each other so well, and they listen to each other so intently, that almost every kick played by the drummer's foot is accompanied by a perfectly timed note from the bass player's fingers. We mix them as one, for a brief full sustain in the low end punctuated by punchiness in articulation—a concoction possible only through the coordinated combination of bass guitar plus bass drum. The casual listener may not notice—and arguably doesn't care—that it comes from two different instruments. They just like the sound, and feel the beat.

Use a similar philosophy for merging a gated room sound with a close microphone snare hit (see Chapter 6 on expansion and gating). As mix engineers, we know that it takes many tracks and a lot of finesse to create the overall metasnare sound, but the end user just hears it—and enjoys it—as a thrilling, moving source of tempo and energy.

Take it even further and combine any instruments into a logical new sound, perhaps an electric bass plus a very low, left-hand piano part plus a bit of lower brass. If the performances are consistent enough to act as one, you can deliberately unite them into a new texture—a single new sound fusion. It might sound like electric bass, but with a twist, an added inexplicable. Or we might take it to an extreme, and the bass sound might now have a tone like no other. It's clearly not an electric bass guitar, nor is it a piano, and it's certainly not a tuba. The bass/piano/tuba metacombination has the aural presence of some sort of synthesized bass sound that is perfect for the tune, even though the casual listener may have no idea what instrument was played.

As mix engineer, we must consciously control this. Some tracks are sculpted, panned, and processed so that they can make their uniquely identifiable contribution to the overall mix. Other tracks are creatively blurred and obscured, tucked into and around other tracks for a more subliminal role.

We seek to please the end user through any and all means appropriate. Delete what they can't hear, convert obscurities into opportunities, and recognize that although we can audition the individual multitrack pieces, the end user hears only the final, artistic whole.

SEIZING THE SPECTRAL OPPORTUNITY

Boosting one track while applying complementary cuts in a spectrally competing track is a proven EQ strategy, and it reveals the logic of another, preferred strategy. The better approach is to listen for and identify the existence of half of this process. That is, once your critical listening skills have gained sufficient confidence, you listen for a spectral opening (e.g., a preexisting lack of energy in the string pad at 4,500 Hz) and seize the opportunity with the mix-enhancing boost at that frequency on the acoustic guitar.

Not every boost on one track requires a symmetric cut on another track. Good arrangers and orchestrators know how to piece all the instruments together for a natural fit. The producer and recording engineer must learn this too, using equalization only as needed.

The mid frequencies are definitely the most difficult spectral region to sort out. It is very competitive space spectrally as almost all instruments have something to say in the mids. When listening to full-range music, it can be the most difficult place to hear in detail. Less-experienced engineers tend to gravitate toward the more obvious low- and high-frequency areas when they reach for the equalizer. On the road to earning golden ears, plan to focus on the middle frequencies as a key challenge and learn to hear the subtle differences that live between 500 Hz and 6,000 Hz. It's much easier to fit all the pieces of the spectral puzzle together when you can exploit the full potential of the lows, the mids, and highs.

3.3.3 Feature

A natural application of equalization is to enhance a particular part of a sound in order to bring out the positive spectral components of the sound. What are the positive spectral components of a sound? That is a terrifying question—at first.

To be a good mix engineer and use EQ to highlight the necessary features of each and every track so that the resulting mix serves the music best, you must have a point of view regarding what sounds good and what sounds bad. It is subjective, to be sure, but it is an opinion informed by deep experience with music, frequent encounters with moments of exquisite sonic beauty, and necessary, accidental exposure to mistakes of gloriously extreme magnitude.

Mix engineers take the time to study their art so that they can navigate the decisions of what "sounds good" and what "sounds bad," and be so good at it that artists and producers hire them to express their opinion and mix their work. A mix engineer needs to remain an avid music fan, no matter how many times the clinical, technical side of engineering rears its head. A mix engineer must seek out and listen to a vast range of recordings, making time for critical analysis of why—to them—one piece of art is better than another. The successful mix engineer grows to rely on a community of equally passionate colleagues with whom they share music, ideas, and opinions. All great engineers must have periods of eating, drinking, sleeping, and living all things audio. It mustn't be allowed to take over your whole life—in fact, family and friends will remind you to turn off your intense audio side every now and then. But if you are ever going to get good at this—really good at this—some controlled obsession is going to help. And it is in those moments of controlled obsession that you will take on the challenge of adding to your ever-expanding, ever-deepening understanding of the spectral components of the most important tracks in the mix.

The detailed instrument-by-instrument discussion and data that follows should be digested slowly. If necessary, come back to this section later, when a session challenge requires you to tackle some specific detail of a particular instrument. Know that this information is typically absorbed over years of practice, not in a single sitting.

No two guitars, no two drums, and no two singers are ever exactly alike. Issues of design, manufacture, tuning, performance, and so on all conspire to make precise spectral knowledge of all types of instruments in all styles of music impossible. Yet some spectral trends and traits are fairly universal, plus or minus a few hundred Hz and plus or minus a few dB.

💡 More important than the instrument-by-instrument discussion and data is the *process*: the idea of building a mental template of key spectral features of the instruments. They don't have to be exact. You always listen carefully to each track as its own unique entity. But the mental template remains a good starting point.

The following information is introduced as a way of thinking about the tracks you wish to EQ, so that you can get your head around the challenge of deciding what attributes need to be featured in your mix.

Start with the instruments you play or the instruments played by the clients you work with most. Work your way out from those instruments to similar ones—from guitar to bass, from piano to other keyboards, from female vocals to male vocals, and so on. Feel no obligation to commit every detail to memory. The goal is to build an easy-to-remember oversimplification of the major spectral features of every instrument you encounter:

- The fundamental frequencies of the playable range of the instrument
- A general sense of the harmonic content that accompanies these fundamentals
- Any unique spectral landmarks associated with the instrument
- The spectral qualities of any mechanical noises or artifacts

- Frequency ranges that might be strongly associated with the beginning and the end of notes
- Any particular special performance techniques that radically alter the spectral content, and more

Starting with these features, you are strongly encouraged to build your own templates.

VOICE

▶ Really low male voices reach a fundamental frequency just below 100 Hz. Female voices start about an octave higher, around 200 Hz. (See Figure 3.5.) Think of vocals as a two-part signal: sustained vowels and transient consonants. The vowels happen primarily at lower mid frequencies (200 to 1,000 Hz) and the consonants happen at the upper mid frequencies (2 kHz and up). With the possible exception of that singer from REO Speedwagon in the early 1980s ("I meant that I love you for everrrrrrrrrrrrrrrrrrrrrrrrrrrr"), most singers sustain the vowels and move rather quickly through the consonants. For a richer overall tone to the voice, manipulate the vowel range. Having trouble understanding the words?

FIGURE 3.5
Frequency range of fundamentals and spectral landmarks (vocals).

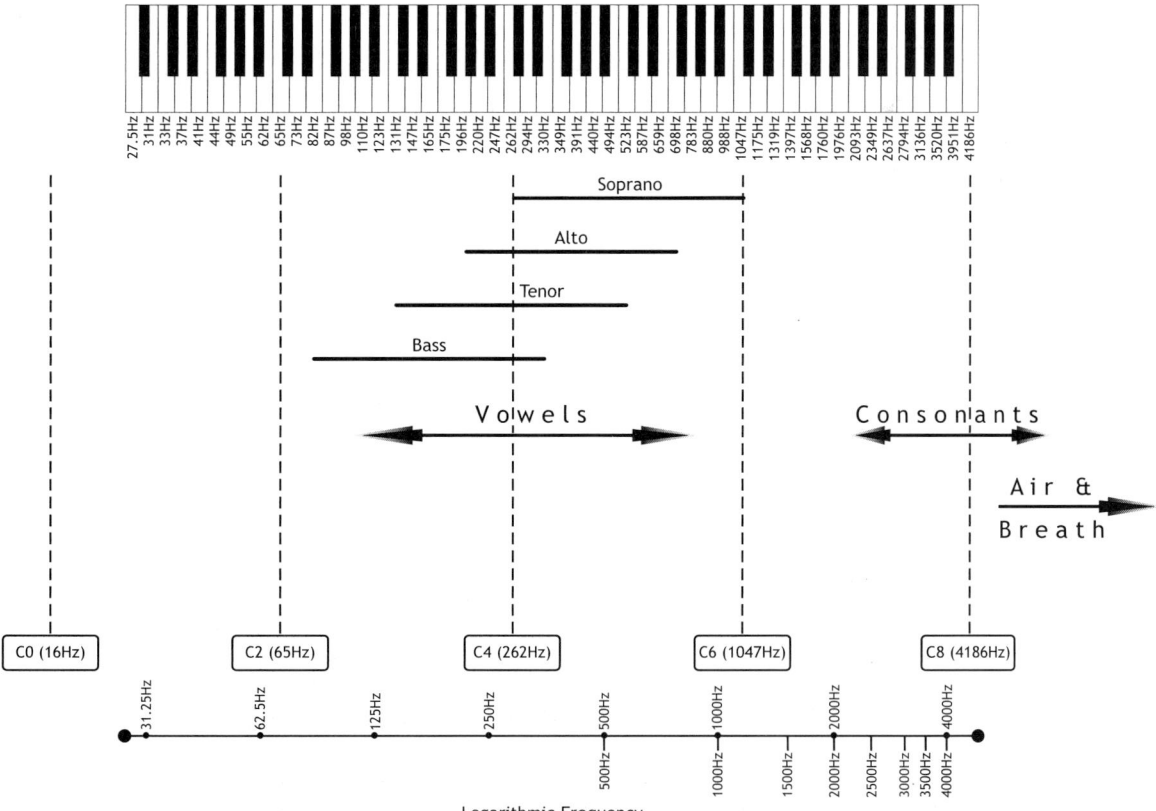

Enhance a bit of the consonant range with a modest, typically less than 6-dB boost, with low Q, in the frequency range that sounds best for that singer, likely somewhere in the 2-kHz to 6-kHz range.

! For added intimacy and human fragility, perhaps bring out some air and breathiness through a lift at the very high frequencies, a 3- to 6-dB boost using a very high-quality shelving EQ at 10 kHz, maybe 12 kHz if the EQ sounds good that high. A side effect of this emphasis of air is the creation of overly sizzling sibilant "S" sounds. In between the "Ss," the voice sounds brilliant with this EQ move. But the high-frequency boost makes each "S" into an aural assault of too much high-frequency energy. The solution is to momentarily attenuate these overly loud "S" sounds by using a De-Esser, a compressor with a bit of side-chain processing (see Chapter 5), leaving us free to emphasize some of the human expressiveness of the singer taking a big breath right before a screaming chorus.

SNARE DRUM

One mental template for the snare drum is to think of it as a sudden, short burst of broadband noise. Snare can be difficult to EQ in isolation, as it reacts to almost any spectral change. The context of the other tracks will be essential when choosing spectral regions for emphasis/deemphasis.

A starting point might be to divide the snare sound into two parts (Figure 3.6). One is the lower-frequency energy coming from the drum itself—the sound the snare would have if it were just a tom, without the snares stretched across the bottom, resonant head. Second is the mid- to high-frequency energy up to 10 kHz and beyond due to those rattling snares. Narrow the spectral possibilities by looking for power in the drum-based lows and exciting, raucous emotion in the upper mid frequencies of the noisy snares. Vocals, guitars, pianos, strings, and reeds are going to want their own various upper-midrange areas to themselves. The snare, with much to offer in the mids, welcomes a complementary set of gentle cuts to make room for these instruments, balanced by a gentle boost in any midrange area that remains so that it can express itself.

🔊 A common area of focus for more aggressive cutting is in the mid frequencies, usually narrow and in the 1-kHz to 2-kHz range. This narrow midrange area often has a sharp sort of ringing "ping" sound which, when attenuated (high Q, greater than 4, with a 6- to 12-dB cut), reveals a more open, easier to enjoy midrange character above. Yes, it is counterintuitive, but a cut here highlights desirable spectral features in the frequency range just above.

KICK/BASS DRUM

Consider reducing this instrument to two components (Figure 3.7). There is the upper-mid-frequency click of the beater hitting the drum followed by the low-frequency pulse of the large, ringing drum. The attack lives broadly up in the 3 kHz range. The tone is down around 80 Hz and below. These are two good targets for tailoring a kick sound.

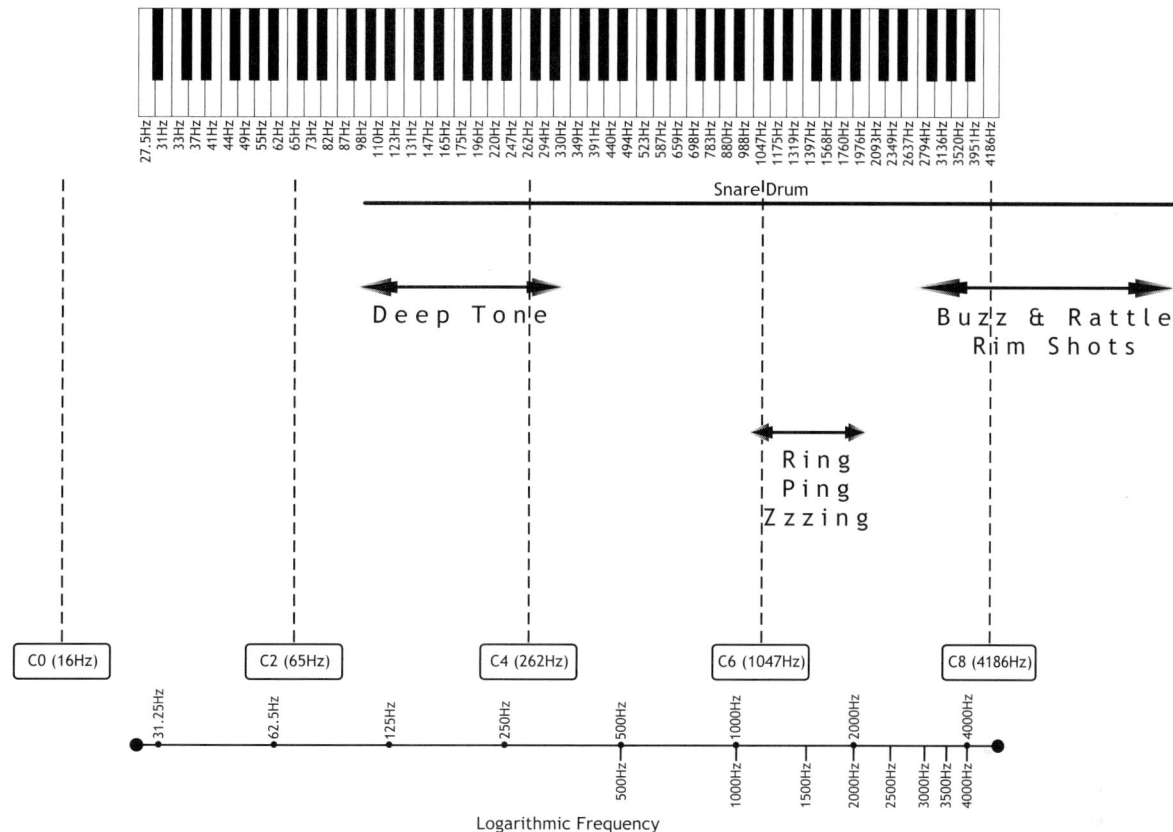

FIGURE 3.6
Frequency range of
spectral landmarks
(snare drum).

▶ If the kick lacks low-frequency excitement, remember this before you dial in a simple low-frequency boost (discussed earlier in this chapter, see "Fix" above.): kick is a common candidate for a narrow cut in that 200- to 400-Hz mud-range (high Q, greater than 4, with a cut of 6 to 12 dB, sometimes more!). Find it by boosting 6 to 9 dB with a narrow Q, and sweep around slowly from just below 200 Hz to just above 400 Hz. Find the ugliest, muddiest frequency range and then suck it out with an aggressive sweep from boost to cut. Listen carefully when you remove the muddiness. This EQ approach is productive if lower-frequency (100 Hz and lower) energy seems to rise into audibility, leading to a punchier and more powerful kick.

Depending on the tuning of the instrument, there is often a signature papery/ boxy sort of sound associated with the front resonant head. Often somewhere in the 300- to 600-Hz range, some kick tracks will benefit from a gentle, surgical cut here (2 to 6 dB of attenuation with a medium to narrow Q at the most problematic frequency).

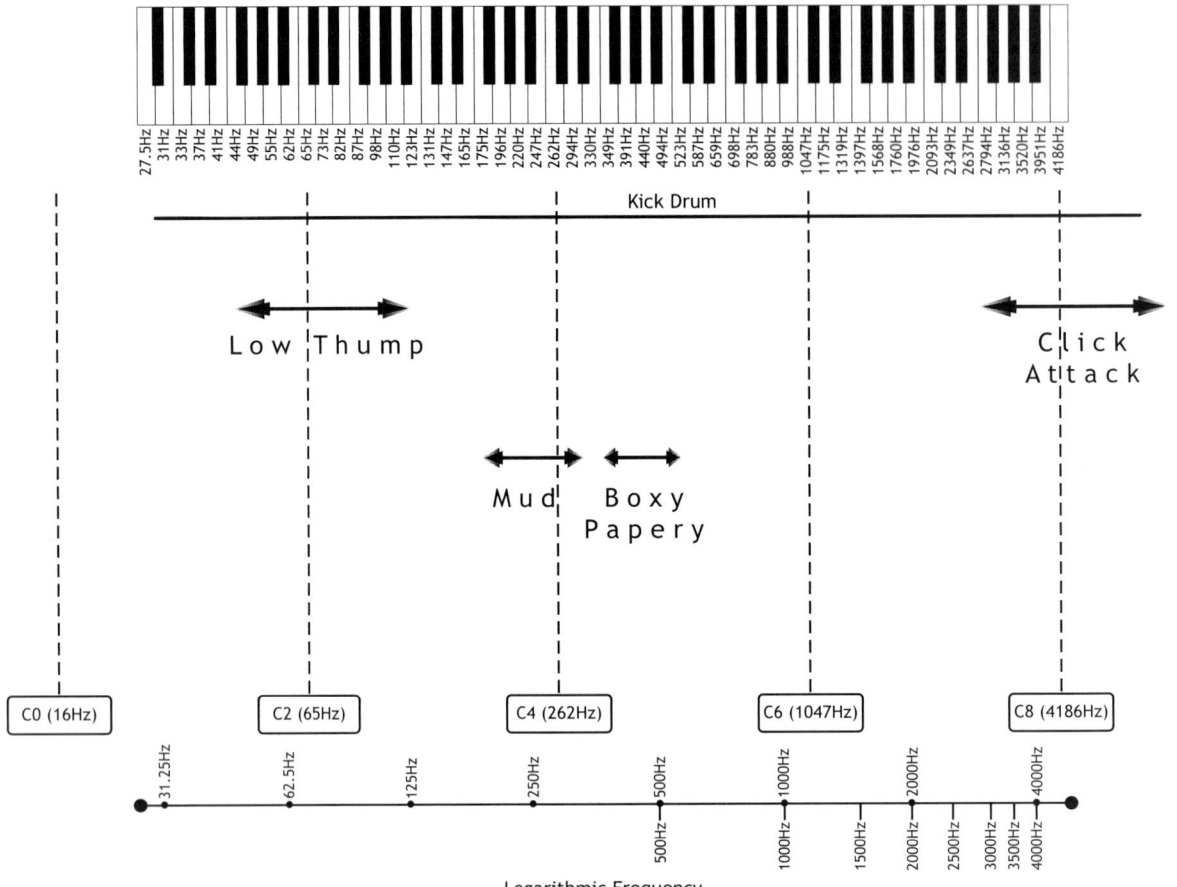

FIGURE 3.7
Frequency range of
spectral landmarks
(kick drum).

ELECTRIC GUITAR

The fundamental frequencies in the playable range of the typical, in-tune electric guitar covers the range from about 80 Hz to about 1200 Hz (Figure 3.8). Of course, the harmonics of every note played represent energy at multiples of these frequencies. Any amp- or stomp-box-induced harmonic distortion adds yet more energy to the sound, from 80 Hz well up to 6 kHz and beyond.

! The equalizer doesn't empower the engineer to separate fundamentals from harmonics, only to access a specific frequency range. So the engineer should note that an EQ setting above 1200 Hz affects only the harmonics of the instrument. EQ processes below 1200 Hz are a bit messier to keep up with, adjusting the harmonic balance for some notes and changing the level of the fundamental for others. Shaping timbre above 1200 Hz is more straightforward than dipping into that 80-Hz to 1200-Hz range, where the engineer's cuts and/or boosts will affect the very phrasing and note-to-note level of the guitarist's performance. EQ changes in the 80–1200-Hz range aren't forbidden. In fact, such settings are

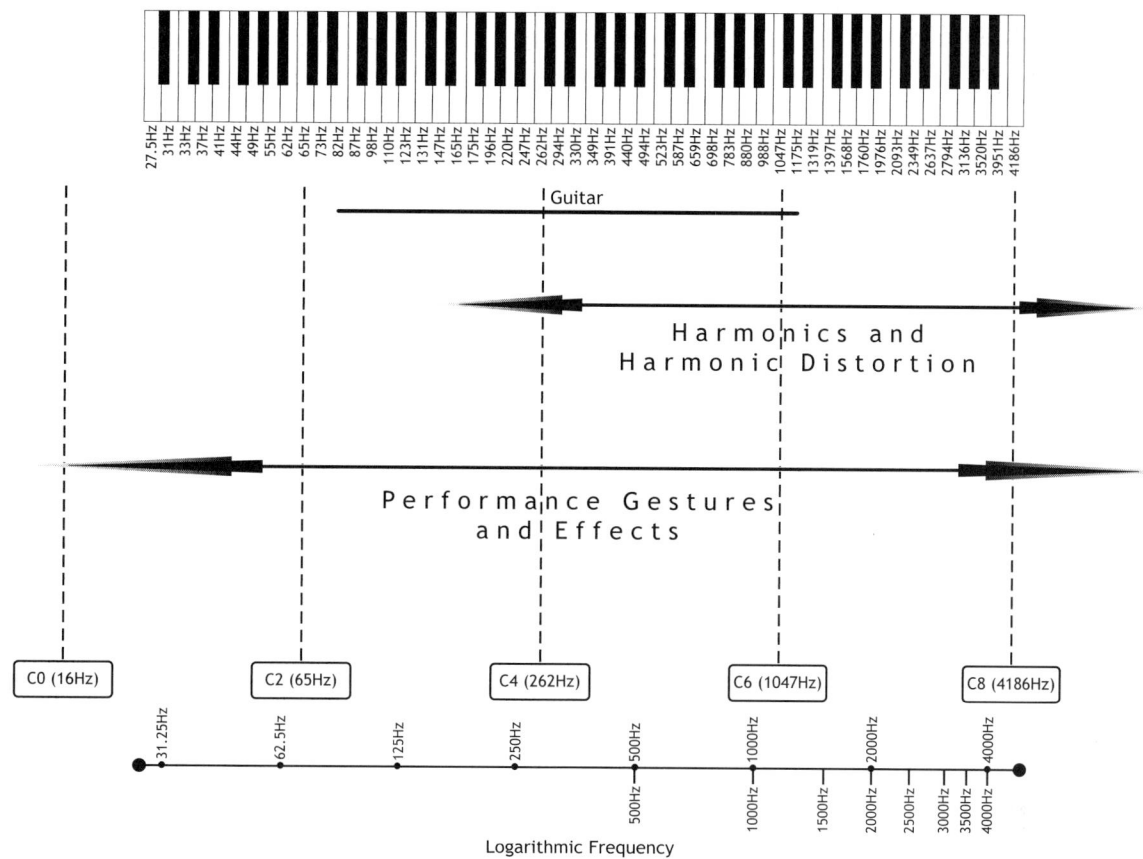

FIGURE 3.8
Frequency range of fundamentals and spectral landmarks (guitars).

quite common. But expect to use more restrained gain settings (less than +/–6 dB) and lower Q settings (less than 2), and pay particular attention to single-note lines in the performance, if there are any.

🔊 Although the lowest sustained pitch for the instrument is that 80 Hz fundamental, this doesn't mean that the instrument has no output below 80 Hz. As with all instruments, there can be audible broad band bursts of energy associated with the onset of every note or chord, creating instantaneous bursts of energy that include energy below the lowest fundamental frequency. Similarly, the ends of notes and chords can have broadband energy associated with them, very much influenced by the performance gestures invoked. This spectral interest at the beginning and end of notes is very important to the electric guitar sound, so pay particular attention below 80 Hz. Playing with a pick, playing with fingertips, playing with fingernails, hammering on and pulling off, palm mutes, and so on. The spectral subtleties of these techniques are most revealed in the upper midrange (2 to 6 kHz), and in the subharmonic range (below 80 Hz).

ACOUSTIC GUITAR

Try separating the acoustic guitar into its musical tone and its mechanical sounds. Listen carefully to the tone as you seek frequencies to highlight. Frustratingly, this covers quite a range from lows (fundamental frequencies start at about 80 Hz) to highs (harmonics for a steel-string acoustic played with a pick reach 10 kHz and above).

In parallel, consider its more peculiar noises that may need emphasis or suppression, depending on your production wishes: finger squeaks, fret buzz, pick noise, and the resonant sound of the wooden box of the instrument itself that thumps along with every aggressive strum. Though their exact frequency location varies from instrument to instrument and performer to performer, look for these frequency landmarks in every acoustic guitar you record and mix. EQ is a powerful way to gain control of the various elements of this challenging instrument. Manipulate the relative loudness of these elements as part of shaping the tone and character of the acoustic guitar.

ELECTRIC BASS GUITAR

The four-string bass guitar reaches an octave lower than a regular guitar, leading to a fundamental near 40 Hz (see Figure 3.8). The highest fundamental is close to 350 Hz. As with regular guitar, mentally distinguish between and among the fundamentals, the harmonics of the instrument, and the harmonics of any distortion device.

🔊   As the bass rarely plays chords and usually plays single-note lines, particular attention must be paid to any EQ moves in that range of overlapping fundamentals and harmonics (40 to 350 Hz). The bass player has likely worked hard to make the level of each note suit the performance intent of the overall tune. The mix engineer's EQ settings shouldn't undermine the music by crudely altering the relative level of the notes.

Landmarks to be on the lookout for include fret buzz (when the string is allowed, usually mistakenly, to vibrate partially against a fret), pick noise, finger squeaks, and the sound of fingers and/or a pick striking the pickups.

At the risk of oversimplifying: the primary role of the bass in most pop music arrangements is harmonic—it lays out the bottom of the arrangement, orienting listeners to the very chord being played and giving essential meaning to the rest of the notes being played and sung by others. The bass plays a significant second role, often underestimated by engineers new to their craft: time. Not only is the pitch of the note played important to the tune, but the instant at which the note is played is critical to the feel and the groove of the tune. Is it just a little late and lazy, lending a laid-back and relaxed quality to the tune? Is it a fraction of an instant early, adding urgency? In either case, the bass player is in league with the drummer to make these things happen. Listen carefully to the interaction of the kick drum and the bass guitar. They usually act as a single instrument, each kick accompanied simultaneously by a bass note. With bass, it's about pitch and time.

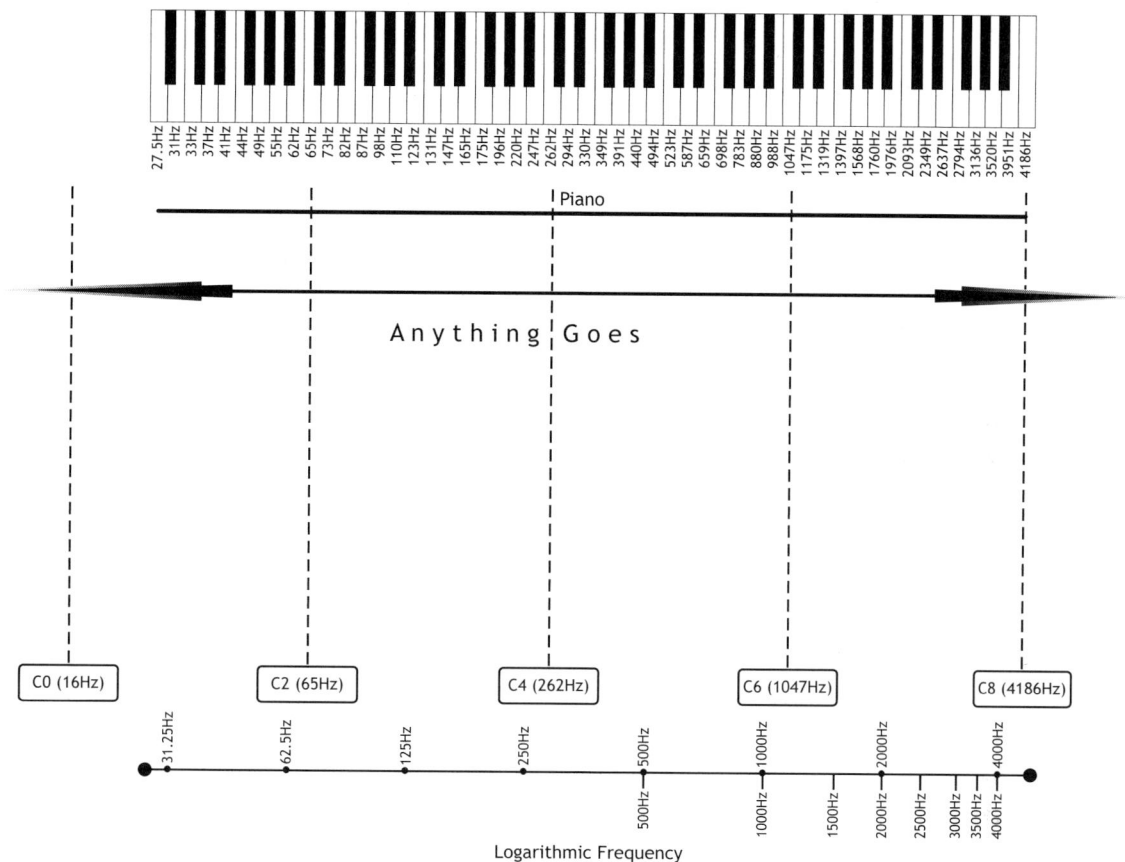

FIGURE 3.9
Frequency range of fundamentals (piano).

PIANO

Ah, piano. The mother instrument. Its 88 keys span fundamental frequencies from just under 30 Hz to just over 4 kHz (see Figure 3.9). It reaches about as low and as high as any instrument we are likely to encounter (the pipe organ is a rare exception).

As if that vastness weren't enough, the piano is also a percussion instrument. That is, the strings are excited into action by the pounding of hammers (the frequent use of tuxedos and fancy gowns seems to make it look less crude in concert). So, like a drum, the notes begin with a broadband, pitchless burst of energy that spans the audible range. The notes sustain—well, the notes sustain with that characteristic rise and fall of any and all harmonics in a complex, ever-changing, unique piano way.

So the sound begins with a bang, and sustains with an ever-evolving harmonic recipe. For many of us, we half-panic when we think of applying equalization to a piano track. EQing a piano is like hugging Jell-O. The harmonic content of the sound keeps moving. Just when we think we've found a spectral region worthy of highlighting or in need of a bit of attenuation, the piano itself allows that harmonic region to swell or recede or both. Prepare to be challenged by this instrument—hopefully, pleasantly so.

For those instruments that a mix engineer plays and records often, it is wise to spend some time examining their sounds with an equalizer. Insert an equalizer, dial in a narrow (Q = 3 or 4) 9-dB boost, and sweep the frequency parameter slowly throughout the audible range. Look for and make note of any defining characteristics of instruments and their frequency ranges. Also look for the less-desirable noises that some instruments make and file those away on a "fix" list. These mental summaries of the spectral qualities of some key instruments will save you time in the heat of a session when the client wants more punch in the snare (aim low) and more breathiness in the vocal (aim high). The ability to successfully feature any aspect of any instrument depends directly on both your general knowledge of the overarching attributes of the entire audible spectrum from 20 Hz to 20,000 Hz as well an intimate knowledge of all instruments in the multitrack arrangement.

3.4 SUMMARY

Equalizers enable the mix engineer to alter the relative levels of the spectral content within any signal. The parameters of frequency select, cut/boost, and Q are variously available on parametric, semiparametric, program, and graphic EQs. Peak/dip, shelving, high-pass, and low-pass filters provide the necessary alteration to frequency content so that the engineer can fix, fit, or feature audio signals as needed to help the mix best realize the artistic intent of the musicians who created it.

CHAPTER 4
Distortion

The most tempting effect: overdrive it, without overdoing it.

MIX SMART QUICK START: Distortion

GOALS

- Avoid unwanted and accidental distortion.
- Fit tracks together by strategically allocating the spectrum associated with harmonic distortion.
- Feature key tracks in need of attention through the addition of a well-chosen flavor of distortion.

GEAR

- Become distortion-literate by learning the unique sound qualities of each: tubes, transistors, transformers, converters, transducers, analog magnetic tape, each and every stomp box, guitar amps and their software simulators— all offer their own unique distortion qualities.

There is something innately appealing about distortion. That extra crackle and purr that occurs when we overdrive a device adds attention-getting adrenaline to a performance. Accidental distortion is a rookie mistake. Deliberate distortion, judiciously applied, hits the thrill button on the listener's home stereo.

Electric guitars are often deliberately driven into distortion, and rock and roll simply would not have survived without it. But drums, vocals, bass and most other instruments sometimes welcome the flattering gritification of distortion. It is an important mix element to be mastered.

4.1 DISTORTION OF AMPLITUDE

It's almost insulting to our art form to admit it, but all of audio is defined along two dimensions: amplitude and time. The waveforms on the screen of your DAW make it clear. The vertical axis is amplitude. The horizontal axis is time. Every

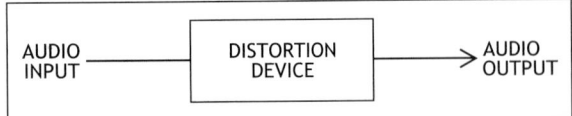

FIGURE 4.1
Any device could be a source of distortion.

track we record and every final mix we upload is nothing more than a specific pattern of amplitude changes over time.

When we monkey about with either time or amplitude, we are distorting the waveform. In this chapter, we choose to ignore time error problems and time manipulation possibilities and focus entirely on distortion along the amplitude axis.

In order to take a closer look at the distortion of amplitude, we just need to compare the level of the signal going into and coming out of the device—any device (Figure 4.1). Though the device might be a stomp box or other effects device designed deliberately for distortion, it might also be an equalizer, a mic pre, or some other signal-processing device driven into distortion. When the input level is so hot that the device simply can't muster the necessary output level, distortion has occurred.

Looking for the amplitude errors of distortion can be tricky, but is made easier by a simple plot of the output amplitude versus the input amplitude. An oscilloscope or software equivalent can easily do this for you. Figure 4.2 shows such a plot for a device that simply passes the signal, without any amplitude change whatsoever—it might be as simple as a cable, or it might be an equalizer not equalizing. For this

FIGURE 4.2
Output amplitude versus input amplitude.

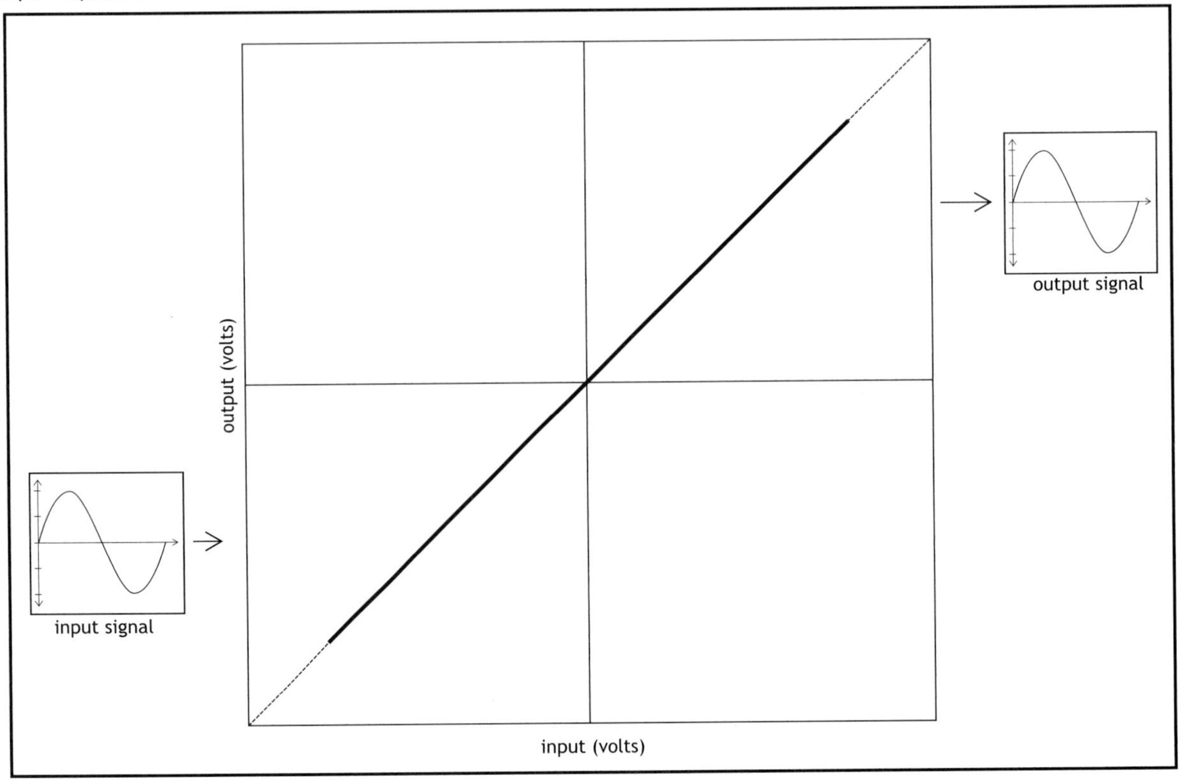

output signal

input signal

output (volts)

input (volts)

device, the output amplitude is identical to the input amplitude at every instant. Because every input value is matched by the exact same output value, this undistorted signal follows a perfect line on this plot, at exactly 45°.

A fader or amplifier that increases the gain without otherwise changing the shape of the waveform remains on a straight line, but deviates from the 45° angle (Figure 4.3). The amplitude is consistently scaled up or down in value.

💡 Distortionless unity gain occupies the 45° line. Devices that increase level without distortion are described by lines steeper than 45°. Devices that attenuate without distortion are described by straight lines shallower than 45°. By preserving the same waveform *shape*, the output versus input plot remains a straight line. For this reason, a device that does not introduce amplitude distortion into the signal is said to be *linear*. Devices that do not offer clean gain, which reshape the waveform along the amplitude dimension, will deviate from this straight-line behavior: they distort. They are *nonlinear*.

4.1.1 Harmonic Distortion

A signal cannot be raised in level without bound. In this world, all devices have only finite capabilities to create amplitude. Each and every hardware device in the recording studio reaches a voltage limit. All plug-ins run out of bits eventually.

FIGURE 4.3
Output amplitude versus input amplitude for linear gain.

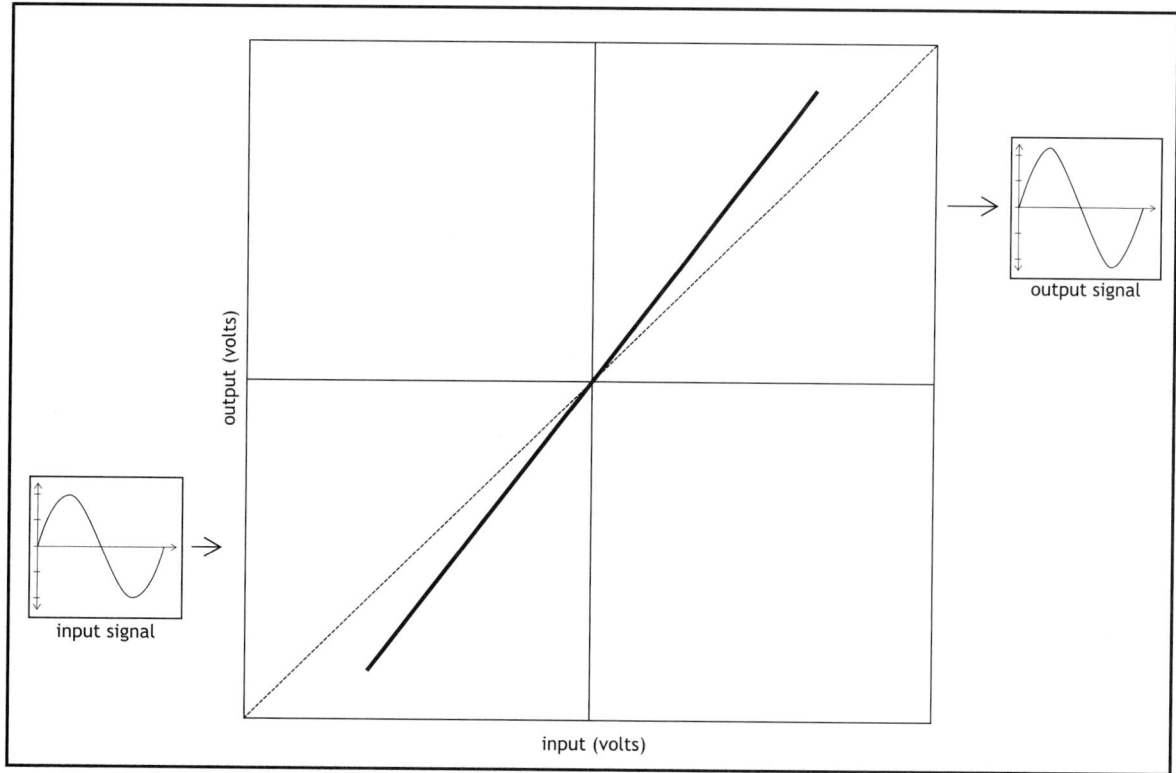

Distortion is what happens when the output level limits are exceeded, changing the amplitude of the audio waveform into a new shape, creating a sound with new spectral content.

Consider a simple domestic light switch—a musical instrument only in the hands of a rare talent. Turn the switch on, and the full voltage supplied by the power company is applied to the lightbulb. Flipping the switch harder or faster does not change this. Dimmers can effectively *lower* the average voltage fed to the lightbulb. Barring the addition of other electronic components, the voltage may never be *higher* than that supplied to the input of the switch. That is the maximum voltage available.

Power amplifiers present a similar situation. The power amp is fed a line-level audio signal. The voltage of that incoming audio signal is to be magnified by the power amp and fed to the loudspeakers. The power amp driving the speakers in our mix room is expected to do this linearly.

That power amp is plugged into the wall. It receives the grid's power and converts it into internal power supply voltages. Those internal voltage values represent the hard-and-fast limit of voltage capability for the audio output. No matter what sort of voltage swing is supplied on the audio input to the power amplifier, at some level, the output from the amp can increase no further. The output no longer tracks linearly with the input. It begins to distort.

! All audio devices, including mixing consoles, guitar amplifiers, and equalizers possess an output voltage limit. It must be understood what happens when that limit is reached.

HARD CLIPPING

Consider a device that abruptly hits its output voltage limit (Figure 4.4). Below the point of distortion, the device behaves linearly, and there is no warning that the device will run out of output voltage capability. The threshold of distortion is crisp and absolute. The audio below this limit is completely undistorted. As long as the input remains below the threshold of distortion, the performance of the device is perfectly linear.

Send an input whose level seeks to drive the output above this limit, however, and the peak of the wave is flattened. A sine wave input exits the device looking as if its peaks had been clipped off. At low amplitudes, the output looks similar in shape to the input. It looks like a sine wave. But at the highest levels, it becomes a different wave shape entirely. Because the device couldn't create the output voltage necessary to preserve the shape, the peaks—top and bottom—are obliterated. The term for this kind of distortion, *clipping*, comes from a visual inspection of the output waveform. As the transition from linear to nonlinear is abrupt, it is classified as hard clipping.

Sine wave inputs below the threshold of distortion remain sine waves on output. Sine wave inputs of ever-increasing amplitude develop more severe clipping. Taken to an extreme (Figure 4.5), the output based on a sine wave input starts to resemble a square wave.

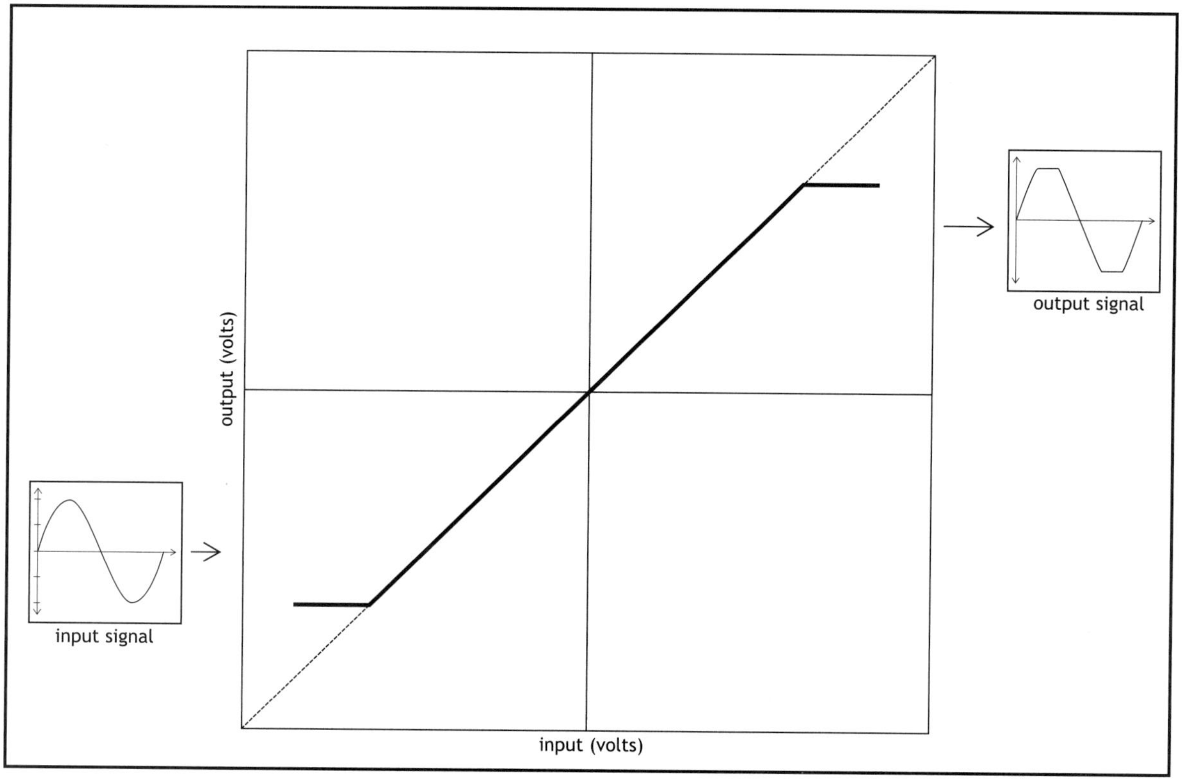

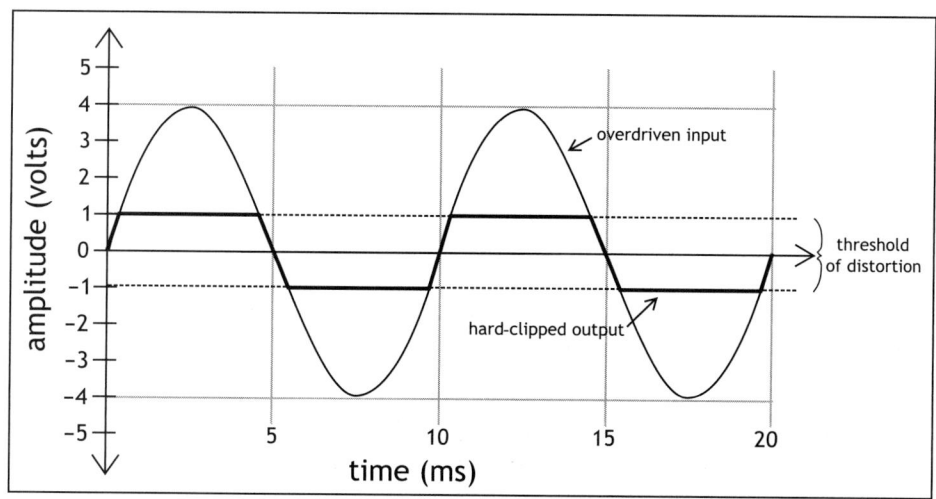

FIGURE 4.4
Output amplitude versus input amplitude for hard clipping.

FIGURE 4.5
Hard clipping a sine wave in to a square wave shape.

🔊   Severely clipping a sine wave until it resembles a square wave has the effect of adding harmonic content to the sine wave signal. That is, send a 100-Hz pure tone through a hard-clipping device and the square-ish wave output is a signal that retains some energy at 100 Hz while shifting the rest of that energy up to 300 Hz, 500 Hz, and so on. The act of clipping a waveform essentially synthesizes harmonics, a phenomenon known as *harmonic distortion*. Hard clipping begins to introduce the harmonics that make up a square wave, which—if you feel like a little math—follows the equation:

$$Y(t) = \frac{4A_{\text{peak}}}{\pi \sum_{n=1}^{\infty} \frac{1}{2n-1} \sin 2\pi (2n-1) ft} \tag{4.1}$$

!   A specific set of odd harmonics makes a square wave square. As the equation makes clear, a true square wave has the impractical requirement that an *infinite* number of harmonics are needed. When a device is driven into hard clipping, beware of the possibility of side effects due to the generation of a broad range of odd harmonics, possibly reaching into extremely high frequencies—even beyond what is nominally audible.

SOFT CLIPPING

Clipping is not always so abrupt. Imagine a device that approaches its amplitude limits more gradually (Figure 4.6). To the naked eye, there may not be a well-defined threshold of distortion that separates the distorted from the undistorted. As inputs are raised in level, the degree of output shortcoming gets gradually worse, too.

This type of distortion also alters the peaks of the waveforms, but in a kinder, gentler way than hard clipping. Call it *soft clipping*. As the output versus input plot makes clear, soft clipping is nonlinear. It is a kind of amplitude distortion. The peaks are squeezed down, not lopped off. Distortion of a different shape is, unsurprisingly, built of a different recipe of harmonics. There is no single shape to soft clipping. You can imagine a slight rounding or a more extreme rounding-off of the high-amplitude peaks. As a result, we are spared the messy equation of the square wave and instead absorb only the conclusion: softly attenuating the high-amplitude portions of a signal is a regular, repeating change in shape possible only through the coordinated use of new harmonic content. The details of the recipe of the harmonics are less important. It is essential to know that they exist and to listen for them whenever amplitude is manipulated.

By way of analogy, physicists are fond of placing a mass on a spring in a friction-less world. The spring remains linear if not overstretched (i.e., no distortion). If stretched too far, the opposing force of the spring starts to rebel and overreact (i.e., soft clipping). If the spring is stretched so far that the mass bumps into the floor, motion is instantly stopped (i.e., hard clipping). Pressure waves in air can reach similar limits. Voltage oscillations in audio equipment suffer the same indignities. Calculations within a digital audio workstation run out of numbers—that is, they reach numbers that eventually become too big for the system. Distortion results, and we can hear it via the introduction of new harmonics within the signal.

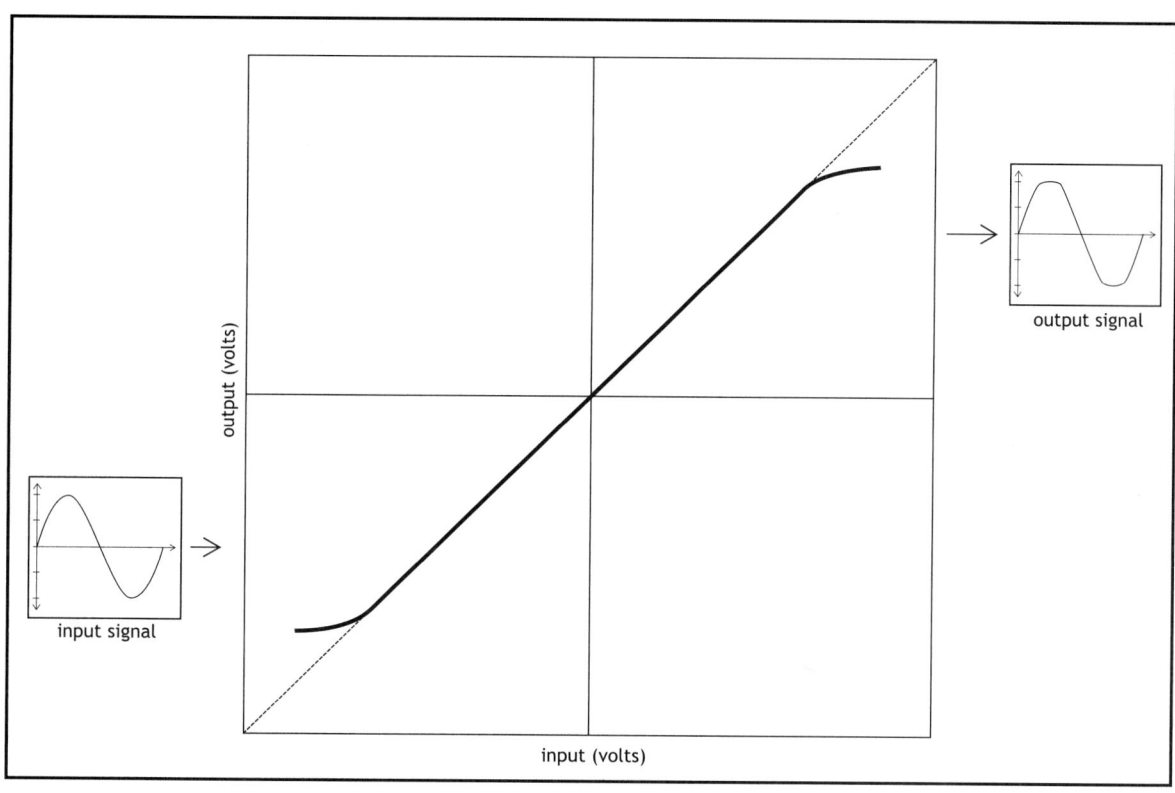

FIGURE 4.6
Output amplitude versus
input amplitude for soft
clipping.

4.1.2 Intermodulation Distortion

The previous discussion, evaluating the spectral changes to a sinusoidal input into a device reaching its amplitude limits, is not the whole story. Nonlinear devices have a more profound effect on complex signals. That is, when an input contains energy at more than one frequency (e.g., vocal, guitar, snare, and pretty much every track we record, except pure tones!), the harmonics generated are not limited to just the odd and even multiples of each frequency component. These harmonics are created, to be sure. But additional frequency components that are not harmonically related to the input spectrum are also created. Called *intermodulation distortion*, energy at frequencies related to the sum and difference of the spectral components appears at the output of the distortion device.

An input consisting of just two frequencies, f_1 and f_2, would develop distortion components at frequencies equivalent to $f_1 + f_2$. Similarly, distortion will be found at the differences that exist between the two spectral components: $f_1 - f_2$ and $f_2 - f_1$.

Intermodulation components between the harmonics may also develop. That is, harmonic distortion of frequency f_1 leads to multiples of f_1, such as $2f_1$, $3f_1$, and so on. Sum and difference distortion products built on these harmonics may become significant, which leads to distortion that is still more spectrally complex, containing terms such as $2f_1 - f_2$, $3f_1 - f_2$, $- 2f_2 - f_1$, $3f_2 - f_1$, and so on.

🔊 When spectrally rich signals—*music* signals—are subjected to clipping of any kind, harmonic and intermodulation distortion must be anticipated. Listen for new harmonic energy when you are trying to figure out if distortion is occurring. Listen for a specific set of new harmonics and their inharmonic accompanists when you are choosing which type of distortion you wish to add to the track.

4.2 DISTORTION DEVICES

To introduce distortion, we most often insert a distortion device. Components of analog circuits such as diodes, transistors, transformers, and tubes each exhibit a characteristic type of distortion. In their simplest application, analog magnetic tape will soft clip, transistors and analog-to-digital converters will hard clip, transformers may hit saturation at low frequencies yet pass higher frequencies without harm, and tubes offer a unique, asymmetric form of amplitude axis anomalies, shown in Figure 4.7.

FIGURE 4.7 Asymmetric amplitude axis distortion from a tube-based gain stage.

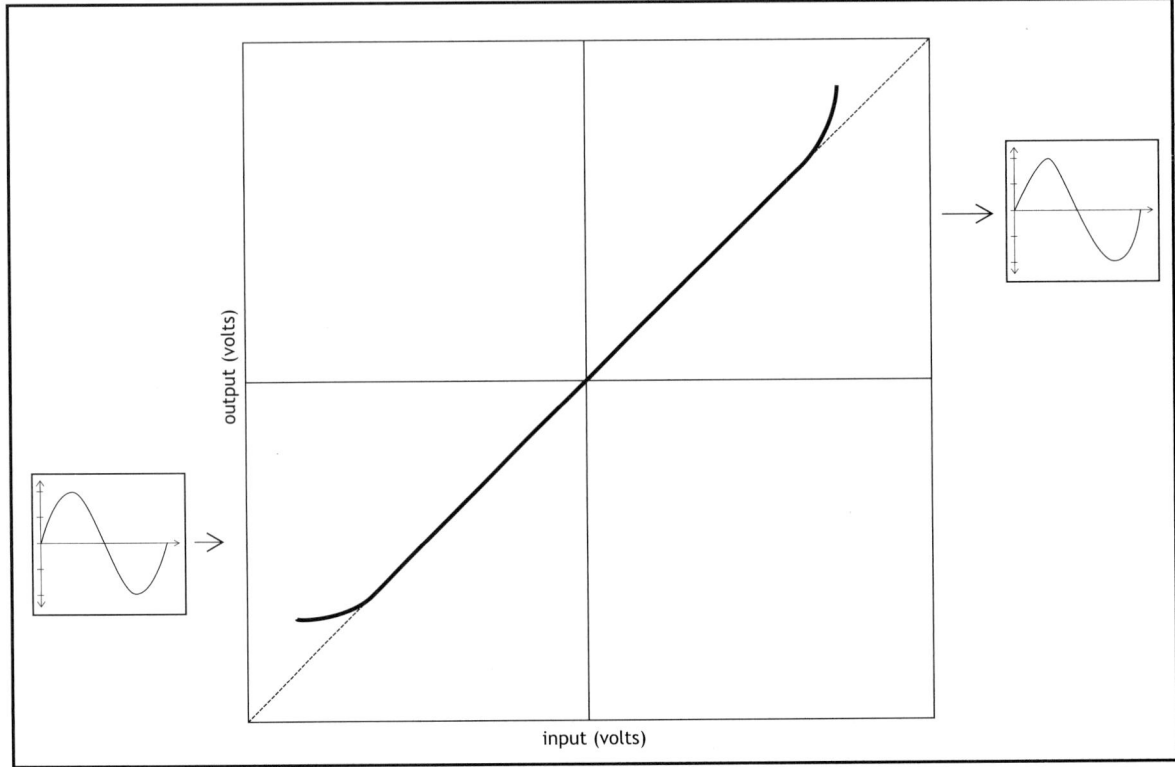

Assemble these building blocks into various circuit designs, and a new type of nonlinearity results: a kind of signature of the system. In the digital domain, algorithms may be written to reshape the signal in almost any way desired as headroom is exhausted. The types of nonlinearities are as diverse as the number of people who write the code and design the circuits. Audio systems utilizing these circuits and software—plus any magnets, coils, ribbons, optical components, and so on—possess an output versus input behavior that grows more complex still.

Think of hard clipping and soft clipping as two icons of harmonic distortion, but recognize that analog and digital devices can develop unique combinations and variations of distortion that are a complex result of the overall design and layout of the components and algorithms employed.

Some audio devices are purpose-built to distort (guitar amps, guitar amp simulators, and distortion pedals), offering a unique flavor that comes from the combination of harmonics associated with a device's output versus input shortcomings.

🔊 Other types of audio equipment possess a bit of distortion more as a side effect. Transducers, compressors, tube gain stages, and analog magnetic tape are among the more notable examples. These devices have other design goals. A microphone must convert acoustic pressure changes into electrical voltage changes, a loudspeaker must convert electrical voltage changes back into air pressure changes, a compressor automatically attenuates audio when the signal is above the specified threshold, and so forth. In addition to this nominal function, many of these devices also introduce harmonic and intermodulation distortions of varying types, which may be taken advantage of by the clever mix engineer to further refine the sound quality of the audio signal at hand.

Musicians think of timbre as that attribute of sound that separates a piano from a guitar when they play the same note. Engineers must sort out what sonically separates a Steinway from a Bösendorfer among pianos and a Les Paul from a Telecaster among guitars, even as these instruments play the same note. Trained and careful listeners can reliably distinguish very fine properties of a sound. The unique distortion traits of each audio device in the studio endow that device with a sort of sound fingerprint related to distortion. Distorting devices (such as fuzz boxes and tube amps) offer their own characteristic qualities of distortion, which is often audibly obvious. Some mixture of hard clipping, soft clipping, asymmetric, and other forms of distortion add up to a distinct and possibly quite desirable sound quality. Any component, pushed beyond its linear limits, offers harmonic and nonharmonic alterations to the signal passing through it that may impart the tasty sonic flavor you seek for that track in this mix.

4.3 MIX STRATEGIES: DISTORTION

The distortion effect isn't just for engineers. Distortion of the right kind is naturally sought out by musicians and music fans alike. Spared the pleasure of reading this text, guitarists and singers and drummers still employ distortion as a method of musical expression. Fans of many—if not most—styles of popular music react positively to an expressive amount of distortion almost instinctively.

When a device is overloaded, something exciting must be happening. Someone is misbehaving. Rules are being bent or broken. Distortion in rock and roll is as natural as salsa in Mexican food. It is the caffeine of music. Or perhaps it is the garlic. And why isn't there caffeine and garlic in my salsa?

Used musically, most people can't resist. Accordingly, we audio engineers reach for distortion as a staple effect. There is a downside. Adding distortion risks adding spectral clutter, confusion, and distraction. As mix engineers, we must guide our instinctive attraction to distortion by strategies for using it effectively.

4.3.1 The Fix

! A good place to start with any signal processor in the studio is to use it to fix any problems. We discuss this goal throughout the book, fixing tracks as necessary with equalizers, compressors, expanders, delays, pitch correctors, and reverb as a first step with any of these effects. Distortion is an exception to this trend. It never fixes a problem. In fact, distortion itself often *is* the problem.

PREVENTION

💡 Unintended, unwanted distortion occurs whenever we fail to effectively manage the dynamic range. Clumsy, accidental distortion is a dead give-away to all who listen that the recording engineer did not know what they were doing. Overload the converters and your signal gets kicked in the teeth with harmonic distortion. Overdrive the input to the equalizer, the compressor, or any other signal processor, and a burst of harmonic distortion pollutes your track. When distortion is the problem, the only fix is to turn down the input level on whatever device is being overloaded.

If you are offered a track with distortion already in it, there is no fix. There is no undistorter. Once the harmonics are part of the complex musical signal or once the detailed shapes of the amplitude are modified, there is no way to restore it fully back to its undistorted shape. We fix distortion best by preventing it.

CAMOUFLAGE

Meantime, there is no common mix problem for which distortion is the solution. It can't fix a problem. At best, it can hide a problem, but hiding is always a second choice to actually fixing. If you have a problematic track that can't be fixed; if it has technical flaws that are distinctly unmusical and no amount of mix surgery with EQ, compression, expansion, delay, pitch correction and reverb can make it go away; if you can't delete or replace the track, you have no choice: hide the problem. Here, distortion can be of help. Using a strategic bit of distortion, you can obscure the flaws, covering over them with the chaotic, distracting harmonics of distortion.

4.3.2 The Fit

Distortion might sometimes be thought of as a creative alternative to the more straightforward spectral effect of equalization (see Chapter 3). The fact that amplitude limitations create spectral complexity means that you can reshape timbre in surprising, complicated, and glorious ways through distortion. Equalization, as discussed in Chapter 3, emphasizes and deemphasizes spectral regions already present in the signal being processed. Distortion fabricates entirely *new* spectral components, fodder for further manipulation in the multitrack mix.

ALLOCATING SPECTRUM

As with EQ, we must deliberately divvy up the available spectrum across all of the tracks in our mix. Extra care is needed with distortion. Rich with the new harmonics of overdrive, the distorted signal might easily obscure the enjoyment of other important sounds competing at similar frequencies.

▶ Consider the following inspiration: you find the vocal benefits from a bit of distortion. Careful! Dialing up distortion on something as important as the vocal counts as a brave mix move. You don't want to make the vocal difficult to understand or exhausting to listen to. If you find that the right touch of distortion on the vocal is appropriate to the mix, don't relax just yet.

It is essential that you review the impact this vocal distortion has on the other tracks in your mix. The harmonics introduced by the distortion may now make it hard to enjoy the spectrally complex snare sound you had fine-tuned into perfection earlier in the session. The new spectral energy that distortion creates may interfere with our enjoyment of the delicate reverb you placed on the ukulele. Distortion on the vocal isn't just about the vocal. The wide sonic footprint of distortion added to any track forces us to reevaluate and rebalance the entire mix.

Exciting distortion casts a potentially large and destructive shadow across the frequency axis. Care and restraint must accompany the thrill and freedom of cranking up some distortion.

CONTRAST

We also tap into distortion to distinguish one signal from another. When two signals on their own are fighting to be heard, held back because they occupy similar spectrum, create some ear-grabbing contrast by adding a tasteful amount of distortion to one track and not the other. If the harmonics of the harmonic distortion are easier to hear, it gives your listeners a chance to better hear the whole track, even in the presence of another track that is otherwise too spectrally similar to enjoy.

4.3.3 The Feature

Nothing draws attention to a track better than distortion. It can be used to feature a track in a mix the way a spotlight illuminates one person on a crowded stage.

INSTINCTIVE APPEAL

When the song calls for excitement, introduce a distorted track to the mix. When the mix needs to soar to still new levels of thrill and defiance, add a dose of distortion to a key track. There is instinctive sensory appeal to hearing music push the boundaries through distortion.

!   Too much distortion makes a mess of your tracks, a tangled ball of barbed wire that drives listeners away. Recognize distortion as a timbral decision rich with opportunity but demanding of musical thoughtfulness. Our emotions must be guided by calm, strategic ideas or we'll lose control of our mix. An engineer carefully chooses distortion—what kind of distortion device will I use, and on which track? The creative process is very similar to that of the arranger choosing a horn section—which horns shall we use, and who plays what? Grounded in sound musical judgment, distortion exacts innate appeal.

GETTING NOTICED

Beyond the visceral, distortion has practical benefits. Distortion can influence your use of the humble fader. Augmenting a sound with additional, spectrally related energy likely makes that sound easier to hear. In this way, distortion becomes a creative alternative to the more obvious approach of simply pushing up the fader to make it louder (see the "Case Credo" in Chapter 10). Any single track of a multitrack project fighting to be heard in a crowded mix can achieve distinct audibility through the harmonic lift that comes from at least a little well-chosen distortion.

Guitarists like attention, so they drive their amps to distortion—and they step on stomp boxes for more. Distortion sonically empowers a single note on a single string of the guitar to rise above the loud cloud of energy created by the drummer throwing hands and feet at their kit. Push distortion effects when a key instrument needs to outdo another.

🔊   Transformers are a common component in many audio circuits and can solve problems related to grounding, noise, impedance and isolation. The best ones are sonically transparent, passing a signal without audible corruption. However, there are times when we reach for a transformer specifically *for* distortion. When a transformer is allowed to distort, it offers a unique kind of distortion. These audio transformers introduce more harmonic distortion to low frequencies signals than high. The sonic result is that it adds growl and texture to tracks with low-end content, while staying out of the way of the middle- and high-frequency part of the signal. Whereas hard or soft clipping introduce harmonic distortion to the entire signal based only on the amplitude of the signal, a transformer reacts more to the low frequencies. The transformer can be used to add enough harmonic energy to distinguish a track while preserving essential upper-frequency detail. Kick drum, snare drum, toms, bass, and left-hand-oriented bass parts on keyboards often react quite well to transformer saturation.

EXCEEDING EXPECTATIONS

▶ There are many moments in music when we deliberately wish to transcend regular life and create something supernatural, better than the real thing, more intense, more exciting, more heartfelt. Mix in some distortion and orchestrate the emotional contour of the tune. When the chorus is supposed to elevate to a higher emotional state, look for a distorted electric guitar. If the tracks offer you none, create one out of a clean guitar. If even clean guitar is not in the offing, process any other clean rhythm track—piano, horn sections, synth lines, whatever you wish. Layer in some distortion to support the intent of the song and enhance the performance of the band. Take your listeners on a ride.

You've got it just right when you surprise everyone—the band, their fans, yourself—with a distorted sound that is a bit more intense than expected. Care is needed. Too much of a good thing can lead to other problems. You must still allocate the spectrum of your mix carefully, as discussed previously. You have to make room for the distortion. You must preserve the integrity of the other tracks in similar spectral zip codes.

It takes intelligence and grace to pull off these moments of distortion intensity, but no one will notice. They'll be too thrilled with the tracks to ever suspect you were the brains behind it.

4.4 SUMMARY

The temptation to add doses of adrenaline and excitement through distortion must be kept in check. Informed by an understanding of the harmonic structure of distortion, the smart mix engineer allocates spectrum carefully, contrasts conflicting tracks strategically, makes sure key tracks get heard, and finds ways to push the limits without undermining the overall aesthetic.

CHAPTER 5

Compression and Limiting

The most misused effect: too much is a turnoff.

MIX SMART QUICK START: Compression and Limiting

GOALS

- Practical motivations include preventing overload, overcoming noise, and de-essing bright vocals.
- Fit tracks together in your mix more easily through tamed loudness, improved intelligibility and articulation, smoothed performances, and cleverly reshaped amplitude envelopes.
- Feature tracks by extracting ambience and artifacts, manipulating timbre, and creating distortion through compression.

GEAR

- Master the quirky set of parameters and features: threshold, ratio, attack, release, makeup gain, hard knee, soft knee, stereo, and multiband.
- Recognize the sonic traits and operational pros and cons built into the topology: tube, optical, FET, VCA, PWM, and digital.
- Acquire an internal sense of the sound qualities and dynamics capabilities of each and every make and model.

Compression is an effect easily misunderstood, and often misapplied. Don't be fooled by the apparent simplicity of the concept: *compression is the automatic reduction in signal level whenever the amplitude exceeds a specified value. Limiting is simply a faster, more extreme constraining of amplitude.* The effect almost seems like a crutch—a process for engineers failing to do their job. Can't we just turn it down when we want to—when we think the level is too hot? Why relegate level adjustments to a machine or a piece of software? We've got faders, and we know how to use them.

5.1 NARROWING DYNAMIC RANGE

There are times when a fader just will not do. Music signals are rarely consistent in level, which is generally a good thing. Each thump of the kick drum, every blat of the brass section, each and every syllable of the vocal produces a signal that—instant by instant—surges up and recedes down in amplitude. Figure 5.1 shows the changing amplitude during a fairly representative two bars of music. Signals like this one must fit within the amplitude constraints of every device in our entire audio chain without damage (see the discussion of "Dynamic Range" in Appendix B): the microphone, microphone preamp, analog-to-digital converters, console/DAW, plug-ins, outboard gear, digital-to-analog converters, multitrack recorder, two-track master recorder, power amp, and loudspeakers. The highest peak must get through these devices without clipping. The faintest detail of our music must pass through well clear of the distracting and detail destroying effect of noise. When we aim for 0 VU (volume units) on our meters, we are targeting an amplitude that just avoids driving any device into distortion when the music peaks, yet isn't lost in noise when the signal recedes to fainter but no less important levels.

With agile fingers and intense concentration, we could constantly adjust the faders in reaction to every kick, horn stab, and lyrical phrase so that the levels move musically without overwhelming the equipment. But we would run out of fingers and find it hard to focus on anything else if we had to ride gain on all our tracks throughout the entire song. We delegate such tedious level control to a machine, an automatic, semi-intelligent fader—a compressor.

Before you get too comfortable, it is essential to notice that the compressor, born from this simple concept, has found applications far, far beyond what anyone might have expected. This same dynamic-range-controlling tool is also the basis for a wealth of other effects—many far from intuitive—that go well beyond the concept of simply controlling dynamics.

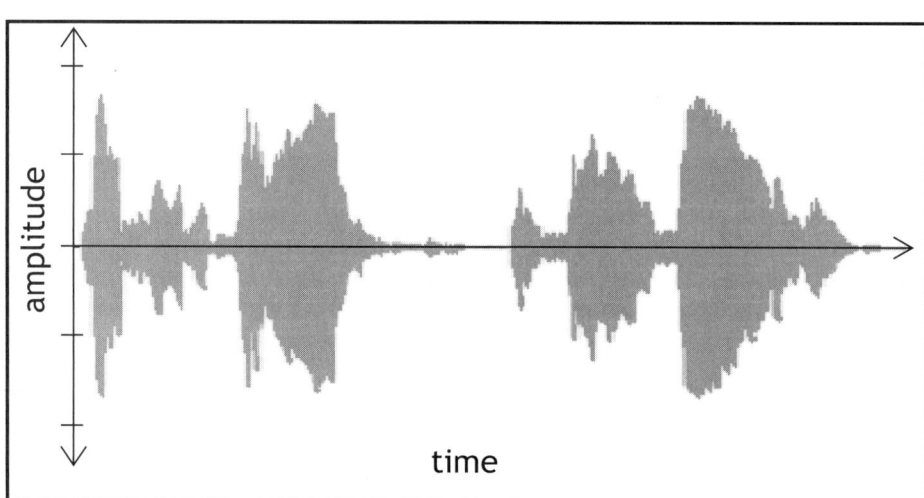

FIGURE 5.1
The ever-changing amplitude of two bars of music.

This chapter reviews the common parameters found on compressors and limiters, summarizes the core technologies used, and details the various production techniques—both practical and creative—built from this device so that we may fix problems in the amplitude domain, fit tracks together so they can be heard in even the most crowded mix, and feature specific qualities of any audio signal as desired to create the mix we want.

5.2 PATCHING AND PARAMETERS

What is the compressor's task? Quite simply, when a signal's amplitude is too hot, the compressor turns it down. What counts as too hot? How much should it turn it down? How quickly? How long? Some compressor controls will be needed. First, we hook it up.

5.2.1 Patching and Plugging In

Compressors are sometimes built into the console or DAW, preconfigured for convenient availability on each and every channel, and are a stock part of any fully featured channel strip.

INSERT

Any additional outboard and plug-in compressors are introduced into the audio signal flow using the *insert* of the mixer or digital audio workstation (see Appendix A). The very action of a compressor—to lower the level of the signal—often makes little sense if the compressed signal doesn't in fact replace the higher-amplitude uncompressed signal.

The insert is the signal flow structure that effectively forces the track through the compressor, altering its tone and giving us no chance to accidentally mix together both the compressed and the uncompressed versions of the sound. Similarly, compression can be patched directly between the multitrack return and the mixer/DAW line input, which effectively inserts the compressor into the signal flow. The use of auxiliary sends or buses is unusual for this type of effect, as we rarely need to send multiple different tracks to the same compressor, and any adjustment of a gain stage in front of the compressor (in the form of a fader or an aux send) directly affects its interaction with the compression algorithm.

PARALLEL COMPRESSION

Although the insert is the general signal flow approach for adding compression, we sometimes set up parallel compression. *Parallel compression* refers to the signal routing strategy in which you monitor both your uncompressed and your compressed signal together. It is an apparent philosophical violation of the correct use of the insert just described: a violation that is sometimes allowed and is unique to compression and expansion. Using the insert send only (without the insert return), or a prefader, unity gain effects send, the compressor is fed the desired signal and returned to any spare strip or effects return on your mixer. This process is described more completely in Appendix A in the section "Hybrid Parallel Effects."

In this way, you have uncompressed and compressed versions of the signal feeding the mix bus in parallel, each contributing to your mix independent of the other. You access both the uncompressed and the compressed elements as you wish. This feature presents a plethora of parallel production possibilities, which are discussed in detail throughout this chapter.

5.2.2 Threshold

A line is drawn separating amplitudes that are too great from those that are not, using the *threshold* control. The threshold setting on the compressor sets the amplitude above which compression is to occur and below which compression is to stop. As long as the signal remains below this threshold, the compressor is not triggered into action. When the signal exceeds the threshold, however, the compressor begins to attenuate the signal, like a fader automatically being pulled down. Once the compressor is attenuating a signal, the signal must fall back below this threshold before the compressor will stop attenuating and return to unity gain.

5.2.3 Ratio

When the audio is above this threshold, the compressor begins to attenuate. The amount of compression is primarily determined by the *ratio* setting. Mathematically, the ratio compares the amount of the input signal above the threshold to the amount of the attenuated output above the threshold. For example, a 4:1 ratio describes a situation in which the input level above the threshold is to be four times higher than the output above the threshold: 4 dB above threshold in becomes 1 dB above threshold out, and 8 dB above threshold in becomes 2 dB above threshold out. A ratio of X:1 sets the compressor so that the input must exceed the threshold by X dB in order for the output to achieve a level just 1 dB above the threshold.

It is important to note that the ratio only applies to the portion of the signal above the threshold. When the input is below the threshold, the compressor is not applying this ratio to the signal. It is in fact in the business of not compressing signals below the threshold. Only that part of the input that exceeds the user-defined threshold is multiplied by the compression ratio.

It is worth making the distinction now between a compressor and a limiter. The websites, catalogs, and showrooms selling compressors often tout these devices as compressor/limiters. This is basic marketing; one piece of equipment seems to have two functions—two for the price of one.

We dedicate an entire chapter here to the thorough discussion of the broad scope of production potential for the compressor/limiter. You'll soon find it easy to tap into more than a dozen different effects, all created by this one device. The sonic output of the compressor is the result of our strategic use of the compressor parameters available. The compressor/limiter in the hands of a talented engineer (you and me) is much, much more than two effects for the price of one.

The key defining parameter that separates compression from limiting is the ratio setting. A ratio below 10:1 indicates compression. A ratio above 10:1 classifies the effect as limiting. Other attributes—fast attack and fast release—are accompanying features for a limiter. It is simply a matter of degree. For very tight control of the amplitude of a signal, *limit* the dynamic range with a high ratio, fast acting dynamics device. For less abrupt modification of the amplitude of a signal, gently *compress* the dynamic range with a low ratio.

As the principal technology for compression and limiting is the same, most compressors are also limiters, and most limiters are also compressors. Therefore, most faceplates and brochures declare the device to be both: a compressor/limiter.

5.2.4 Attack

It is perfectly appropriate to think of the compressor as a fader—a smart fader able to do some analysis and apply some intelligence when deciding when and how fast to adjust the level. If you've seen moving fader automation on a mixing console or digital audio workstation, you have a good visual analogy for what goes on inside a compressor. A machine is pulling down and pushing up a virtual fader in reaction to the amplitude of the signal.

The speed with which the signal is attenuated is a function of the *attack* setting. Attack describes how quickly the compressor can fully kick in after the threshold has been exceeded. Audio signals change level constantly. Attack time defines the agility: the reflexes of our compressor as it reacts to these changes. The compressor watches a signal exceed the threshold we specified and then applies the necessary gain reduction to satisfy our ratio setting. When 5 dB of attenuation is called for, attack time describes how quickly that 5-dB decrease in level is introduced, from instantaneous to gradual.

The importance of attack time is not limited to that moment of transition when our signal passes up through threshold. In fact, the attack character of the compressor is in play for any signal at or above the threshold and increasing in level.

Recall that with a ratio of 4:1, an input signal 4 dB above the threshold will be attenuated by 3 dB so that it is just 1 dB above the threshold on output. I had to reread that sentence, and I just typed it. Make sure that the numbers so far, leading to a needed compressor action of negative 3 dB, make sense. As the input signal increases to an amplitude that is 8 dB above the threshold, attenuation must increase from 3 dB to 6 dB in order to honor our 4:1 ratio and achieve the target of 2 dB above the threshold output. Attack time governs this further reduction in gain, even as the signal remains above the threshold.

Attack describes the speed of the imagined fader as it turns down the signal whenever attenuation is required. Fast attack times enable the compressor to react promptly and pull the fader down quickly; slow attack times are more lethargic. Both have production potential, as we discuss shortly.

5.2.5 Release

The speed of the imagined fader within the machine as it moves back up toward unity gain is determined by the *release* setting. When the amplitude of a compressed signal above the threshold starts to head back down in level—for example, during the decay of a snare hit or at the end of a sax note—the compressor must begin *un*compressing. Whereas 10 dB of attenuation might have been called for an instant ago, maybe only 8 dB of attenuation is expected now, based on applying the ratio to the now-lower portion of the signal above the threshold. The release time setting governs the speed of this reduction in attenuation. That is, the release parameter sets the speed that an attenuating compressor can turn the gain back up toward unity. When the amplitude of the music returns to a level that is below the threshold, the compressor must stop compressing. This same parameter—release—still applies, setting the speed of the compressor as it returns from negative gain (i.e., attenuation) to unity gain change (i.e., zero attenuation).

5.2.6 Makeup Gain

Finally, because the compressor attenuates signals based on the four parameters mentioned above, it is often desirable to turn the signal back up by a fixed amount using *makeup gain*. Managing the dynamic range of the entire signal chain is made easier if one can insert and remove a compressor without a significant change in average overall levels. In the course of a recording session, we often need to compare our compressed and uncompressed signals in order to verify that the effect envisioned is being achieved, without unwanted side effects. Hopefully, the signal is in fact better, not worse, with our particular compression settings. Such comparisons—with and without the effect—benefit from level matching so that we may react to the overall quality of the signal without being distracted by loudness differences. As the compressor is generally attenuating, on average, it is common to turn the output level up a bit overall—a final bump in gain. Makeup gain is the parameter that raises the overall level of the signal by a single fixed amount.

These five primary parameters—threshold, ratio, attack, release, and makeup gain—enable the device to carefully monitor and make fine adjustments to the amplitude of a signal automatically. Now we can concentrate more on other things: is the piano in tune? Does the pizza have enough sauce?

5.2.7 Options and Features

Your understanding of compression requires mastery of the five key parameters discussed in the previous sections. Additional features are found on some compressors: auto attack and release; hard knee versus soft knee; and look ahead.

AUTO

Clicking or pressing the auto button on a compressor, if such a feature is provided, tells the device to ignore your settings of attack and release times and to apply its own. In *auto*, the compressor has the ability to analyze the incoming signal and apply attack and release times that seem most appropriate. Typically, auto circuits speed up for transients and slow down for steady state sounds.

The goal of the auto feature is to have a transparent compression effect, offering level control with minimal audible artifacts. If you are going for an obvious, aggressive effect, you likely won't want the auto feature. If it is not in your personality to trust such important decisions to the ghost in the machine, consider the following: many auto circuits are very good. They can be quite musical, reflecting the care and cleverness of the designer.

More importantly, automatic attack and release times are rarely constant attack and release times. The time constants flex in response to the signal being applied. Take the compressor out of auto mode, and you must dial in the attack and the release time that is best for the signal. Your adjustments to the attack and release parameters define a relatively fixed (or close to it) reaction time for the compressor. You might tweak until you like what it does to the transients yet find that the sustained notes don't sound quite right. Modify attack and release until the sustained portion of the signal improves, and you undo everything you just achieved on the transient sounds.

When specifying the parameters yourself, you may have to compromise between two or more goals. Engaging the auto feature typically causes the compressor to apply changing attack and release times, treating transients and sustained sounds to different reaction times. That approach may be the only solution for a challenging track.

Complex, variable, and important tracks, like vocals, bass lines, and many solo instruments, often have transient spikes of energy comingled with legato long lines. These are good candidates for the auto feature.

KNEE

The transition from unity gain to a state of attenuation and back again—necessary each time the signal exceeds and then falls back below threshold—may be abrupt or gradual (see Figure 5.2). Hard knee invokes compression crisply; soft knee applies compression with a bit more cunning and coaxing.

We regularly use compression to reshape the amplitude envelope of our musical waveforms, as explained in more detail later in this chapter. This knee feature directly drives that reshaping. The application of compression with a hard knee, in which there is a kink in the transfer function, creates a small discontinuity in the slope of the waveform: a deliberate and often audibly apparent act of harmonic distortion.

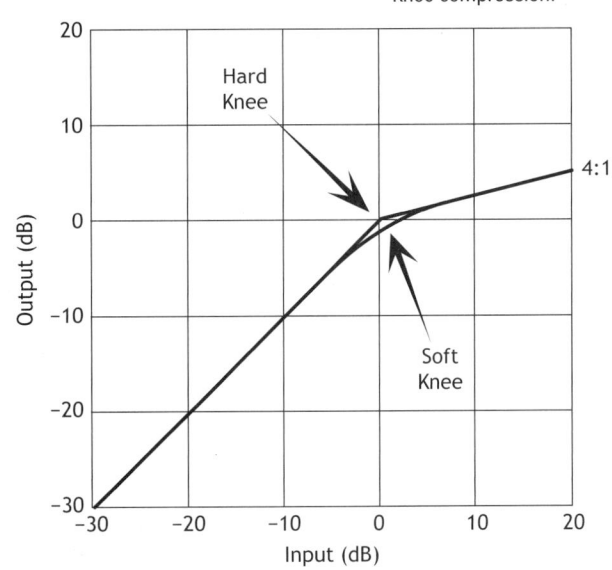

FIGURE 5.2
Hard knee versus soft knee compression.

Soft knee compression is a more transparent, sonically purer, application of compression, leading to a gentler reshaping of the waveform with less-noticeable distortion artifacts.

For each and every use of compression you implement, one key decision must be made: apparent effect (hard knee) or subtle effect (soft knee).

LOOK AHEAD

When your signal processor is digital, it finds opportunities not easily realized in the analog domain. For compressors, the most important feature accessed through digital signal processing is the ability to look at and analyze the signal before it compresses it. An analog compressor must react in real time to the signal it is compressing. Look-ahead compressors don't wait so long. They can read your hard disk well ahead of when you actually hear the associated audio. They look at what is coming and prepare for the necessary gain change. Look-ahead compressors have the ability, if desired, to shorten attack and release time concepts down to zero, compressing and uncompressing instantaneously because they saw what was coming. In fact, *negative* attack and release times are possible; the compressor begins compressing in advance of a threshold-exceeding event and releases before the signal falls back below.

The most obvious development attributable to look-ahead compression is loudness maximization (see the "Loudness" subsection under "The Fit" later in this chapter). Ever more sophisticated algorithms are being developed that evaluate the level, spectrum, and envelope of the signal so that it can be attenuated instant by instant and the overall signal turned up—often rather aggressively—with minimal impact on the quality of signal, particularly in the form of harmonic distortion.

! Any change to the shape of a waveform leads to harmonic and inharmonic distortion. In this way, all compressors are in fact distortion devices. Digital compressors offer the ability to implement gain changes in ways that minimize the audibility of unwanted distortion artifacts. As the look-ahead analysis of the compressor grows more sophisticated, and as the signal-processing code implementing the compression grows more elegant, it seems our tracks have the option to grow ever louder.

5.2.8 Caution

▶ The functional logic of a compressor is explained by the compressor parameters and features discussed previously. If you are new to compression, you owe yourself a break from this book now so that you can get to work in the studio and digest the clumsy logic of it all. Using a compressor is odd at first, as the parameters aren't exactly intuitive, and some surprising sounds are likely to result. After a bit of experience, return here for more advanced study. In order to use compressors successfully—particularly in more advanced applications—a finer understanding is helpful. We must get beyond this top-level understanding of parameters and start to explore some of the subtleties.

NUMERICAL NONSENSE

The actual values of the attack and release parameters aren't particularly universal and interchangeable. Transferring the full set of parameter values from one compressor to a different make and model—faithfully matching the threshold, ratio, attack, release, and makeup gain values—does not lead to the creation of the exact same effect on the new compressor. The sound of a compressor is not fully defined by these parameters.

When we drive a car, we can reasonably rely on the speedometer to accurately quantify the speed of the automobile, no matter who made the car or what year it was manufactured. But knowing that two cars are driving the same speed doesn't tell you anything about how they got there. Did they accelerate quickly or slowly, through a four-speed automatic or a six-speed manual transmission?

💡 Attack and release values—typically based on steady-state sine wave test signals rather than dynamically complicated musical signals—are an incomplete description of the reaction time of the compressor. They are best understood as indicating a relative range, from fast to slow, for the attack and release rates on that particular compressor. On an absolute scale, they are essentially meaningless to the practicing sound engineer, serving only as shorthand to indicate general trends. You needn't obsess over the numeric values of the attack and release parameters, only their rough order of magnitude. When you are using a different compressor, made by a different company, you can ignore the specific data and instead focus on the directionality of your settings—are you pushing to a faster time, or a slower one? Experienced engineers track attack and release through the words fast, medium, and slow, not through any specific numbers.

PARAMETER INTERDEPENDENCE

❗ Wait, there is a bit more confusion: attack, release, threshold, and ratio are actually more variable and more dynamic than the discussion so far admits. Sometimes one parameter changes its value based on the settings you've specified for the *other* parameters. Changing the threshold parameter can influence the attack time of the algorithm. Altering ratio may shift the threshold. I'm not making this up.

PROGRAM DEPENDENCE

Just as earnest engineers new to the field start to get their heads around the twisted logic and sonic surprise of compression generally and attack and release specifically, yet another layer of complexity appears. Most, if not all, compressors have some amount of program-dependent behavior. That is, even as you work to find the most effective attack and release values on a compressor for the track at hand, the compressor then has the audacity to flex those attack and release parameters based on qualities of the signal being compressed.

It is not unusual for compressors to treat transients with faster attack and release times than more sustained, steady-state signals. And it makes these adjustments to attack and release times even when we haven't directly modified the attack or release settings.

Moreover, the attack and release trajectories are sometimes a function of the amount of gain change currently implemented by the compressor. Slow settings as the compressor initiates compression are often desirable. However, when a compressor is in the midst of compressing a signal, already applying some 6 to 12 dB of attenuation, the attack and release times might start to increase. More intense attenuation begets quicker reaction. Meantime, the attack and release parameters themselves, as set by the engineer, remain misleadingly fixed on the faceplate of the device.

The attack and release settings we specify amount to starting points driving the general behavior of the compressor. They rarely represent rigid values strictly followed by the compressor at all times. Listen closely, and be open to surprise.

This interaction among parameters and waveform dependence are frustrating at first and can slow our ability to master compression. Have patience and stick to it, because these seemingly random qualities usually do improve the usefulness of the compressor. With experience, you'll find things formerly mysterious prove musical. What at first is frustrating in the end is artistic.

PUMPING AND BREATHING

The most obvious artifact of too much compression is *pumping and breathing*. Pumping and breathing refers to the audible, unnatural level changes associated primarily with the release of a compressor. The audio signal contains material that changes level in unexpected ways. It might be steady-state noise, the sustained wash of a cymbal, or the long decay of any instrument holding a note for several beats or bars. Listeners expect a certain amplitude envelope: the noise should remain steady in level, the cymbals should decay slowly, and so on. Instead, the compressor causes the signal to get noticeably louder in a way the sound itself never would. This unnatural increase in amplitude occurs as a compressor turns up gain during release.

Of course, compressors always turn up the gain during release. Pumping is the generally unwanted, audible artifact of compressor release revealed by the slow or unchanging amplitude of the audio signal being compressed. The fun part is that the release value isn't necessarily too slow or too fast. The problem may be because the release parameter on the compressor is too, well, *medium*.

A much faster release time could remove pumping by having the compressor release immediately and unnoticeably after the compression-triggering event. A snare drum hits. The compressor attenuates. The snare sound ends almost immediately. The compressor releases instantly. If this all happens very quickly on each snare hit, the snare sound itself makes it difficult, if not impossible, to hear the quick releasing action of the compressor. The result is a believable sound, despite compression. The release happens so soon after the snare hit and occurs with such a steep release slope that pumping is not easily heard. In this way, fast release settings help prevent unwanted pumping and breathing.

A slow release time might also be a good solution for removing this artifact. Slow the release time substantially so that the level change of the compressor during release takes a second or two longer and the gradual increase in gain by the compressor is slower than the gradual reduction in amplitude during the decay of the sound. The result is a natural-enough-sounding decay of the audio signal, even as the compressor slowly turns up the level. The cymbal takes longer to decay, but it still sounds like a naturally decaying cymbal. Very slow release is another solution for unwanted compressor pumping and breathing.

! Beware of pumping whenever a steady-state, near-steady-state, or sustained sound occurs within the signal you wish to compress: cymbals, long piano notes, tape hiss, electronic or acoustic noise floors, reverb tails, synth pads, whole notes, double whole notes, or longer. Signals that have a predictable, slowly changing amplitude envelope can inadvertently reveal compression, forcing us to abandon compression altogether, back off the degree of the effect (raise the threshold and/or lower the ratio), or at least modify the release settings.

On the other hand, note that the pumping artifact may on occasion be an interesting effect worth highlighting. If each snare hit causes the cymbals to pump, the result can be an interesting, unnatural envelope in which it sounds as if the cymbals have been reversed in time. If the performance on the congas causes the piano to wobble and tremolo in time to the music, it might amount to a new, synthesized piano-like texture that contributes a unique texture to your mix. Used judiciously, such an effect can add great interest to certain productions.

5.2.9 User Interface

Like any great musical instrument, compressors come in many forms. Some are easier to play than others. Any given compressor may prove better at some jobs than others, very much analogous to a musical instrument. A guitarist reaches for one type of guitar for one song and a different guitar for another. We develop the same subtle sonic finickiness for compressors. Expect to spend some time learning the ins and outs of a long list of makes and models, but it helps to know that they all possess the same core capabilities.

PARAMETER SHORTCOMINGS

Although these five parameters are always at work, they are not always on the faceplate of the device; that is, they are not always user-adjustable. There are compressors at all price points that leave off some of these controls. For example, attack and release may not be user-adjusted. They are embedded in the design, determined by the manufacturer. The particular attack and release characteristics are an unmodifiable part of that compressor's sound.

Other compressors offer full adjustability of all the parameters yet also offer presets. The presets reflect someone else's careful tweaking of the parameters for a satisfying sound in specific applications. Certain presets might be recommended for bass and others for drums.

💡 Sometimes the presets simulate the ratio, attack, and release characteristics of other, vintage, collectible, famous sorts of compressors. You should certainly spend some time with the fully adjustable types of compressors for exploring and ear training, but don't choose against compressors just because they offer only a few knobs and sliders. A lack of user-adjustable parameters does not mean a lack of sound quality. These simpler user interfaces can often get the job done quickly and with terrific sonic results. A lack of adjustable parameters may not be a shortcoming; indeed, it might be a signal-processing virtue.

MULTIBAND COMPRESSORS

The user interface grows profoundly more complex with multiband compressors. Though it is richly complex and a challenge to master, it isn't difficult to understand how multiband compression works. Integrate two, three, four, or more compressors together in parallel into a single processor and divvy up the audio band among them. The lows go through one compressor, with its own settings for threshold, ratio, attack, and release. The highs go through another compressor, with its own unique settings. Need more? Divide the spectrum again. Send the mids through another compressor, and the presence range (upper mids) through yet another. The outputs of all of these individual, band-dedicated compressors sum together, reconstituting your signal back into full bandwidth, now made of spectral regions treated to different forms of compression.

As you get smart working with single full-band compressors, you'll develop the ability to tailor different compression ideas to different spectral regions, perhaps treating the lows to slow releasing but high ratio-level control, while the mids get faster-acting, adrenaline-inducing fast attack and release compression, while the highs get … your imagination is the limit.

🔊 Listen carefully: first, because hearing different compressors acting on different frequency ranges of the same signal is an acquired skill. Second, and likely more important, note that you are sending the entire signal through some rather severe filtering. The signal is carved up along the frequency axis and fed to different compressors. The outputs from all the compressors are then summed together, reuniting all frequency components of the signal back into a single, full-band signal. The decision to run any signal through a bank of filters and the requisite summing back together that follows should not be taken lightly. The filters themselves are a significant amount of signal processing and are often far from transparent, sometimes introducing phase distortions to the signal so severe that transients are blurred, clarity is diminished, naturalness is reduced, and beauty is robbed.

💡 Before you use a multiband compressor, it is recommended that you first pass the desired track through the compressor without compressing. Listen for any side effects due simply to passing through the filtering and the summation process. Important tracks like the vocals, bass, drums—or your entire mix—may not sound as good having passed through the filters, through no fault of the various compressors. Fragile, natural signals full of transient detail and spectral subtlety, such as strings, piano, and other acoustic instruments, may start to sound unnatural just for being subjected to the filters on the input to the multi-band compressor. Make sure the multiband compressor passes this test first: does the track still sound good simply passing through the processor, with all compression out of the way (set the threshold so high that the audio never exceeds it and the ratio to 1:1 so that it still doesn't compress it)? If you don't like the hit in quality that occurs *without* compression, it is unlikely you'll ever be happy with the sound *with* multiband compression.

ENVELOPE MANIPULATORS

A variation on the compressor theme is the family of envelope manipulators that offer us threshold-independent ways to emphasize or deemphasize the transients on the front of a waveform and to lengthen or shorten the decay at the end of the waveform. Best suited to discrete, isolated, consistent waveforms like drum hits or simple walking bass lines, envelope manipulators track the rate of change of the amplitude envelope of the signal and let us adjust the level up or down during note onset and offset. Parameter names for this relatively new approach aren't standardized. Instead of threshold, ratio, attack, and release, you'll see parameters like attack, sustain, length, or similar. The attack parameter turns up or turns down the waveform instantaneously during the beginning of the note. The sustain/length parameter allows for increases or decreases in level at the end of the note.

5.3 TECHNOLOGIES

The signal flow through a compressor (Figure 5.3) consists of an audio path through a gain stage, with a level-detector circuit controlling the level of the gain change component. The level detector receives the same signal as the gain change

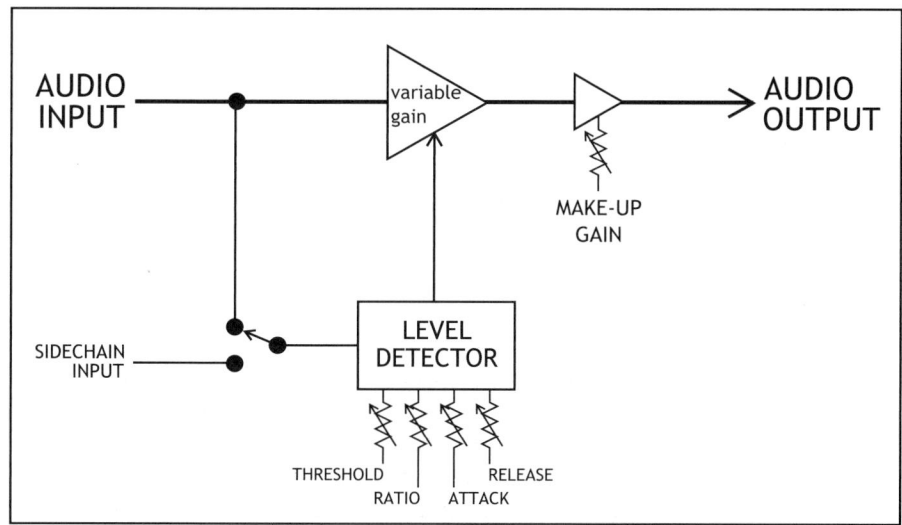

FIGURE 5.3
Signal flow through a compressor.

element in most applications. A side chain input makes it possible to feed a modified or entirely different signal into the level detector. This enables a great range of signal-processing effects, as discussed later in this chapter. The parameter settings (threshold, ratio, attack, and release) directly influence the sound of the compressor, but the type of gain change device employed in the design of the compressor and the level detector that assesses level and applies your settings make a significant sonic contribution as well. You'll learn when to use which compressor on any given track by keeping track of the basic technologies within. Just as we divide the microphone closet up based on the components and technologies within (moving coil, ribbon, condenser, tube condenser, transformerless, and so on), we separate compressors based on the type of gain stage and level-detector circuit used. Knowing a bit about how they work gives you some intuition about what to listen for and enables you to draw some general conclusions that apply to other compressors using similar technologies.

5.3.1 Tube Compressors

Once upon a time, tube-based (valve-based) compressors were the only choice for machine-modifying gain. The resulting sound is often so desirable that we still go to a lot of trouble and expense to use tube compressors to this day, despite all the innovations that have followed.

The level-detector circuit sends a voltage to the tube that directly drives the gain of the tube. Fairchild and Altec compressors were early devices built on this principle. Today, Manley, Summit, and a handful of others make a state-of-the-art update of this approach. You are probably very familiar with the often-desirable qualities of tube-based components. The sometimes musical and appealing distortion introduced by a tube-based gain stage is part of a tube compressor's sound (see Chapter 4). In addition, there is the reaction time of the tube as its gain is changed by the level-detector circuit. Any decrease or increase in gain is not absolutely instantaneous, giving tube-based compressors their own unique attack and release time properties. As we discuss in detail in the following section, the rate at which the gain is increased and decreased has a major influence on the overall quality of the compression effect.

5.3.2 Optical Compressors

Optical compressors use the interaction between a light source and a light-sensitive resistor to influence gain. The level-detector circuit uses the audio signal to illuminate a light. That light—which might be an incandescent bulb, a light-emitting diode (LED), or an electroluminescent panel—shines on a photocell. The amplitude of the signal through the compressor is driven by the resistance of the photocell in reaction to the amount of light shining upon it. The entire LA series of compressors (LA-2A, LA-3A, and LA-4A) formerly by Urei and now made by Universal Audio—as well as some Joe Meek compressors, among others—leverage this principle.

Switch on a light, and there is a time element to how long it takes the light to achieve full illumination. Switch it off and it goes dark, losing brightness at a rate characteristic of the type of light source. LEDs can snap on and off nearly instantly. Incandescent bulbs turn on more slowly and linger briefly with a little glow even after they are turned off. Luminescent panels exhibit a memory effect in which, if they have been bright for a period of time, they remain illuminated even after the illuminating voltage has been removed.

The particular type of photoresistor similarly contributes to the optical compressor's attack and release characteristics—there is a reaction time to the electrochemistry of the cell as its resistance increases and decreases in reaction to the light intensity. An optical compressor's attack and release characteristics are ultimately determined by the complex interaction of the light source and light receiver chosen by the equipment designers.

5.3.3 FET and VCA Compressors

Solid-state compressors might leverage a field effect transistor (FET) or other transistor-based voltage-controlled amplifier (VCA) to change the gain of the compressor. The Urei (now Universal Audio) 1176LN compressor is a particularly famous use of an FET gain stage. The dbx compressors (old and new) take advantage of high-quality audio VCAs for compression. Using transistors, the detector circuit is empowered to drive the gain via a control voltage. Compared to tube and optical compressors, FETs and VCAs are typically capable of much faster attack and release times, and the gain stage is usually capable of more extreme swings. Although transistors can be very linear when not overdriven, their distortion character is markedly different from tube-based gain stages—typically edgy and easy to hear. This distinct sonic coloration is often the reason for using a FET or VCA compressor—but don't tell the purists.

5.3.4 PWM Compressors

Pulse width modulation (PWM) offers a more abstract approach. The level of the signal drives the frequency of a very-high-frequency—in the megahertz range!—pulse stream. That pulse stream now turns on and off based on the level of the signal and acts as a very agile form of level detection feeding a FET or VCA (typically). This approach is an elegant form of level detection, acting very much like a class D amplifier. Look for PWM compressors to offer fast, transparent gain reduction across a wide dynamic range.

5.3.5 Digital Compressors

Compressors in the digital domain are liberated from the constraints of real-time analog circuit components. The qualities of digital compression are as varied as the companies that design them. When audio becomes a string of numbers, digital compressors become a bank of calculations.

One might be tempted to conclude—numbers being numbers—that all digital compressors sound alike, which is simply not the case. Humans write the code that governs the behavior of a digital compressor. There is no right answer and no single solution to compression.

A digital compressor represents one algorithm for achieving compression through calculations, an algorithm that is the result of countless decisions, trade-offs, and moments of inspiration by the creative software engineers who write the code.

! Software might be written with the goal of trying to emulate old analog compressors. With so many desirable analog compressors, it makes sense to attempt to capture some qualities of tube, optical, or solid-state compressors digitally. The very quirks and personality traits that you must learn associated with all of the analog technologies are simulated in these digital compressors. Of course, they are just simulations. When you click the "optical" button on a digital compressor, it doesn't make the digital compressor optical—it just modifies the algorithm within the plug-in so that it takes on some of its traits. This result can be a bit confusing. As you learn optical compressors, don't let digital emulations of optical compression confuse you. They'll sound slightly different. The best way to learn the sound of tube compression is to use tube compressors. As you begin to internalize the unique sonic attributes of tube compressors, it then tells you what to listen for when you use a plug-in that has a tube compressor check box. A picture of a tube in the plug-in doesn't give it the sound of a tube. Careful circuit modeling by clever digital signal processing engineers can impart many appealing characteristics of a tube through digital means. You are encouraged to file digital simulations of reference analog designs in a different place, mentally. You must internalize the qualities of tube compressors, optical compressors, FETs and VCAs, and digital versions of each. They are different enough to qualify as a different compressor technology, even though theses digital tools are highly influenced by their analog predecessors.

Another valid approach to digital compression is to create entirely new algorithms that essentially reinvent the compressor—leveraging opportunities not achievable through analog topologies. Loudness-maximizing-type compressors are a critical example of this. When the compressor is a plug-in, it has the ability to read the sound file well ahead of when the sound is actually played. Analog compressors react to the audio the instant it happens. Look-ahead digital compressors can prepare themselves for each and every spike and transient in the waveform well before the spike or transient occurs. The attack and release and knee characteristics can be adjusted in any way the digital designers see fit—obtaining loudness with fewer artifacts, for example. Compressing for loudness is usually best saved for the mastering session and doesn't much belong in a mix session. Mastering engineers invest a great deal of time and resources finding ways to enhance loudness without damaging the many layers of complexity and the fragile weaving of tracks and effects that you, the mix engineer, created.

💡 Your challenge, with compressor technologies, is to be aware of the type of design being used within every compressor you own and—over time—to learn what the sonic advantages and disadvantages of each are. As you accumulate experience, you develop a sense of the subtle flavors and features of the different compressor types, similar to how a guitarist knows the differences between makes and models of guitars—both in terms of how they sound and how they feel in the hands of the guitarist. In this way, you'll develop essential intuition about when to reach for which compressor.

5.4 MIX STRATEGIES: COMPRESSION AND LIMITING

The compressor is used to create a range of effects, each pursuing different goals artistically, through different means technically. The most intuitive use of compression is to reduce—sometimes slightly, other times radically—the audio dynamic range of a signal. Less obvious, but no less important, we use compression to improve intelligibility, sharpen or dull the onset of notes, emphasize ambience, highlight breaths, alter timbre, add distortion, and more: so many possibilities from this one, humble processor.

As happens throughout our work in this book, we organize the rich production possibilities into a logically organized set of strategic motivations: we fix, we fit, and we feature qualities of the signal best obtained by using compressors and limiters.

5.4.1 The Fix

Smart recording engineers anticipate any potential problems in the amplitude domain and insert compression and limiting to make any needed corrections first. Preventing problems is the best form of fixing them.

SAFETY COMPRESSION

▶ Performance dynamics can vary tremendously. When a guitarist gets confident and excited, he or she may start playing louder. A sax player taking a solo steps it up, but as part of a horn section, tucks back in. If any of these great performances overloads the analog-to-digital converters or the analog tape machine, the track becomes unusable. Without some amplitude protection, a magic performance is easily lost to distortion. Safety compression to the rescue. Be ready with some gentle (around 4:1 or less) compression across the track, medium to fast attack and release, with the threshold set a good 6 dB below the threshold of distortion. Applying a 4:1 ratio 6 dB below distortion converts that last 6 dB of headroom into 24 dB of usable input level! Between this safety compression and your manual riding of gain during expressive takes, you should be able to prevent any unwanted overload while capturing an expressive performance. As a result, the signal racing through our picky, rather inflexible equipment will not sour when our musicians soar. Aided by compression, we are able to successfully capture the work of musicians who have found that special, life-changing moment when they are performing as they have never performed before—technological precision captures artistic expression.

PEAK LIMITING

There may be no more physical form of music making than playing drums. And during the course of a song, some snare hits are understandably struck harder than others. During the chorus, the best drummers physically live and physiologically experience the dynamic changes they play. When the music gets going, they start slamming. As a result, the drum levels during the chorus might be substantially hotter than the delicate, ghost-note-filled snare work of the bridge.

When we encounter very different performance levels, it isn't unusual to subject the track(s) to some dynamic range–reducing constraints. A limiter is employed to attenuate the extreme peaks and prevent nasty distortion—distortion that comes from overloading any gain stage along the signal path, especially the converters and/or analog tape recorder.

A limiter is nothing more than a compressor taken out to rather extreme settings: the ratio is high—greater than 10:1—so that any signal that breaks above the threshold is severely attenuated, kept very close to the level of that threshold; the attack is very fast so that nothing above threshold sneaks through, even briefly, without limiting; the threshold is high so that it only affects the peaks, leaving the rest of the music untouched; and finally, release is as fast as it can be without causing audible artifacts.

If you find it necessary to do some peak limiting but can't seem to dial in any settings that work without causing a noticeable hiccup or an instant of disappointment in the sound at each peak *and* if you have a very high quality multiband compressor, evaluate the spectral nature of the offending peaks and do the limiting only on the offending band. If the other bands get through without gain reduction, you may be able to preserve the overall excitement of the performance and richness of tone while letting a limiter solve your technical offense: distortion due to too much level in some spectral zone.

Figure 5.4 gives an example of peak limiting, the sort of processing used to prevent distortion and protect equipment. Fitting a signal on tape without oversaturating or broadcasting a signal without overmodulating requires that the signal never exceeds some specified amplitude. Peak limiters are inserted to ensure these amplitude limits are honored.

! In live sound, exceeding the amplitude capability of the sound reinforcement system can cause feedback, overdrive amplifiers that may in turn damage loudspeakers, and turn happy crowds into hostile ones. Limiters offer the solution again. They guard the equipment and listeners downstream by confining the signal amplitude to safer levels, no matter how much the artists might wail on stage. In this way, an unpredictable artistic performance is squeezed into an unforgiving amplitude processor.

Analog-to-digital converters, radio transmitters, and tape recorders can be particularly unpleasant when overdriven. Compressors and limiters are a regular process to add to the signal chain, giving us a protective buffer before troublesome distortion ruins our music.

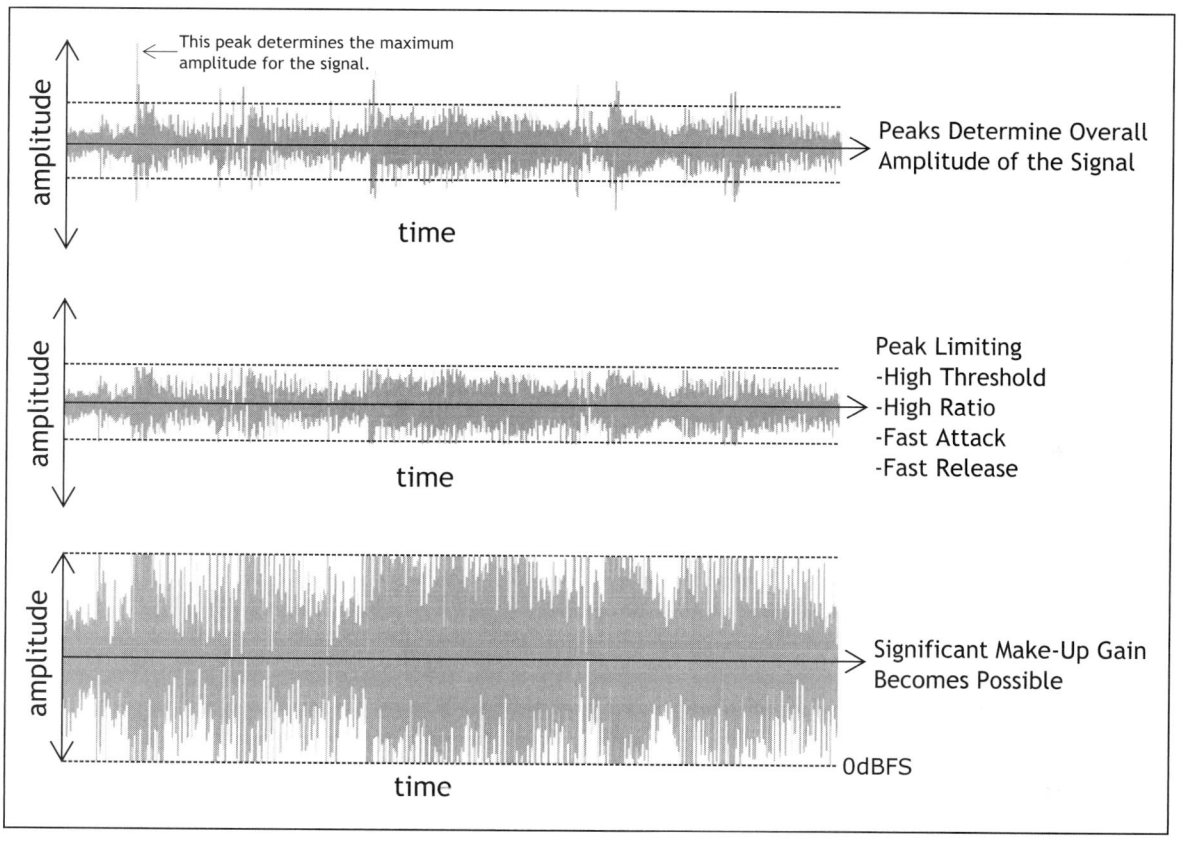

FIGURE 5.4
Peak limiting.

OVERCOME NOISE

Sometimes it's music versus noise. Music recordings must fight the noise floor of the storage medium (e.g., analog tape) or the listening environment (e.g., an automobile). Compression offers much-needed help, reducing the higher-amplitude portions of a signal to a level that is a little nearer the lower-amplitude parts. With the loud bits turned down, we then turn the whole signal up. That is, with the high-amplitude parts somewhat attenuated, we are able to turn the overall level of the signal up some without subsequent risk of overload distortion, which makes the softest parts of the performance a little louder, lifting them up out of the noise floor.

Some engineering finesse is required, as this application of compression generally does not intend to change the *musical dynamics* of the signal. The dynamics intended by the composer and the performer should not be undermined. Overcoming noise with compression requires the *audio dynamics* to be narrowed without harm to the musicality of the piece.

It should be noted that overcoming noise with compression is, at first, a bit of a logical conundrum. Compressors attenuate. How can turning our music down help us avoid noise?

! The key is to recognize that the noise-suppressing compression happens to the signal before it encounters noise. Compression is used to reduce the high-amplitude portions. Makeup gain is used to then turn up the entire signal, and *then* the signal is allowed to hit a bit of noise. Because our compression effect had net makeup gain, the softer portions of our musical signal are now higher in amplitude, better able to outshine any distracting noise. If your signal is already noisy, the damage is done, and no compressor settings will help. Expansion and gating, however, might do the trick (see Chapter 6).

DE-ESSING

Pop-music standards push our productions to have vocal tracks that are bright, airy, larger-than-life, better-than-the-real-thing, in-your-face, over-the-top, only more so—and that's a bit of an understatement.

🔊 Our vocals must rise above sizzling synths, gritty guitars, shimmering cymbals, resonating reverb, and everything else we throw into the mix. Needless to say, it is common to push vocals with a heavy dose of high-frequency hype (via EQ; see Chapter 3). We pretty much get away with this aggressive equalization move, except for those instances in which the vocal was already bright to begin with: during hard consonants like S, Z, and F (and, depending on the vocalist and the language, even X, T, D, K, ch, sh, th, and others). These sounds are naturally rich in high-frequency content. Run them through the desired EQ-induced increase at high frequencies, and the vocal track now zaps the listeners' ears with pain on every S-like sound. It's a real problem when everyone in the room blinks each time the *sssinger sssings sssome essssssesss*.

This problem is also tricky to solve. Tracks on the radio have wonderfully bright and detailed vocals. When an inexperienced mix engineer tries this for the first time in their studio, they are stuck. With added high-frequency hype, the sound of much of the vocal track may be improved, but the esses are too loud—so loud they hurt.

All we need is some clever, well-timed compression to solve the problem.

In the compressor applications discussed so far, our settings of threshold, ratio, attack, and release have been based on the same signal that is actually being compressed. The level detector reads the same signal being passed through the gain change element (see Figure 5.3). It makes sense. When compressing a vocal, shouldn't the compressor evaluate the incoming vocal track?

What if the compressor modified the gain of one signal while "looking at" another? What if the signal feeding the level detector portion of the compressor were different from the signal running through the gain change device?

Specifically, consider our aggressively equalized lead vocal. Rather than compressing the vocal based on the vocal track itself, let's use a different signal to govern the compressor's behavior. Using a *side chain*, a different vocal signal is fed into an alternative input—an input that feeds the level detector only. The vocal itself is what gets compressed, but the behavior of the compressor—when,

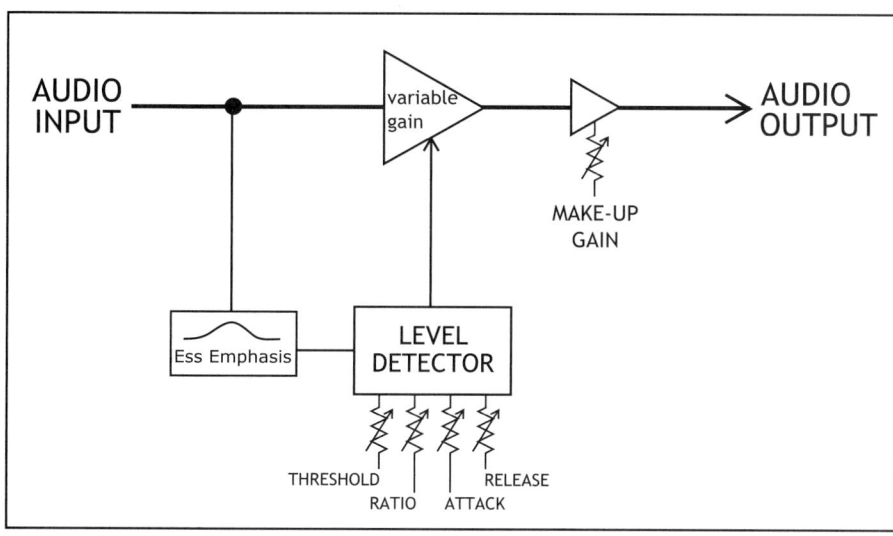

FIGURE 5.5
Signal flow of de-esser.

how much, how fast, and how long—is governed by the side chain signal. To get rid of esses, we feed a signal into the side chain that has enhanced esses. That is, the side chain input is the vocal track specifically equalized so as to exaggerate the esses, and deemphasize the rest (see Figure 5.5).

This S-emphasized version of the vocal never makes it into the mix. We don't want to hear this signal. Listeners at home never hear it. Only the compressor does. When the singer sings an "S," it goes into the compressor's level detector extra essy. And the extra-essy esses break through the threshold and trigger the compressor into action. The side chain signal is the vocal with a high-frequency boost, maybe a +12-dB high-frequency shelf somewhere around 2–8 kHz, wherever the particularly painful consonant lives for that singer. The side chain vocal can be cut at other frequency ranges; the lows aren't needed. The compressor is set with a mid to high ratio, very fast attack, and very fast release. The threshold is adjusted so that the compressor operates on the loud "S" sounds only, and ignores the side chain vocal the rest of the time.

The result is a de-essed (as in, no more loud letter "S") vocal. The compressor instantaneously attenuates the great-sounding vocal every time an S is sung, just enough to take the edge off. You can still hear the vocal perfectly fine. You can still hear that an "S" was sung. You can still understand the vocal. In fact, the vocal sounds perfectly natural, and perfectly intelligible. In between the "S" sounds, the compressor doesn't change the level of the vocal at all. During each "S," the compressor clamps down.

Now the vocal can be made thrillingly bright without fear: use EQ to boost the highs, knowing that each "S" sung will be reliably pulled down in level just the right amount by the de-esser.

There is a multiband compressor variation on this approach. Some of the best de-essers have the ability to de-ess only the high-frequency portion of the vocal, passing the low-frequency portion uncompressed. If the filters separating low from high are of high quality, a better vocal sound might result. This extra processing makes it possible for the de-essing process—the instantaneous reduction in level during sizzling consonants—to affect only the relevant spectral portion of the signal, leaving the rest of the vocal track completely unmolested.

▶ If you lack a de-esser but have a multiband compressor, consider bypassing all bands but the one tuned to the "S" range (search between 2 kHz up to 8 kHz, rarely higher) and dial in a high ratio, fast attack, fast release on that band only. Adjust the threshold so that it only grabs the zingy zaps of energy and otherwise passes the vocal, and you've got an elegant solution to the sizzling "s" situation.

5.4.2 The Fit

With challenges presented across some 48 or more tracks, compression can be one of the most effective tools we have in the studio for helping us piece together our mixes.

LOUDNESS

Balancing a mix (see Chapter 1) is a fundamental and rather challenging part of mixing. We iteratively turn up and turn down the various tracks until an effective blend of all tracks is found. Faders, mutes, and pan pots are the essential tools of the balancing task; compression offers another. Compress to raise the average level of a track and it becomes louder—a variation on the act of pushing up a fader.

As with raising faders, care and artistic maturity are needed. Increasing the loudness of *all* your tracks via compression will solve few problems and likely cause more. If all the tracks are made louder, the level conflict will remain: some will still drown out others. The mix will still be out of balance. Balancing finesse is still very much needed. Selective use of compression on some tracks to make them more easily heard is the right approach.

We should make an essential distinction between level and loudness. Proper use of level is a full-time job for engineers. We are forever strategically navigating the dynamic range constraints of all of our audio systems—too loud and it distorts, too soft and noise becomes a problem. Engineers watch levels at all times; meters are studied at each and every stage in the life of a signal. Managing level simply means that when noise is a problem, we seek ways to turn our signal up, out of the way of the noise; when distortion is a threat, we seek ways to turn our signal down, preventing overload and its associated distortion. Use compression as needed to narrow the dynamic range of your signal and place it within the finite dynamic constraints of your studio hardware and software.

A handy side effect of compressing to control level—reducing the overall dynamic range of the signal—is that now it can be turned up. As discussed above, we do this to overcome noise. But turning it up offers a seductive bit of loudness. And on first listen, we might think that louder is better, and we might then want a little more, and then a little more still, and then a lot more. When we expand the scope of the discussion from level to loudness, we need to be philosophically prepared.

A commitment to watching and controlling level does not give us universal permission to chase associated loudness. Nurturing our signals through careful checks on level is a technical necessity, done constantly. Hyping our signal through loudness is a very different effect, implemented tastefully, only when needed.

Figure 5.4 demonstrates the sort of compression that raises loudness. With the peaks attenuated, the overall signal is raised in level (see the lower part of Figure 5.4), with the result, when desired, of an increase in the overall loudness of the signal. The average amplitude is higher, the area under the amplitude curve is increased, and the perceived loudness is raised.

! There is no doubt that this effect, compressing for loudness, is overused. It is often the coarse signature of inexperienced or low-quality engineering talent. You and I must use mature musical judgment to find the appropriate amount of peak limiting and associate loudness maximization, without taking it too far. We should try to create productions that invite repeated listening by passionate fans without distortion-driven fatigue while still sounding exciting and competitive in the market.

What works on the vocal track and the finger cymbal should also work on the overall mix. Why not patch in a loudness processor on the mix bus? Maximizing the loudness of the entire mix is too often taken to rather radical extremes where mixes are absolutely crushed (i.e., *really* compressed; see also squashed, smooshed, squished, thwacked, spanked, slammed, and so on). The hollow goal: compress for enough extra loudness that your song sounds louder than all the other songs on the radio dial or shuffled playlist.

Selling music recordings is a competitive business. Loudness does seem to help music sales figures, at least in the short term. And so it goes. You make your mix loud today, and someone else makes their mix louder tomorrow. Often the music suffers in this desperate search for loudness and hope for sales.

Sorry: only one of us gets to be loudest. Everyone else is just loud. Trying to make your mix the loudest mix ever is a worthless goal. Allow yourself to make your mix loud, but recognize that look-ahead compressors adding loudness do so with some trade-offs, distortion being chief among them. Don't subject your recorded art to unnecessary compression artifacts in search of that last fraction of a decibel of loudness. Let someone else win the "loudness mix of the universe" award this week and instead chase other, more important sonic attributes.

Artist, producer, and engineer must make this trade-off carefully: long-term musical quality versus the short-term thrill of loudness. One must not forget that the consumer has the ability to adjust the level of each and every song they hear by operating the volume control on their playback system. If listeners want the song louder, they can make it so. Loudness might help initial sales, but deep artistic quality is the better way at it. Loudness won't fix a weak performance or correct poor songwriting. If you want to up the odds of commercial success, focus on musicianship and artistic merit at every step of the production rather than exaggerated loudness at the end of the process.

If you are a music fan, you likely listen to recordings that are more than a decade old, recordings made before this race for loudness empowered by look-ahead digital compressors had begun. Perhaps you even listen to recordings from the last century—the 1990s, the 1980s, the 1970s, the 1960s, and earlier, back through the history of music. The iconic recordings that we find worth listening to decades later had no benefit from loudness processors; they hadn't been invented yet. As a result, they suffer none of the associated overdone, ear-exhausting distortion.

🔊 If you play a contemporary track and a vintage track from the 1950s, side by side, you'll likely need to turn up the old record a bit. Once you've done so, however, there is no reason the recording from the 1950s can't sound every bit as exciting as the current production. Loudness is a misleading kind of enhancement. When we compare a louder recording to a softer one, it is easy to prefer the louder song. But when we compare a loudness-maximized recording to a less-compressed version, and use the humble volume knob to raise the level of the less-compressed version to approximately match the level of the processed mix, we are no longer distracted by loudness and listen instead to other qualities in the recording, and these are magic "other" qualities: detail, depth, width, timbre, spaciousness, bloom, contrast, airiness, and so forth. It is often the less-processed mix that offers better sound quality on all of these essential, nonloudness dimensions. Any kid can make a mix loud. Only the experienced, committed, smart, creative, clever engineer can make a mix art.

The only way to make a fair comparison between two recordings is to level-match them, and then listen carefully. Most mix engineers who go through this careful comparison find that the transient toppling, crusty film of distortion covering the entire song caused by too much loudness processing is an unacceptable indignity for a mix that reflects many hundreds of hours of their best work. Louder is not better, but it can sure seem that way at first. And on second, careful listen: louder is often worse!

As mix engineers, we must be good stewards of all qualities of the recording, including the most subtle musical qualities that might attract listeners decades from now. Chasing track loudness can deafen us to these essential other qualities. Track by track, choose loudness effects only when you need it, and don't fall for its seemingly universal initial appeal. Then, save overall mix loudness for the mastering session. Mixing smart means anticipating the contribution of a great mastering engineer—their ears, their loudspeakers, their room, and their compressors deliberately designed for increasing loudness with minimal side effects. Here endeth the loudness lecture.

IMPROVE INTELLIGIBILITY AND ARTICULATION

The vocal isn't much use to the mix if listeners can't understand what they are singing. There is real artistry and technique in singing and speaking for recording and broadcast. That singers need vocal lessons and practice is accepted. Actors and voiceover artists do too. Part of what is studied and rehearsed is diction. It is essential in almost all styles of music and forms of speaking that each and every word be plainly understood. The intelligibility of words, spoken or sung, depends in part on the ability to accurately hear consonants. Vocal artists are trained to control their vowels and make distinct each and every consonant. Pronouncing a word, the letter B must not be confused with the letter D; Ns and Ms must be differentiated.

Imagine looking at a meter on the mixing console while a singer sings "bog" and "dog." The meters would reveal no visible difference between these two words. Imagine looking at the waveform of each word on the screen of your digital audio workstation. They appear, to the naked eye, incredibly similar. There is precious little difference between these two words, differences that aren't easily measured or confirmed by any device available in the typical recording studio—except our hearing. The incredibly subtle difference is extracted from the signal our ears pick up through the work done by our hearing physiology and the analysis performed by our hearing neurology.

Add in hiss, hum, static, and other noises, and distinguishing "bog" from "dog" becomes more difficult. Add drums, keys, and ukulele to the mix, and the intelligibility is further threatened. Add delay, reverb, and other effects, and hearing detail in the vocal may become more challenging still. "Bog" or "dog?" Is the song about a soggy swamp, or a furry creature? Music fans want to know.

▶ Compression can help fit a vocal in among all this distraction. Adjusting the gain of the vocal, syllable by syllable and word by word, compressors help the engineer ensure that no important part of a vocal performance gets swallowed by competing sounds. Set the threshold low enough that it is below the average amplitude of the vocal. Reach for a fast attack time and medium to fast release time, and raise the ratio as needed; 4:1 is often enough, but you may have to push as high as 10:1. The effect is not usually transparent. The vocal is clearly being processed. Done well, though, it can be worth it. Intelligibility rises.

Compression is used often—particularly on the dialog portion of film, television, and game sound tracks—to tame the loudness inconsistencies of the vocal performance so that no portion of the track is difficult to hear. Every syllable of every word is tamed in amplitude by the compressor so that the intellectual meaning of every syllable of every word is communicated clearly. In these situations, vivid intelligibility often trumps natural believability.

The concept of intelligibility isn't limited to vocals. The articulation of each and every note by a melodic instrument can also carry significant musical value, yet risks being drowned out by other elements of a crowded mix. The intelligibility and the articulation of a horn line or a guitar part can be every bit as important as the intelligibility of the lyrics. Many styles of music, from jazz to progressive rock, make their artistic living based on communicating the exact phrases—the timing and the pitch—of sometimes several instruments at once. Careful control of dynamics through compression may be, at times, a required part of the production. We use compression to improve intelligibility for vocals and articulation for instruments.

As with so many effects, tread carefully. This compression strategy alters the musical dynamics of the track processed. If the sax player wanted some notes to be softer, with a more impressionistic and expressively vague form of articulation, this kind of compressor effect will be unwanted. We must have the musical good sense to apply only as needed, where artistically appropriate.

SMOOTH PERFORMANCE

Performers aren't perfect. Experienced studio musicians augment their musical chops with total performance control, but microphones are unforgiving contraptions.

Lead singers are obligated to be expressive. It is insufficient to just sing the words. They must communicate the associated feelings through all audible attributes of their presence in the recording. They do it through fine adjustments to pitch, level, timbre, phrasing, breathing, and more. Listeners can't see them, so singers have to get the feeling out using only sound. It's a miracle it ever works. Not surprisingly, some phrases, some words, and some syllables are too loud and others are too soft. Constant fader riding of key tracks like the vocal will be essential. Gentle level riding via compression may bring essential help.

When the acoustic guitar player gets nervous, he or she might lean to and fro on the guitar stool, leaving us with the challenge of a static microphone aimed at a moving target. A compelling performer, nervous in the studio, still deserves to be recorded. Without the constant gain riding of a compressor, listeners will hear the guitarist moving on and off the microphone. A little gentle compression might just coax a usable recording out of an inexperienced studio performer.

Beware of the vintage bass guitar. When the bass player pulls out that wonderful, old, collectible, valuable, sweet-sounding, could-sure-use-a-little-cleaning-up, aren't-those-the-original-strings, couldn't-stay-in-tune-for-eight-bars-if-you-paid-it gorgeous beast of an instrument, we can be sure that—even in the hands of a master—some strings are consistently a little louder than other strings. The instrument isn't balanced, a sad fact that might be revealed by abrupt level changes in the bass line being played. Of course, one solution is compression. Careful, note-by-note, precision amplitude adjustments must be made to the signal or the very foundation of the song becomes shaky.

We humans are very sensitive to changes in bass levels. The perceptually volatile low-frequency portion of our music requires particular focus on levels. On bass, mix engineers almost always invoke the careful level adjustments of gentle compression. Multiband compressors are called on in particular to smooth the low-frequency part of a track. The low band gets a bit higher ratio and a slower release time, with a threshold set specifically to ride the low-end content in your track. Smoothing the performance dynamics of the low end of your audio allows your mix to have the stability that only comes from always-present, never-too-loud, never-too-soft, low-frequency energy throughout the piece.

When a drummer's foot gets (understandably) tired during what may be several takes of the same song, the kick drum performance can become a little ragged. Some individual kicks are noticeably louder than others. Use compression to help make the performance more consistent.

It is important to note that smoothing a performance with compression as a way to better fit various signals together in a multitrack mix should never really be needed—we hope. That is, it is always best to get a musically useful blend of the musicians at the time of performance, if at all possible. Teach the singer microphone techniques in which they back off the microphone for louder parts of the performance and lean in slightly on the quieter portions. Find a way to get the guitarist to remain in a relatively steady position by the microphones. After all, the quality—not just the level of the sound—is changed when the acoustic guitar moves toward and away from the microphone.

Hire bass players who are so sensitive to their instruments and performances that their playing techniques adjust for level differences, note by note, string by string. Great players can create a performance with musically expressive and technically sound dynamics across the entire range of their instrument.

If the drummer is getting tired, coordinate with the producer to have strategically timed breaks so that a consistent performance is recorded to the multitrack and needn't be chased with smoothing compression. Also, if the drummer gets a pizza, they'll often give a slice to the mix engineer. There is no doubt that a kick drum hit hard has a different spectral content than a kick drum hit softly. Having a compressor match levels of various kicks will not lead to a believable timbre.

Compression is the last step in reigning in the performance dynamics. After getting the best performance possible, we first pull the dynamics of the performances in line through careful, active fader riding, helped out by automation. The goal is not to find a single fader position that is best for the whole tune. The goal is to obtain the appropriate level for each and every track, note by note, phrase by phrase, verse by chorus, no matter how often we have to move the faders. Automation (see Appendix C) is essential. Key, expressive tracks like vocals, solo instruments, and bass get particular attention, but mixing most songs usually includes a constant amount of fader finesse—up a bit here, down a bit there—across all tracks throughout the tune. You add just a bit of additional polish to your mix through additional level riding, performance smoothing compression, narrowing the level variability just a bit more to a range hard to achieve through automated fader manipulation alone.

Parallel compression offers a different kind of smoothing. It is done only rarely, as it requires more work, more time, more equipment, and very careful listening, but a more convincing recording might result. Instead of forcing a track through compression to smooth a performance and living with any unnatural side effects, consider layering the uncompressed track with the compressed track. The compressed track offers a smoothed performance in which loud parts and soft parts are brought closer together in level. The uncompressed track is mixed in with it, adding back a bit of believable, hopefully musical level variation. You get the tactical benefit of level riding in the compressed track and the expressive benefit of the rawer, less-compressed performance. Moment by moment, throughout the song, the louder of the two tracks dominates. Phrases, words, syllables, or notes that were performed at a louder dynamic will be primarily presented to the listener via the uncompressed track. Softer moments are too quiet in the uncompressed track and listeners will instead hear the compressed, smoothed version. In this way, a track with a variety of performance dynamics can be made to fit in the mix. But from loud to soft, there remains a human amount of variation in the performance, so that the performer's emotions are still communicated, not squeezed out of the track.

RESHAPING AMPLITUDE ENVELOPE

Compression can be used to change the amplitude envelope of the sound, making it possible for the mix engineer to sculpt sounds for a better fit. The envelope describes the "shape" of the waveform, how gradually or abruptly the sound begins and ends, and what happens in between. Envelope is a step back from the fine-level, cycle-by-cycle detail of the waveform, looking instead at the more general signal amplitude fluctuations as in Figure 5.4. Drums, for example, have a sharp attack and nearly instant decay. That is, the envelope resembles a spike or impulse. Synth pads, on the other hand, might ooze in and out of the mix: a gentle envelope on both the attack and decay side. Piano offers a combination of the two. Its unique envelope begins with a distinct, sharp attack and rings through a gently changing, slowly decaying sustain. All instruments offer their own unique envelope.

▶ The compressor is the tool used to modify the envelope of a sound, and doing so lets us fit together pieces of the mix puzzle with more precision. A low-threshold, medium-attack, high-ratio setting can sharpen the attack. The sound begins at an amplitude above threshold (set low). An instant later (medium attack), the compressor leaps into action and yanks the amplitude of the signal down (high ratio). Such compression audibly alters the shape of the beginning of the sound, giving it a more pronounced attack.

▶ This approach can be applied to almost any track. A good starting point for this sort of work is a snare drum sound. It's demonstrated in Figure 5.6. Here's how. Be sure that the compressor's attack-time setting isn't too fast or the compressor is likely to remove the sharpness of the snare entirely. Set the ratio to at least 6:1, and gradually pull the threshold down. This type of compression has the effect of perceptually morphing a transient spike onto the front of the snare sound. Musical judgment is required to make sure the newly formed click or crack of the sharper snare hit fits with the remaining decay of the snare. Adjusting the compression attack time while trading off a low threshold with a high ratio

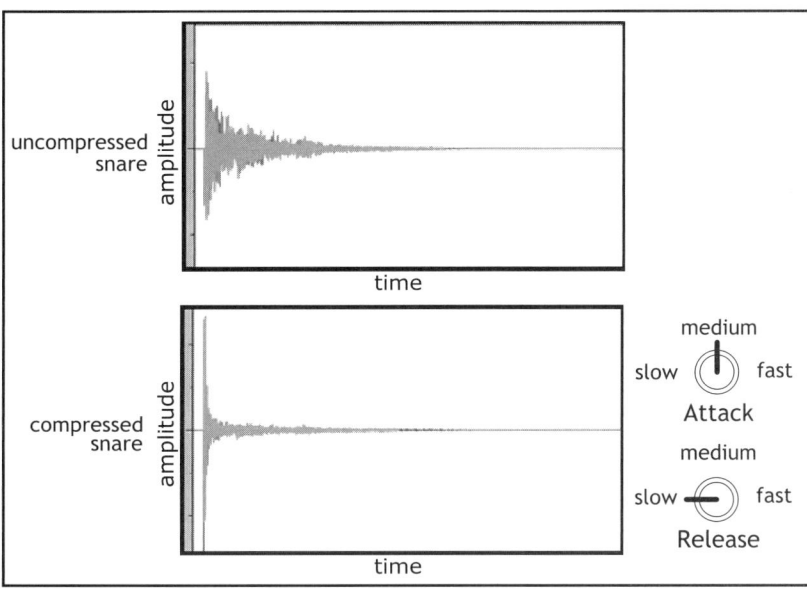

FIGURE 5.6
Compress to sharpen attack.

helps us obtain more control over the shape of the more aggressive snare sound. This compression effect works well on any percussive sound from congas to pianos. It can also sharpen the amplitude onset of less percussive tracks, from the saxophone to the electric bass. Done well, this method creates a more exciting sound that rises above a crowded mix and gets noticed.

A further reshaping of the envelope can be created through the compressor's release parameter. A fast release time instructs the compressor to pull up the amplitude of the sound even as it decays, as shown in the snare example of Figure 5.7. Notice the raised amplitude and increased duration in the late, decay portion of the waveform. Dial in a fast enough release time, and the compressor can raise the volume of the sound (i.e., *un*compress) almost as quickly as the track decays. This approach increases the audible length of the snare sound, making it easier to perceive each drum hit in and among the distorted guitars and ear-tingling reverberation. The drums and guitars achieve a better fit. Altering the decay of instruments can be taken to radical extremes. Applied to piano, guitar, and cymbals, the instruments can be coaxed into a seemingly infinite sustain, converting them into bell- or chime-like instruments in character while still retaining the unmistakable sound of the original instrument.

Multiband compression, in which you sculpt a new envelope through different attack, release, threshold, and ratio decisions across multiple audio bands allows you to further synthesize an unusual sound.

File it under "Special Effects" and let it be odd. Or do it all with parallel compression and add back a bit of the uncompressed, more natural, more believable

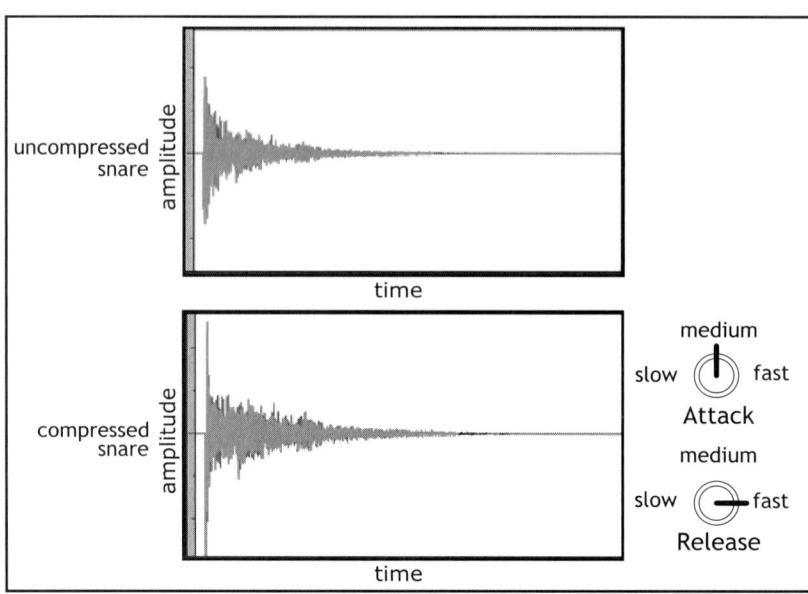

FIGURE 5.7
Compress to lengthen sustain (attack remains sharp).

sound. It can be just the sort of slightly unusual effect a track needs to get noticed, making it possible for you to fit together a more complicated arrangement of tracks with artistic success. Unwanted chaos and murkiness becomes tantalizing richness and complexity.

5.4.3 The Feature

Yes, now the fun part: compression is one of the most powerful ways to alter a sound in the studio. The user interface can prove clumsy, but with experience, you'll feature aspects of any track through these advanced applications of compression.

AMBIENCE AND ARTIFACTS

A coordinated adjustment of compressor parameters (threshold, ratio, attack, and release) enables us to manipulate—even indirectly—the shape of the amplitude envelope toward still other goals.

▶ Consider some extreme compression: high ratio, low threshold, with a very fast release time. If the compressor pulls down the peaks of the waveform and then quickly releases the signal as it decays, listeners may be able to hear parts of the sound that were previously inaudible. Through fast-release compression, the mix engineer effectively turns up the later parts of the sound, revealing more than just the decay of a snare. This kind of compression upgrades the barely audible to the unmissable with artistic benefits. It reveals the expressive breaths between the words of a vocal, the ambience of the room in between drum hits, the delicate detail at the end of a sax note, and so on. It is often the case that the compression settings have to be pretty aggressive, pretty radical, in order to lift up the formerly hard-to-hear, in-between-notes portion of the signal.

💡 Such extreme compressor settings may highlight emotional features at the expense of some other essential attribute. Reach for multiband or parallel compression whenever the unwanted side effects are too much.

For example, multiband compression can enable you to dial in some high-ratio, low-threshold fast-acting compression in the mids (and/or highs) for ear-tingling liveness while leaving essential lower bands less processed for a tight, stable foundation in the track. Conversely, if you find a thunderous low-end bit of ambience worth underlining, apply the heavy compression only to the associated bottom band and spare the rest of the timbre from the process.

Another way to back off the weirdness of some extreme compression is to hook it up for parallel compression. With the uncompressed track on one fader (or *less* compressed track in which a compressor is inserted in the usual way in service of any of the other production goals discussed elsewhere in this chapter) and the intensely compressed track on another fader, you can combine them to taste: a generally believable and beautiful sound courtesy of the less-processed track *plus* added ambience and other artifacts in between the notes by way of the heavily compressed track. Ride their relative levels to tailor the emotion of the sound to the song form, maybe it gets extra ambience in the chorus, but returns to a more natural sound in each verse.

Cues about space, emotion, and performance intensity often live in the ambience and other artifacts between the notes. Where musically appropriate, bring them forward through fast, extreme compression for powerful sensory impact.

TIMBRE

Timbre is one of the most misunderstood properties of a sound. We think of it, casually, as the spectral content of the sound, the distribution of energy among the harmonics that make up the sound. That is an incomplete definition. In fact, timbre is pretty much everything in a signal not captured by loudness and pitch.

Consider the sonic differences among a few different instruments as they play the same pitch, at the same loudness: imagine a piano, trumpet, voice, and ukulele simultaneously playing middle C. When they all play the same note at the same loudness, how can we tell which instrument is playing which sound? The answer is timbre.

It is the timbre that enables us to identify each instrument, unaided by loudness or pitch. There are obvious differences in the spectral content of these instruments; they have a different tone, even as they play the same note at the same level. But at least as important is that each of these instruments begins and ends the note with its own characteristic envelope—its amplitude-over-time signature.

Did a hammer hit the string, was it bowed, or was it plucked? These are unmistakable, yet equally indescribable timbral cues. When wind on plumbing creates the sound, we can still tell if the user interface was flute, trumpet, sax, or other. Ignoring loudness and pitch, we still hear many great details in a sound, without much effort or training. We hear, and we enjoy the timbre.

In all of the earlier discussion about using compression to reshape the amplitude envelope of a track, altering its onset and its decay, we are also affecting timbre. So use compression to sharpen and lengthen a signal with the practical goal of fitting things together better. But recognize also that the same compression gestures are sometimes desirable because, well, we just like the way it sounds. We are taking audio signals and altering the timbre in ways that make it more attractive, that accentuate existing good qualities of the instrument, or that simply make it unique, weird, even ugly on purpose.

Do this with a multiband compressor, manipulating the amplitude envelope band by band, and you are profoundly reshaping timbre, if your ears can follow all the changes. Parallel compression lets you fabricate a new net timbre by layering different timbres together.

Timbral adjustments through EQ are valid but can be rather static. Timbral manipulations through compression, though abstract, are an alternative way to feature any instrument in your production, making your mixes strong, different, or somehow attention-getting. Compressing for timbre is perhaps the highest form of compression, difficult to hear and rather counterintuitive to fine-tune. Through practice and careful listening, you'll find compressors to be powerful timbre-modifying sound synthesis tools.

DISTORTION

It is generally most accurate to think of compression acting at a time scale that alters the amplitude envelope of the signal, not the fine, cycle-by-cycle detail of the waveform. Picture the waveform when the digital audio workstation is zoomed out looking at a few bars of music, not zoomed in looking at a few individual cycles of an oscillation. All of the compression effects discussed previously happen at the envelope level of resolution. But sometimes compression is made to attack and release so quickly that it chases individual peaks and valleys of a sine wave (Figure 5.8).

FIGURE 5.8
Distortion through compression.

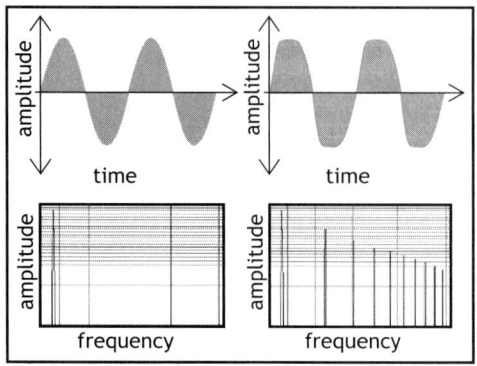

With a threshold set below the peak amplitude of a sine wave and an attack and release setting faster than a fraction of the period of the sine wave, the compressor will attenuate during each peak (positive or negative) and uncompress in between. The output from the compressor is clearly no longer a sine wave. If the ratio is high enough (as was used in the creation of Figure 5.8), the sine wave starts to turn into a square wave possessing the same fundamental frequency as the uncompressed sine wave but containing all the additional harmonics known to exist in a square wave. For this effect, compression is used to generate related harmonics, hopefully featuring a flattering kind of harmonic distortion (see also Chapter 4).

With multiband compression, we have more sophisticated opportunities. You can add some ear-tingling distortion, but preserve any spectral regions that are critical to the track. On a vocal, you might allow the bottom band to have fast compression-induced distortion but keep the middle- and upper-middle-frequency regions essential to intelligibility out of the distorting process. This approach uses compression to fabricate a kind of distortion sometimes found—and sought out—in transformers (see Chapter 4). On a bass guitar, you might choose to *not* allow low-band distortion so that the bottom of your mix stays strong and stable. Meanwhile, the upper bands of the bass guitar might get fast-acting, distortion-inducing compression so that fret buzz and finger slaps get decorated with a bit of harmonic ear candy. This method turns the transformer distortion signature on its head.

If you feel like you've pushed it too far, consider layering in a bit of the undistorted (i.e., uncompressed) track using the parallel compression approach. Now you have an authentic sound plus a variable amount of it pushed into distortion. Vary the clean-to-distortion mix in a way that supports the song form—maybe more distortion in each chorus, less in each verse, and none in the bridge—and you take your listeners on an orchestrated emotional journey driven by distortion.

It takes imagination, as there are no limits to the possibilities. You can apply a strong dose of compression—to an individual track or the entire mix—for the well-loved effect of distortion. There is something visceral and stimulating about the sound of distortion that makes the music more exciting. The distortion typically dialed in on most electric guitar amps adds an unmistakable, instinctively adrenaline-producing effect. Compression—with settings that deliberately modify the detailed shape of the waveform—offers us another option for creating distortion while still communicating an intense, on the edge, pushing-the-limits sort of feeling.

5.5 LEARNING TO HEAR COMPRESSION

This single device, the compressor, is used for a wide range of very different-sounding, sometimes difficult-to-hear effects, using counterintuitive parameters that are in turn a function of the type of signal being compressed, the amount of compression occurring, and the type of components or algorithms within the device. Compression is a difficult effect to master. Savor the challenge.

5.5.1 A Learned Skill

There is nothing built into our hearing system that makes it particularly sensitive to the sonic signatures of compression. We humans are generally pretty good at hearing pitch, easily identifying high versus low notes in music. We react instinctively to volume. Without practice, any listener with healthy hearing can separate a loud sound from a sea of quieter sounds.

Our hearing mechanism can do deeper analysis. Without conscious thought, we have no trouble identifying the source of a sound—a bird, a plane, a snare drum. Without additional effort, we typically have a strong sense of the location and approximate distance of that source. Our hearing system deduces the angle of arrival of sounds by analyzing, among other things, differences in amplitude and differences in time of arrival at the two ears; turning our head as we listen often gives us the ability to extract still more information. Our hearing system tells us which way we need to turn our head in order to face the sound source. We don't do any conscious math. We just know which way to turn our heads.

Human hearing extracts additional, subtle (and darn difficult to measure with a meter or oscilloscope) information from a signal, such as the emotional state or personal identity of a sound source. When we hear a dog bark, we know if it is big or small, near or far, threatening or playful. And all of that additional information is presented to us at the same time as the bark itself. We don't hear the bark and then analyze it. We hear the bark and know it. Identifying pitch, volume, arrival angle, distance, emotion, and identity are all easier than riding a bicycle; humans do it naturally, instinctively, preconsciously and with no deliberate, intellectual effort, training, or training wheels.

! Despite these amazing accomplishments, our hearing simply can't make much intuitive sense out of compression. There is no important event in nature that requires any assessment of whether the amplitude of the signal heard has been slightly manipulated by a gain-manipulating device. Identifying the audible traits of compression is a wholly learned, intellectual process that we audio engineers must master. It is more difficult than learning to ride a bicycle. It is possible that most other people, including recording musicians and avid music fans, will never notice—or need to notice—the sonic fingerprint of compression. We mix engineers have this challenge to ourselves.

Early in our careers, most of us make a few compression mistakes. Too much compression is a common problem. This effect that is so difficult to hear must be overdone in order to become audible. We do this with EQ and reverb too, but compression most of all.

We try compressing until it is clearly audible and then backing off, so as not to overdo it. Not a bad strategy. At the early stages of developing our critical listening skills, focusing on compression, we suffer the following frustration: overcompression may not be noticeable until the next day. How can we mix when our evaluation of the effect has that sort of reaction time?

The sonic traits of compression range from quite subtle to quite obvious. Spending all day mixing one song, with ears wide open, can make it hard to remain objective. A fresh listen to the mix the next day can be an effective education. Try to learn from these mistakes, but cut yourself some slack. It isn't easy.

Wait, it gets worse: It is often our goal to set up the compressor so that the effect is inaudible. For many types of compression, we wish the effect to be sonically *transparent*. We oh-so-carefully adjust threshold, ratio, attack, and release until we can't hear it working. Don't let anyone tell you otherwise—tweaking a device until it sounds so good that you can't even hear it isn't easy.

> *You:* Did you hear that?
> *Them:* No, not at all.
> *You:* Exactly.

5.5.2 Imitation

Learning to use compressors and knowing with confidence when to reach for a specific make and model is a nearly impossible challenge, very much like microphone selection.

At first, it makes sense to imitate other engineers whose work we admire. Microphone X on snare drum, microphone Y on female vocals, and so on. We take in the vast complexity of the sounds that result over many recording and mixing experiences. Then, when we reach for a different microphone in a familiar application, the sonic contribution of the microphone becomes more obvious.

So too with compressors. We observe the successful use of certain compressors for vocals, others for snare drums, and still others for electric bass. As a starting point, we can do the same. Once we think we are getting satisfying sonic results using these specific compressors for these specific applications, we can start to branch out and explore, reaching for different compressors in tried-and-true applications. The sonic fingerprint of the compressor begins to be revealed. Over time, we start to internalize the general sound qualities of every compressor we own, and every make and model of every studio we hire. That is our task as advanced users of compression.

Compression is not a single effect. As we've seen, it is used to prevent overload, overcome noise, de-ess vocals, increase perceived loudness, improve intelligibility and articulation, smooth a performance, alter amplitude envelope, extract ambience and artifacts, modify timbre, and add distortion. In order to correctly learn from the way other engineers use compression, we need to know which effect they sought to achieve. There is no such thing as a vocal compressor or a snare drum compressor. When engineers reach for a specific compressor for a specific instrument, they are not demonstrating the universally correct compressor for that instrument. They have a specific strategy. They are trying to fix a problem, fit one track in and around others, or feature some appealing quality that they like in the track. When we learn by watching and listening to others, we need to somehow know—or at least have a good guess—what they were trying to accomplish from a production point of view: which result do they seek and why. We learn by working with others, assisting the best, most experienced engineers we can, whenever possible. But it is rarely appropriate for an assistant engineer to interrupt the productive and creative flow of a session and ask this directly. The wise assistant silently watches, guesses, experiments, and eventually learns. If you have a good relationship with the seasoned engineer, you might be able to have a discussion on a specific compression strategy in a calm moment after the session. Try to find what they were going for when they patched up that particular compressor on that particular track at that particular moment. This knowledge can be the seed for your eventual success with that kind of effect on that kind of compressor.

5.5.3 Multiple Personalities

Obtaining a complete understanding of compression grows more complex still. Every compressor with adjustable parameters offers a great range of sound qualities. Compressors don't have a single personality for us to discover and codify. In fact, for every make and model of compressor that we wish to wield skillfully, we need to develop an understanding of its unique capability across the adjustable parameters. Memorize each compressor for low, medium, and high ratios, across fast, medium, and slow attack times, using fast, medium, and slow release settings. There is a frustratingly large range of permutations to work through. Experienced engineers have done this already. You are encouraged to choose a couple of compressors and develop this kind of deep knowledge for yourself. With experience, you'll find this task less daunting. Then, as you seek to become proficient with additional compressors, you'll assess their behaviors and sonic traits relative to the compressors you already use comfortably.

💡 This learning is not difficult to do, but it is a slow process. There is no substitute for experience. Any serious pianist can tell within an eighth note whether the piano is a Steinway, Bösendorfer, Baldwin, Falcone, or Yamaha. Any serious guitarist can identify a Stratocaster, Telecaster, Les Paul, and Gretsch. Seasoned guitarists can also make good guesses as to the type of pickups, the gauge of the strings, and whether an alternative tuning is being used. Audio engineers need to develop a similar instinctive ability to identify specific compressors. The distinctions, indescribable at first, become second nature.

5.6 SUMMARY

Dialing in the right settings on the right compressor as needed in the course of a multitrack production isn't easy. Experience is essential, but that's about as fair as requiring experience for a job before you can get a job.

> *You:* I'd like a job.
> *Them:* Do you have any experience?
> *You:* No.
> *Them:* We can't hire you until you have some experience.
> *You:* How can I get any experience when no one will hire me?

Compression serves up a similar logical cul-de-sac. How can you know which compressor to engage when you don't yet know the sonic differences between all the makes and models you have available? Experience is essential, but you won't be particularly good at it, at first. Nobody is. Armed with a crisp motivation to fix, fit, or feature your track in some way, assert yourself with confidence, using parameter settings that you think will get the job done based on the discussion in this chapter. Listen carefully to the results, try different compressors, and compare your mixes to recordings you admire. Assist more experienced engineers whenever you get the chance. Expect to make some glorious, audible errors. Savor your early successes. You get the necessary experience by simply trying, listening, researching, learning, and enduring.

The process of using a compressor begins with vision and a strategy: which type of effect do you want? There are many to choose from. Target a specific effect or two from the long list of possibilities: distortion prevention, noise suppression, de-essing, loudness, intelligibility and articulation enhancement, performance smoothing, amplitude envelope alteration, ambience and artifact exaggeration, timbre manipulation, and distortion.

Once you've determined the type of compression effect that the mix needs, choose the make and model that is most effective at achieving this. This choice is all but impossible at first. As with microphone selection, you'll get better at this over time. It can feel like a trivia contest. You have to learn the subtle features and quirks of all these plug-ins and all those boxes. Through experience on many different compressors, though, this step eventually becomes easy and intuitive.

Next, set the parameters to settings appropriate for the type of effect. With an understanding of the overall strategy for the effect, threshold, ratio, attack, and release can be preset to values near where they need to be before any audio is run through the compressor. For example, sharpening the attack portion of the amplitude envelope requires a low threshold, high ratio, medium attack, and medium to slow release.

With the compressor selected and the parameters preset to the appropriate coarse levels, listen carefully to the resulting compression and adjust the parameters until you've fine-tuned the effect to what is needed. Glance at the meters on the compressor to make sure you that are aren't causing unwanted overload. Outboard gear and individual plug-ins can easily be overloaded, causing unintended distortion. As compressors are gain-adjusting devices, we must pay attention to levels both before and after the compressor. Accidental distortion is rarely helpful to the track and to the mix. In addition, it is a good idea to look at the meters on the compressor to confirm the intended amount of gain reduction is occurring. Over time, in your compressor trivia contest, you start to learn each compressor's sweet spot, the amount of gain reduction that the compressor is capable of achieving without too much distortion or other sonic artifacts or with *exactly* the desired amount of delicious distortion and appealing artifacts.

If no amount of parameter adjustment leads to the desired effect, switch to a different compressor. As with microphone selection, the initial choice of equipment determines the range of possible results. Getting this wrong can make certain goals unachievable. Especially early in your career, expect compression to be a highly iterative and sometimes frustratingly slow process. As you become more fluent with the compressors you own, learning their personality traits from a sonic and operational point of view, your mixes get better. Total mastery of this effect is essential to making your mixes sound more professional. The artists may not know why; they'll just know that your mix sounds competitive, exciting, with all of the pieces dovetailed together. By overcoming the quirky interface, developing the ears to hear it, and knowing the many strategic motivations behind the effect, compression and limiting can become one of your key sources of mix success.

CHAPTER 6
Expansion and Gating

The most underutilized effect: hear what you are missing.

MIX SMART QUICK START: Expansion and Gating

GOALS

- Practical motivations include noise reduction and leakage reduction.
- Fit tracks together in your mix more easily through cleverly reshaped amplitude envelopes and supportive gated reverbs and ambience tracks.
- Feature tracks by extracting exaggerated ambience and artifacts, manipulating timbre, and tremolo.

GEAR

- Master the unusual set of parameters and features: threshold, slope, attack, release, fade, decay, hold, and range.
- As you learn compression/limiting, you get better at expansion/gating, and vice versa.

With expansion and gating, the concept of a compressor gets logically inverted. *Expansion is the automatic reduction in signal level whenever the amplitude falls below a specified value. Gating is simply a more absolute application of the effect in which the signal is not merely attenuated but is in fact fully muted whenever the signal falls below that level.* Whereas compressors turn down the hotter portions of our ever-changing audio signals, expanders turn down the low-level portions. This simple concept leads us to a new family of effects.

6.1 INCREASING DYNAMIC RANGE

Recalling the significance of audio dynamic range (see Appendix B), Table 6.1 compares and contrasts the logic of our dynamics devices: compressors, limiters, expanders, and gates. If you are new to mixing, this subject may feel frustrating at first. Keeping track of which device boosts and which device cuts based on whether the signal is above or below the threshold seems a bit trivial. The musical

Table 6.1	Modifying Audio Dynamic Range		
Dynamics Effect	Gain ↑ Boost/Attenuate ↓	Threshold ↑ Above/Below ↓	Dynamic Range ↑ Increase/Decrease ↓
Compressor	↓	↑	↓
Limiter	↓	↑	↓
Upward expander	↑	↑	↑
Downward expander	↓	↓	↑
Noise gate	↓	↓	↑

potential of these devices is not at all captured by Table 6.1. However, knowing these rules is essential to your musical use of these mix tools.

As discussed at length in Chapter 5, compressors and limiters turn the signal down when the music drifts above the threshold—a simple concept capable of a terrific variety of effects. Compressors and limiters *reduce* dynamic range. Expanders and gates, on the other hand, *increase* dynamic range.

By definition, the upper limit of the audio dynamic range is the peak level obtained before distortion (see the section "Dynamic Range" in Appendix B). The lower bound of dynamic range is the noise floor. If the peak amplitudes of the audio signal are increased without distortion or the level of the noise can somehow be decreased, then the dynamic range is stretched. One might say it is *expanded*.

There exist two types or two philosophies of expanders. *Upward* expanders increase the amplitude when the signal exceeds the threshold in an attempt to raise the level of the signal without distorting. *Downward* expanders decrease the amplitude when it falls below the threshold in an attempt to lower the level of the noise. Both approaches increase dynamic range.

▶ That's the theory. In practice, it is quite rare to find an upward expander in the racks of gear and menus of plug-ins available at any great studio. Most devices labeled "expander" are in fact downward expanders. Upward expanders are more difficult to control. Think about it. Taking the higher-amplitude portions of a complex signal and *increasing* the amplitude further still makes the chances for distortion much more likely. No safety cushion is built into this process. It challenges the very idea of headroom. It is inherently unstable. Clipping distortion (see Chapter 4) is a likely result if the upward expander is not used very carefully.

Downward expansion is a bit better at taking care of itself. Downward expansion—working the other end of the dynamic range limit—attenuates the lower amplitudes. If accidentally overused, the lower amplitudes get too low. Distortion is not the result; silence is. By minimizing the audibility of its own overuse—it turns it down—downward expansion is the more stable, easier to use effect. Downward expansion is therefore the more common type of expansion.

As is customary in the recording industry, when we use the term "expander" in this book without indicating upward or downward, we are referring to a downward expander.

Gating is to expanding as limiting is to compression. A gate is simply a downward expander set to more extreme settings. Rather than slightly attenuate that part of the signal that falls below the threshold, it aggressively attenuates. In the case of gating, it doesn't just attenuate the signal: it fully *mutes* the signal when it falls below the threshold.

6.2 PATCHING AND PARAMETERS

Comparing expanders and gates to compressors and limiters reveals that they have many of the same controls and use the same basic idea, but achieve a very different volume effect (Figure 6.1).

Audio flows through a gain element. This variable gain stage boosts or attenuates based on instructions from a level-detector circuit. The level detector might look at the very audio signal being processed. As with compressors, a side chain input permits the level detector to look at one signal while the gain change element processes a different signal. This setup provides the flexibility needed for more creative signal processing.

6.2.1 Patching and Plugging In

Like compressors and equalizers, expanders are sometimes built into the console or DAW, preconfigured for convenient availability on each and every channel, and are a stock part of any fully featured channel strip.

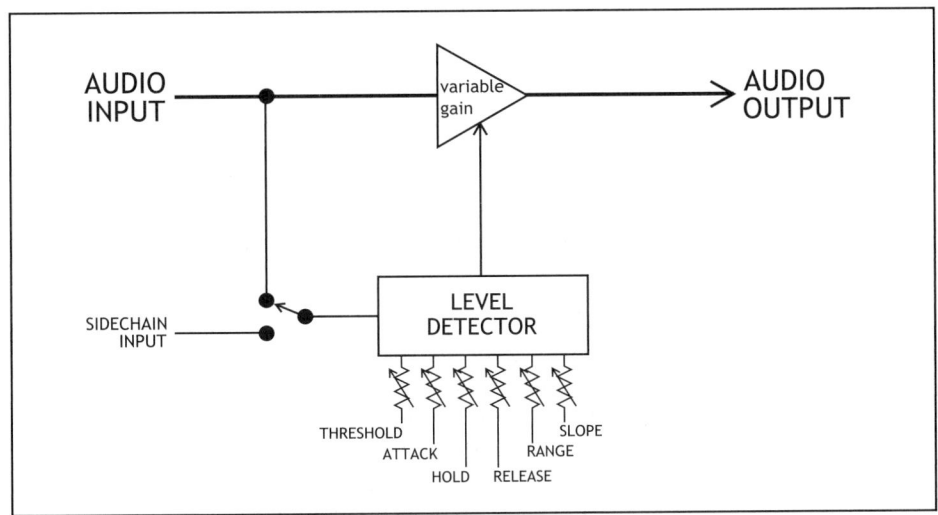

FIGURE 6.1
Signal flow through an expander/gate.

INSERT

Any additional outboard and plug-in expanders are hooked up using the *insert* of the mixer or digital audio workstation (see Appendix A). The intended goal of the expander—to increase the dynamic range of the signal by making the quiet portions quieter still—often makes little sense if the expanded signal doesn't in fact replace the higher-amplitude unexpanded signal. It's hard to hear how quiet you made it if the nonquiet part is drowning it out in the mix.

The insert offers us the desired signal flow logic: it forces the processed track through the expander or gate, altering its level, its noise floor, and its tone and giving us no chance to accidentally mix together both the expanded and the nonexpanded versions of the audio. As with compression, expander/gates can be patched directly between the multitrack return and the mixer/DAW line input, effectively inserting the device serially into the signal flow. The use of auxiliary sends or buses is unusual for this type of effect, as we rarely need to send multiple different tracks to the same expander, and any adjustment of a gain stage in front of the expander (in the form of a fader or an aux send) directly affects its interaction with the expansion algorithm.

PARALLEL EXPANSION

Although the insert is the more common signal flow approach for adding expansion and gating, some specific applications require us to set it up as a parallel effect. Parallel expansion refers to the signal-routing strategy in which you mix the expanded signal together with the nonexpanded version of that same track. This mix is achieved by using an insert send without the insert return, returning to a spare mix input instead. Alternatively, use a prefader, unity gain effects send and return it to a spare mix input. Both of these parallel expansion signal flow structures are described in detail in Appendix A in the section "Hybrid Parallel Effects."

Parallel expansion effects lead to an additional set of mix possibilities not possible using inserts alone, described in the following section.

6.2.2 Threshold

The expander must sort out the portion of the waveform that is to be attenuated and the portion whose amplitude is to remain unchanged. For downward expansion, the threshold control determines the amplitude below which attenuation is triggered into action. As long as the signal remains above this threshold, no expansion is initiated, and the gain stage of the expander stays at unity. When the amplitude of the signal sinks lower than the threshold, however, the expander begins to attenuate the signal, very much like a fader being pulled down automatically. The lower amplitudes are made lower still; the dynamic range is expanded.

Once the expander is attenuating a signal, the threshold identifies the amplitude where the level has returned to a high enough value that the expander should stop attenuating and return to unity gain.

In the unusual case of upward expansion, the threshold control specifies the amplitude above which the gain is to be further increased. The signal above the threshold is to be amplified. When the signal is below the threshold, no upward expansion is needed. The device returns to unity gain.

6.2.3 Slope

When the audio falls below the specified threshold, the downward expander begins to attenuate. The degree of expansion is determined by the slope setting. The *slope* compares the level below the threshold of the input to the level below the threshold of the output (Figure 6.2). For example, a 1:2 ratio describes

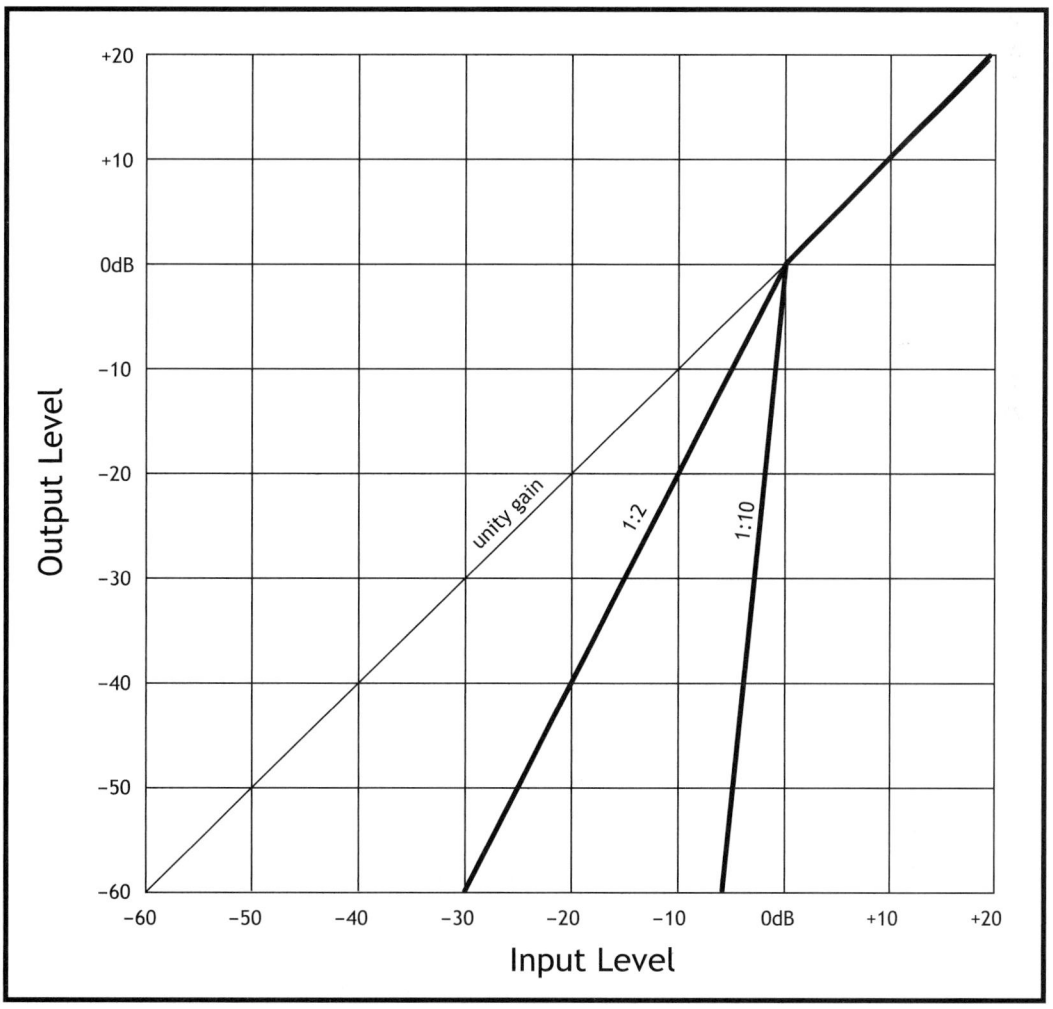

FIGURE 6.2
Comparison of output versus input for a downward expander.

a situation in which the output level below the threshold is two times lower than the original input below the threshold: 1 dB below threshold in becomes 2 dB below threshold out, and 10 dB below threshold in becomes 20 dB below threshold out. A ratio of 1:X sets the expander so that the output must fall below the threshold by X dB whenever the input is just 1 dB below the threshold.

💡 It is no coincidence that the math here sounds reminiscent of the ratio setting on a compressor. The logic is quite similar. In fact, some expander manufacturers choose to label this parameter "ratio" instead of "slope." The terms are interchangeable. Recording tradition has it that the ratio for an expander is less than 1 and the ratio of a compressor is greater than 1.

The ratio (or slope) applies only to the portion of the signal below the threshold. When the input is above the threshold, the expander is not applying this slope to the signal. The expander is specifically designed not to change the gain for signals above the threshold. The hope is that the expander is sonically transparent and that the audio sound is completely unchanged when its amplitude is above the threshold. Only the part of the input that falls below the user-defined threshold is processed according to the expansion slope. The expander applies its attenuation based on the decibel difference between the amplitude of the audio and the threshold.

It is often the case that the slope is not user-adjustable. Although many compressors offer this control, many expanders and gates do not offer the engineer access to this parameter. It is determined internally. Though essential for compression, many expansion and gating processes do not require a user-adjustable slope. The other parameters are more important to how this device is used.

6.2.4 Attack

How fast the signal is *un*attenuated is a function of the attack setting. "Attack" describes how quickly the expander can return to unity gain after the amplitude of the signal has passed upward through and above the threshold. As with compression, attack describes the speed of the imagined fader as the signal increases in level. Fast attack times will enable the expander to react very quickly, snapping nearly instantly to unity gain. Slow attack times are lazier, sneaking the level up gradually to the unity gain of the unexpanded signal. As with compression, the attack parameter does the most to influence the onset of the signal being processed, the beginning of the threshold-exceeding notes.

6.2.5 Release, Fade, or Decay

The speed of the imagined fader within this machine as it moves down is determined by the *release* setting (identified by some manufacturers as the fade or decay control). When a signal falls below the threshold, for example, during the brief sustain of the snare or at the last breath of trumpet note, the expander must begin attenuating. The release time setting governs the speed of this reduction in level. That is, the release parameter sets the speed that an attenuating expander can turn the gain down from unity toward the level required by the slope setting.

Attack and release parameters for an expander are similar in concept to the attack and release parameters of a compressor. Attack describes the agility of the dynamics device when the amplitude of the audio is increasing. Release refers to the speed of action while the amplitude of the audio is decreasing.

When expanding or gating, the parameters listed previously—all of which appear in very similar form on a compressor—leave the engineer wanting a bit more control. Enter two additional parameters with no counterpart on the compressor: hold and range.

6.2.6 Hold

Threshold is not the sole determinant of whether an expander should attenuate the signal. The *hold* setting is a minimum length of time that the expander/gate must wait after the signal crosses the threshold before any attenuation is allowed to occur.

Consider a sound with a long decay, such as a single, sustained piano note. If the piano is recorded in a noisy environment, it might be helpful to expand the signal. When the piano isn't playing, the expander could attenuate the signal, pulling noise out of the mix. When the piano does play, however, the expander needs to return to unity gain and let the piano performance through. The noise hopefully is not distracting while the piano is playing—the piano pleasantly covers up the noise. The threshold setting separates desired signal level versus unwanted noise level. We carefully adjust the threshold so that the noise is below the threshold, causing expansion of the noise down to an unobtrusive level. The threshold must also be set so that the piano is above the threshold, allowing the expander to get out of the way when the piano plays.

During the sustained portion of any piano notes, the desired piano sound will eventually fall below the threshold. The piano sound is still audible, even as it falls down to a level equivalent to or even below the level of the noise. Humans have no trouble hearing music even when it is swimming in noise. Listening to music in the car and having a one-on-one conversation in a noisy bar are common examples (at least for your author) of enjoying a signal even when it is below the amplitude of road noise and cocktail chatter by others.

The trouble is that an expander will start attenuating according to the setting of the slope parameter, and at the rate specified by the release setting, as soon as the piano amplitude falls below the threshold. Any portion of the piano note with an amplitude below this threshold will be attenuated thusly. The hold parameter lets us outsmart this behavior. Setting a hold time that is perhaps a second or more in duration instructs the expander to wait at least this long before expanding, even if the signal falls below the threshold.

In this way, audio signals of all kinds can remain convincingly realistic full of low-level detail and nuance as they decay naturally, yet still be processed by an expander to conveniently remove unwanted parts of the signal between the notes and phrases of the performance.

6.2.7 Range

! Sometimes it is necessary in audio to avoid extreme amplitude limits. When all of the settings above instruct the expander to attenuate by 60 dB or more, the signal at the device's output is essentially silent. When the expander starts to unexpand later, increasing gain, the expanded signal eventually becomes audible again. That transition—from inaudible to audible—might be noticeable and distracting to both recording engineers and casual listeners.

It is not always necessary or desirable to fully mute the level in these sections of parameter-determined gating. Lowering the level by 20–30 dB instead of the more extreme 60 dB might be enough to make the undesired sounds unobtrusive. One solution is to set a maximum amount of attenuation. In this way, the expander attenuates only as far as we specify. When the parameters of threshold, slope, attack, hold, and release have instructed the expander to attenuate, that attenuation hits a stop at the level determined by the range setting. Specifying a milder amount of expansion using the range control enables us to avoid the awkward transition from true silence to just audible. The range control prevents the unwanted artifacts associated with expanding into and out of deep, dark silence.

6.3 MIX STRATEGIES: EXPANSION AND GATING

The expander/gate is used in a number of related ways, discussed next. We organize the production possibilities in the usual way: we fix any problems that need to be solved by this tool, we fit the multitrack elements together for minimal conflict, and feature those elements best revealed by expansion and gating in support of our mix goals.

6.3.1 The Fix

Although the production potential is far greater, the expander/gate was probably invented specifically to fix the problems of unwanted noise and sound leakage across tracks.

EXPANSION

Listen to an audio track when the musician is not playing and plenty of potentially undesirable sounds might be heard: hiss from microphones, preamps, and analog tape; the whoosh and rumble of the heating, ventilation, and air conditioning system; the purr of a refrigerator chilling beverages in a nearby room; the sounds of traffic, construction, neighbors, or residents present during the recording session; leakage of sound from other instruments in the studio playing during the same take; the noodling, practicing, and pitch-finding that happens between parts; foot-tapping, clothes rustling, and even knuckle cracking. The list goes on.

Working in well-designed professional recording spaces, working with experienced musicians, and using good recording craft, these noises can be minimized and made essentially inaudible. Working in lesser studios with talent new to the studio in noisy urban environments, these sounds between performances may become a nuisance.

▶ An expander that is capable of turning such sounds down minimizes the problem. For example, recording a horn quartet using close mics (one each on trumpet, alto sax, tenor sax, and trombone) will no doubt lead to at least a little leakage. The trumpet microphone picks up mostly trumpet, but it can also have a good amount of sound energy from the other horns. Using isolation booths, gobos (moveable, sound isolating baffles), and clever musician and microphone placement, the engineer can minimize this.

It is likely that the section will want to play together as a group standing side by side or perhaps in an arc or circle so that they can see and hear each other easily. An intimate physical layout generally helps create a tighter ensemble performance. Moreover, making the musicians comfortable is much more important than getting clean tracks. The frustration is that recording them physically closer together also leads to more leakage.

Leakage may not be an acoustic problem when they are all playing at once. Good microphone placement probably enables the tracking engineer to get a great section sound—in part *due* to the leakage across the multiple microphones. That is, with care and experience, one hopes to engineer the performance so that the sound of the trombone in the trumpet microphone helps the overall sound of the section.

The trouble starts when the section stops and someone plays a solo. Then the sound of the trumpet alone in the trombone microphone may not sound so good. Use an expander to further separate the volume of the wanted signal (the target instrument of the close microphone) from the volume of the unwanted signal (the trumpet solo leaking into the nearby trombone microphone).

On the trombone track, patch up an expander and set the threshold so that when the trombone plays, it is well above the threshold, and when the trombone doesn't play, the lower-level trumpet leakage into this track is below the threshold. When the trombone plays, the gate stays open and the signal isn't attenuated. When the trombone stops and the trumpet solos, the expander steps in and turns down the unused trombone track: automatic leakage reduction. The trumpet solo develops the necessary clarity to hold its own and change our lives with an inspired solo.

NOISE GATING

More extreme cleaning up of tracks might be warranted, for example, on close-microphone pop/rock drums. Here, a single musician plays multiple instruments (kick, snare, hi-hat, toms, etc.) all at once, in close proximity. It is inevitable that there will be some snare sound in the tom mikes, for example. When they aren't being played, attenuate the tom tracks a little (expansion) or a lot (gating) by plugging in an expander/gate across them.

🔊 The decision whether to expand or gate is a creative one. You must do whatever sounds best for the tune at hand. Through persistent tweaking of the threshold, attack, hold, and release settings, you are able to make the gate turn the tom track down when the toms aren't being played. The gate then leaps out of the way to let each and every tom hit come through. The kick drum and snare drum can be automatically removed from the tom microphones by this kind of signal processing.

Through gating, a tighter drum sound can result, but you should plan to spend a fair amount of studio time getting the gates to cooperate. It takes some experience and good recording technique to get the gates to open and close when they should. Filtering a side chain input helps persuade the gate to respond only to the intended track.

Just because we *can* do it, doesn't mean we *should* (see "Case Credo" in Chapter 10). Gating leakage out of every track is not necessarily a problem in need of fixing. As always, we must pay attention to the music. Some tunes welcome the clarity such fine-tuned gating brings to a drum kit. Some songs, on the other hand, sound better with the leakage offering a foundation of wonderfully messy chaos. The snare may sound best when heard not just through the close snare microphones, but also through all the other microphones that happen to be in the room at the time. A gate across those other microphones would mute that beneficial leakage.

Good engineering has a lot to do with anticipating what a given microphone selection and placement on one instrument will do to the sound of a nearby instrument. So gate not out of habit, but out of necessity. Only gate away the leakage when the music calls for it, and have the courage to not gate if it sounds good without the gate. Each tact has merit; we must create a sonic environment appropriate to the song we are mixing. It is a powerful creative decision.

Taken to an extreme, you can use gating to make the quiet parts dead silent. An overdubbed track without leakage might still have distracting sounds within. Some low-level noises (e.g., tape hiss and amp buzz) can be automatically shut off by a noise gate. The threshold is set so that it is lower in level than all the music, which prevents the gate from trying to attenuate the signal while someone is playing. But the threshold must also be carefully set so that it is above the amplitude of the noise you are trying to remove.

The music stops. The signal falls below the threshold. But instead of hearing the faint hiss, hum, or buzz of the track, the noise gate sneaks in and gracefully turns it down. That'll fix you, noise.

As soon as the music resumes on that track, the gate opens up, gets out of the way, and lets the sound of the instrument back into the mix. Sure, the hiss, hum, or buzz lingers within the track the music is playing, but it is overpowered by the much louder sound of the instrument. With the noise gate turning things off in the quiet parts, and the music cranking the rest of the time, we fix any noise problems.

As with expansion, the threshold when gating is set to an amplitude above these unwanted noises, but below the level of the musical performance. In this way, the gate removes the nuisance sounds while leaving the actual performance untouched.

🔊 Attack time must be carefully set to taste. Very quick attack-times are often desirable so that the gate snaps open immediately in the presence of threshold-exceeding music. However, the very act of the gate opening can create an audible "click" or other sound artifact. This click most likely occurs when gating signals that contain very low frequencies. The abrupt opening of the gate during the slow cycling of a low-frequency signal radically reshapes the waveform. Such a click might be helpful to the music. More often, however, it is perceived as a distracting nuisance. Remove the click by slowing the attack time down. A slower opening of the gate makes for a smoother waveform.

Musical material with a fairly slow onset—strings or vocals—generally benefits from a slow attack time. Sounds with highly transient onset—drums, percussion, or piano—tend to respond well to fast attack times. In fact, the click associated with a gate opening quickly might help intensify the sharpness of the transient in a way that is beneficial to the mix. You dial in an attack-time setting that is slow enough to prevent unwanted clicking or thumping, but fast enough not to miss the natural onset of the musical waveform. As always, some finesse, patience, and musical judgment are required.

Release time is influenced by the nature of the decay of the musical signal. Slowly decaying sounds—strings, vocals, and piano—sound most natural with slow release times, often aided by a bit of the hold parameter, if available. Percussive sounds can tolerate or even be improved by fast release gating.

💡 It should be reemphasized that the noises that occur between musical passages are not always a negative—they aren't always begging to be fixed. The leakage from other instruments can benefit the overall sound of the mix. Applying a gate across each and every track so that pristine silence exists between every musical note may make a beautiful musical statement, revealing detail and subtlety. On the other hand, it may do damage musically, creating a sound that is too sterile and lacks a sense of cohesion or musical ensemble. You must use good judgment in determining whether noise gating is appropriate to the style of music, the band, or the song being produced.

6.3.2 The Fit

Getting the many tracks of the multitrack arrangement to play together without conflict is a constant challenge. Expander/gates offer some help.

GATING THROUGH EDITING

In a digital audio workstation, noise gating shifts to a different paradigm. If you're willing to spend the time, you could work through each and every track and manually edit out any and all unwanted noises. This is called *gating through waveform editing.* Find the parts when the singer isn't singing and simply cut them out—cut between verses, between phrases, between words. Zoom in as much as you dare.

Better yet, let the computer do all the work. Once the audio has been recorded into a computer, let your digital audio workstation do all the work, through an algorithm that simply crunches the numbers and nondestructively deletes all audio below a specified threshold. As a result, a formerly continuous single waveform with unwanted noises in between musical phrases is chopped up into separate pieces—each a musical phrase with total silence in between.

The algorithm is made more powerful if it can impose a fade-in and fade-out at each transition, analogous to attack and release time settings on a gate. In this way, the computer effectively removes the undesired low-level portions of any track. The patient audio engineer can then manually go through each event and tailor the fade-in and fade-out as desired. The result is a playlist full of audio events separated by pristine silence.

Analog noise gates are forced to work in real time: The audio goes in. The audio is immediately processed. The audio goes out. The digital audio workstation has the luxury of being able to read the audio files well ahead of playback. This feature can inform the actions of any signal processor, such as a gate. In this way, processes that simply could not work in the analog domain become feasible in a digital audio workstation; new creative capabilities follow.

In addition, the digital audio workstation has the ability to work "offline." That is, it can implement signal processing such as noise gating across the entire file ahead of time and offer us the chance at any time to accept, reject, or refine the result.

Gating through editing helps us get elaborate, complicated mixes under control. Such precise removal of unwanted portions of a signal gives us room to get the other mix elements heard. We can fit more pieces into the arrangement.

GATING

Noise gating, as discussed earlier, focuses intensely on keeping a natural, realistic sound while removing unwanted noise. Sometimes noise isn't the issue. Gates may also be used to manipulate the amplitude envelope of the sound: reshaping, for example, the reverberant decay of the room. As this is no longer about the avoidance of noise, it is simply called *gating*. The approach is similar to noise gating, but we seek consciously to alter the attack and decay envelope of the signal for an ear-tickling effect. Getting a track noticed makes it easier to tuck into a crowded mix. Nebulous clouds of tracks are coaxed into recognizable form. More layers and more complexity successfully fill our loudspeakers.

Consider the room tracks that are often recorded onto separate tracks during the recording of a drum kit (see Chapter 9). Even in a great-sounding room, those room tracks can be very difficult to use effectively in a mix. Too often, while the room tracks posses a thrilling snare drum sound and a thunderously large tom sound, they are undermined by a loose kick drum decay and a messy wash of cymbals. In some projects, there is no placement in the mix for these room tracks that works. Place them at a level that sounds right for the snare drum, and the kick drum and cymbals will become disappointing, washed out by the room sound. Turn the room tracks down so that the kick and cymbals aren't undermined, and the snare and toms receive no super-charged ambient benefit.

How do we fit the room tracks in and around the snare hits, without obscuring the rest of the kit? Gating is the answer.

A gate is inserted into the room tracks. The plan is to have the gate open on each snare hit but remain closed in between. In this way, the gated room tracks can be placed loud enough in the mix to enhance the snare without any loss of clarity in between snare hits. A few complications, which are easily solved, arise.

In stereo mixes, room tracks are often recorded via a matched pair of microphones that seek to capture a stereophonic image of the room. In surround productions, room tracks are four, five, or more tracks as needed for the surround format. Gating these tracks requires as many gates.

Stereo gating simply uses two identical noise gates, each inserted into one of the tracks that make up the stereo pair. In the room tracks example, we plan to have the two gates open and close together with each snare hit. Matching the settings on the two gates can help achieve this.

If the gates open at different times, listeners will localize toward the track that opens earlier, and then towards the one that closes later. As the snare drum is almost always panned dead center in a stereo mix (see Chapter 2), gated pairs of tracks that do not open and close in lock step are problematic. The all important location of this fundamental track starts to move, unbalancing your mix. A *stereo link* feature is available on many gates. The stereo link ties the gates together so that both are driven by a single detector circuit. Each gain element of each noise gate therefore receives identical instructions, causing the two gates to open and close in unison, even as the audio passing through them is different. The snare image remains centered. Similar linking is needed for surround.

KEYED GATING

Getting room tracks to open on each snare hit is easier when the close microphone on the snare feeds the detector circuit of the gate (Figure 6.3). To be clear, it is the room tracks that flow through and are processed by the gate's gain change element. It is the room tracks that are attenuated by the gate.

But decisions about when to open the gate, how quickly to open and close the gate, and by how much to open the gate are based on the settings of all parameters and the gate's analysis of the close-microphone snare track, not the room track. Keyed gating is the process of gating one track while looking at the amplitude of another. The snare drum side chain signal drives the gating of the room tracks.

ROOM TRACKS

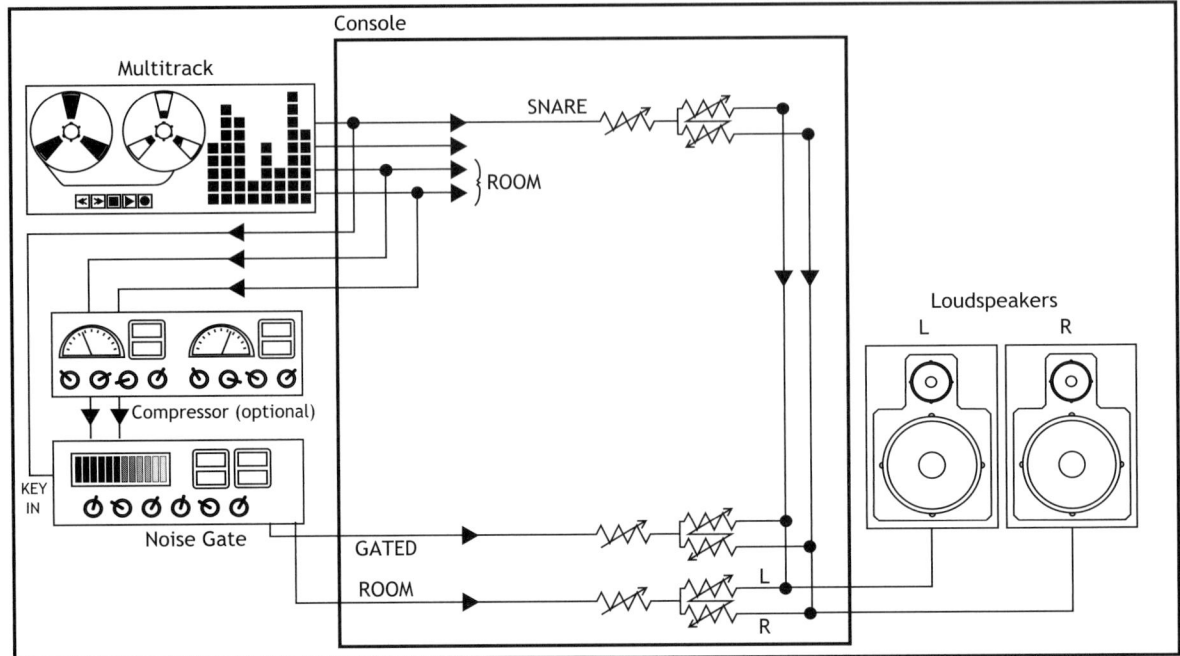

FIGURE 6.3
Signal flow for gated room tracks.

As with the microphones placed out in the room some distance from the drum kit, the microphone placed close to the snare drum will receive plenty of sound leakage from other drums and cymbals. However, unlike the room tracks, the sound of the snare drum will dominate, which gives us a better opportunity to dial in the threshold to a level that is below the amplitude of the snare hits but above the amplitude of the various other elements of the drum kit leaking into the snare track. The gate can be made to open only on snare hits, and to remain closed no matter what happens on the kick, toms, or cymbals.

Filter the side chain input for even more control (Figure 6.4). A high-pass filter could attenuate the low-end-dominated kick drum leakage. A low-pass filter could attenuate much of the cymbal leakage, further sharpening the ability of the gate to detect the snare—and only the snare—ignoring the rest.

! This keyed gating effect can be made more pronounced if the room sound being gated is compressed first (see Chapter 5). The compression can radically alter the room decay associated with each snare hit, exaggerating it in level and duration. However, such compression also makes the problems of a sloppy kick drum sound and a chaotic cymbal wash even worse. Keyed gating becomes essential when the ambience tracks are aggressively compressed.

Although the gates might be patched across the insert of the room tracks, it is often desirable to route them as shown in Figures 6.3 and 6.4. Here, the room

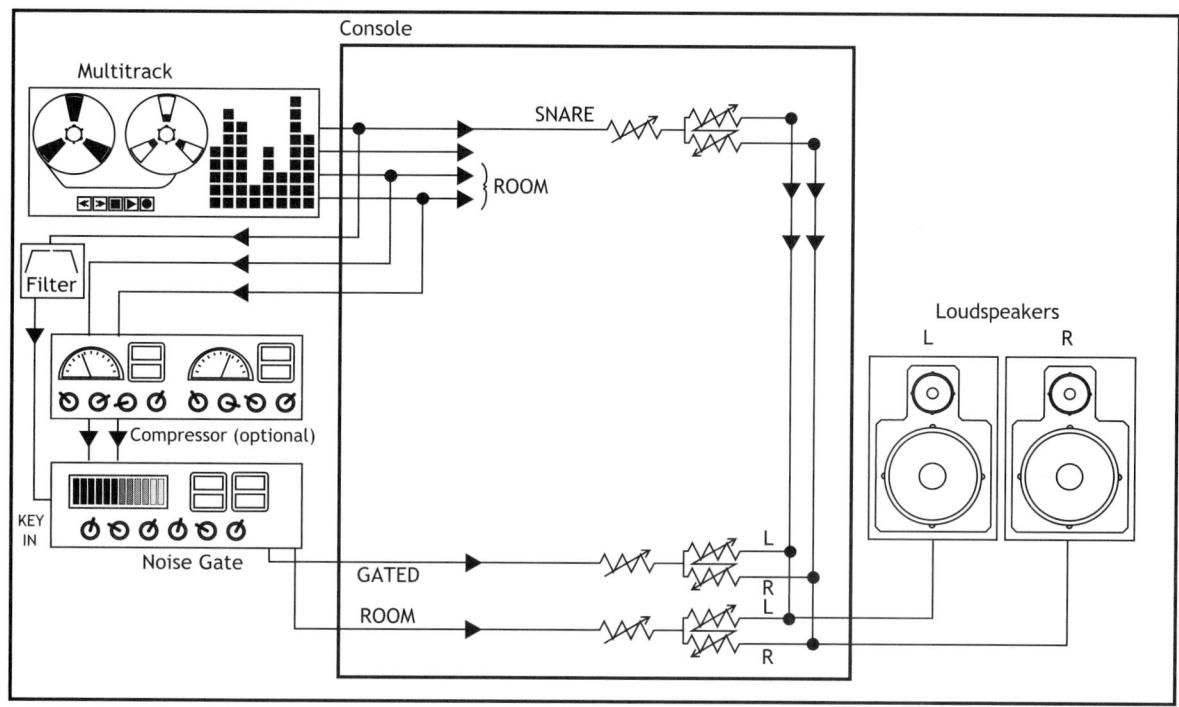

FIGURE 6.4
Filtering the key input for gated room tracks.

tracks make their way to the mix, ungated, for use in any way you desire. Maybe a low-level bit of honest room sound benefits the tune.

In addition, the room tracks are split and sent through an additional path. This additional path is compressed and gated (keyed open by the close microphone on the snare) and introduced to the mix on separate faders—a parallel effect. The mix now places many useful room track production variables at your fingertips. The overall drum sound in your mix is made up of the close microphones, the overhead microphones, the leakage from various other microphones, the room microphones, and the gated room microphones. You must evaluate the pros and cons of all of these elements carefully and orchestrate an overall drum sound suitable for this production.

What works for room tracks likely works for reverb as well (see Chapter 9). Gate and (optionally) compress the reverb returns in exactly the same way, using a close-microphone key input for controlled opening and closing of the gates, and a new sound is created (Figure 6.5). This sort of sound just doesn't happen in nature. It is one of many creations that only exist for the music created in the studio and enjoyed through loudspeaker playback.

The inexperienced engineer may view keyed gating as a heavy-handed effect, appropriate only to certain styles of music or revealing of specific eras of music. It is true that many forms of dance music and electronic music make more

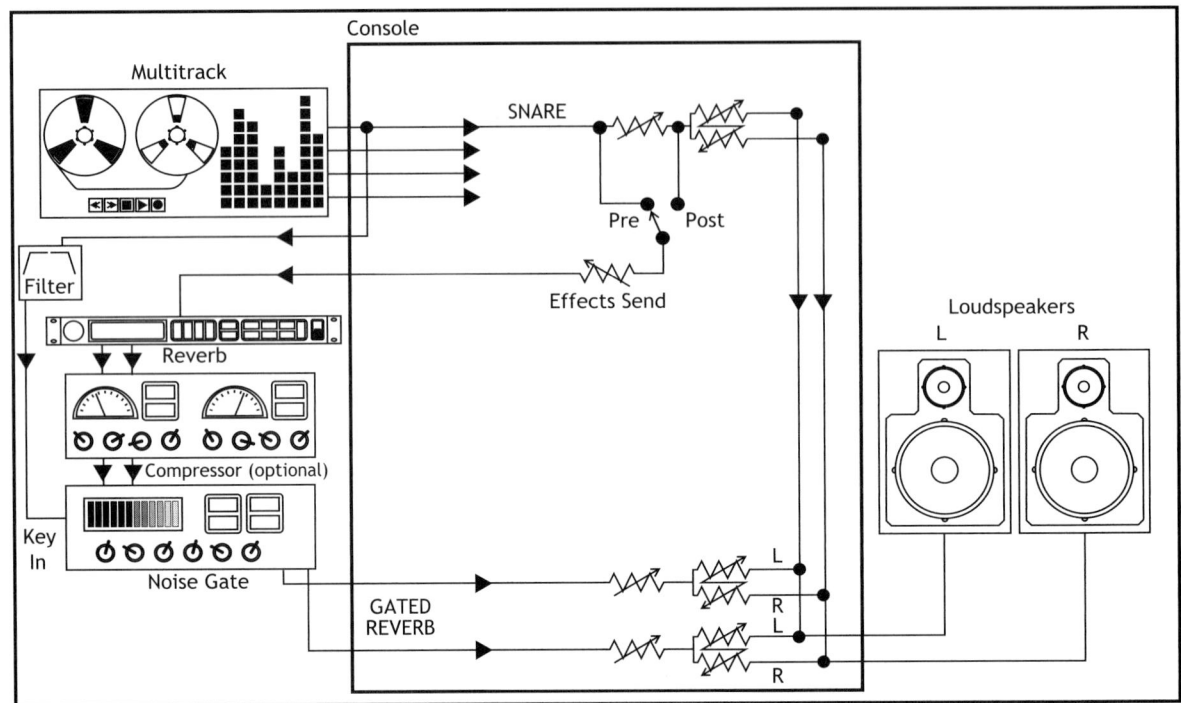

FIGURE 6.5
Signal flow for keyed gating of reverb.

aggressive use of this type of effect than others. It is also true that a pronounced gated snare effect is reminiscent of the synth-pop music of the 1980s. To reduce keyed gating to these musical trends and styles misses many opportunities. This effect may be made quite subtle; it exists in more recordings than you might at first expect.

Using the snare to key open room tracks or reverb returns might be considered a method of enriching the sound of the snare without altering it in a way that is noticeable to the untrained ear. The snare becomes easier to hear in a crowded mix if each hit of the snare has a bit of stereo width associated with it. Room tracks and reverb returns gated by the close microphone on the snare can sound quite natural if the attack and release times are slowed down a bit. Push the gated sounds up in the mix until just audible, then back off a little. Muting and unmuting the returns, your goal is to place them at a level that almost indescribably improves the sound when the gated ambience is on versus off. The snare becomes a little easier to hear without being louder. The duration is a little longer. The stereo image of the snare is broadened; it stays centered, but feels wider than the single close-microphone track. The snare fits into a crowded mix more successfully.

The sound of the snare drum captured by the overheads often sounds better than the sound captured by the close microphone. Split the overheads off to an additional pair of inputs. Gate them with the snare feeding the key input, and layer this into the mix in parallel. It is a standard multitrack production gesture: gated room, reverb, and/or overhead tracks, keyed by the close-microphone snare. Do this to an obvious degree where desired, or explore the benefits of using it subtly.

Snare is but one obvious example among many for this mix strategy. Gate the ambience tracks (recorded in a room or created in a reverb device) associated with the toms, congas, claves, hand claps, or any impulsive sound for a unique change in quality or a subtle enhancement in audibility. Used on less-impulsive sounds, a vocal can be made to sound more live, a piano can be made to sound wonderfully unnatural, and so forth. Keyed gating rewards the mix engineer willing to explore, offering an attention-getting sound that better competes with all the other pieces, or a slight augmentation and alteration to a sound to give it a stronger identity.

RESHAPING AMPLITUDE ENVELOPE

As we are trying here to put together tracks with minimal conflict and maximum musical impact, we should exploit the gain changes to refine the amplitude envelope in any way we desire. The attack-time parameter of the gate governs the reshaping of the onset of the note. Fine-tune the expander's attack time to sharpen, smooth, or otherwise modify the early part of the audio waveform in anyway you see fit. Very fast attack times make the gate snap open, causing the signal to adopt a sharper transient at the front of the waveform. Slower attack times may lead to rounder, punchier sounds.

Similarly, the hold and release parameters govern the reshaping of the later part of the amplitude envelopes, offering ear-grabbing alterations to the sustain and decay of the notes within the signal being processed.

DUCKING

Using a gate with key inputs makes another effect possible. *Ducking* turns down one track by a specified amount in the presence of another signal. Voiceovers provide a good example. Any radio jingle with background music likely wants quality, attention-getting, memorable music. But the music is subservient to the content of the words. The voice speaking over the music must be perfectly audible, always. The music is not permitted to drown out any syllable of any word. As with pop music, perfect vocal intelligibility is compulsory for advertising jingles.

A sensible goal, then, is for the music bed of the ad to be pleasingly forward, loud enough, stylistically appropriate and appealing, and attention-getting. At the instant when the voice speaks, however, the music must immediately be attenuated. Then, when the voice stops, the music should rise up quickly to grab the listeners' attention and keep them interested in the fragrance, food, or fancy frock being fobbed. The music is ducked under the voice.

A noise gate with a ducking function is the solution. The music bed to be ducked is patched through the expander/gate. The voice—the far more important track—in addition to feeding the mix and all other vocal effects, is supplied to the key input of the gate. Whenever the voice speaks, the jingle music is attenuated. Adjust attack time to be very fast, so that the music snaps out of the way as soon as the voice begins. Adjust the release time to as fast as possible without sounding unnatural, so that the jingle doesn't sag with quiet moments, losing the listener's attention.

! The alert reader may notice that a compressor with a side chain input should achieve nearly the same thing. The ducking process described previously is attenuating a signal when the threshold is exceeded—the very goal of a compressor. Background music is attenuated when the voice goes above the threshold. The trouble with using compression for this effect, and the reason that some expanders and gates provide this feature, is the presence of a critical parameter: *range*.

Not available on most compressors, range sets a maximum amount of attenuation, useful in many expansion and gating effects. In the case of ducking, it is likely that the music must be turned down by a specific, fixed amount in the presence of the speaking voice. Regardless of the level of the voice, the music should simply be attenuated by a certain fixed amount based on what sounds appropriate to the engineer, perhaps 10–15 dB. Compression would adjust the level of the music constantly in reaction to the level of the voice feeding the side chain input. The amplitude of the music would modulate constantly in complex reaction to the level of the voiceover. Compression does not hit a hard stop because compressors do not typically possess the range parameter. Therefore, look to noise gates for the ducking feature.

▶ Ducking isn't just for jingles. Engineers occasionally find applications in music productions as well. It is rarely if ever appropriate to have the lead vocal duck the entire rest of the mix. Ducking takes a different form in multitrack music. Consider the situation in which an engineer has recorded drums and finds the leakage among the many drum microphones to be generally beneficial to the production. The sound of the kick drum in the overheads helps the power and excitement of the kick drum. The sound of the cymbals and the room reflections in the tom microphones helps keep the overall drum kit sound convincing, organic, and consistent with the live feeling that the band wants. As we've discussed, there is no rule that says this leakage into neighboring microphones is bad and must be noise-gated out. In fact, it is the presence of this leakage that can help push away the clinical sound of the studio and artistically reduce the precision of multitrack production. Jazz and folk music often benefit from this looser, more integrated sound. Many styles of pop music do, too.

One very loud instrument in the drum kit may make this strategic use of leakage a problem: the snare drum. The drum kit, recorded through several close microphones, may sound great with all the microphones up and open, ungated—except that every snare hit may be a murky mess. Duck the tom tracks on each snare hit, even as little as 6 dB, and the clarity of the snare can be preserved even as leakage elsewhere flatters the rest of the drum sound.

6.3.3 The Feature

Finally, expansion and gating offer us ways to feature a track, drawing attention to it and refining it in flattering ways.

AMBIENCE AND ARTIFACTS

The extreme compression strategies that chase ambience and artifacts (see Chapter 5) can be made more extreme still through gating. When a gate is going to snap off the end of the decay with attention-getting unnaturalness, it may mean that you can get away with pushing the compressor even further.

This fast-releasing, high-ratio compression with distinct gating artifacts to reshape the onset and decay of the sound is often performed as part of a parallel compression/gating effect (see "Hybrid Parallel Effects" in Appendix A). Applied, for example, to the snare, you may find it helpful to retain the unprocessed or at least *less* processed snare in the mix as well. Then the mix contains a believable snare sound plus your highly reconceived compressed and gated version.

! This is not an invitation to add needless additional compression and expansion silliness. You must recognize that synthesizing radically new sounds in this way isn't appropriate for all tunes across all styles of music. The performing musicians and the tracking engineers likely had a vision for the sound of each track, reflected in the qualities of the tracks you've been given to mix. Heavy-handed compression plus gating may not please them. Disconnect yourself from the fun of taking a track in a bold new direction, and have the artistic conviction to evaluate it on its merits. Fun for us when we mix it doesn't mean fun for the band, or their fans, when they listen.

TIMBRE

Our discussion of using compression to alter timbre in the prior chapter applies here to expansion. The gain riding of the expander as it opens and closes creates a modified amplitude envelope, contributing to a new timbre. Very fast attack on a gate introduces a sharpening click or pop to the timbre. Slower attach settings help round out unwanted transients. Apply this effect in parallel and you can layer it with alternative treatments of the same source track. The new timbre is the sum of the component timbres you design.

Coordinate the use of compression with expansion, and you've got a terrific timbre-transforming apparatus.

ENVELOPE FOLLOWING

Courtesy of the side chain input, the amplitude envelope of one sound can be made similar to the amplitude envelope of another through a process called *envelope following*. Insert a gate across a guitar track, but key the gate open with a snare track. The result is a guitar tone burst that looks a lot like a snare, yet sounds a lot like a guitar.

💡 This kind of synthesized sound is common in many styles of music, particularly those in which the sound of the studio—the sound of the gear—is a positive, such as trance, electronica, and many other forms of dance music (always an earnest advocate for studio technologies).

A low-frequency oscillator, carefully tuned, can be keyed open on each kick drum for extra low-end thump. Sine waves tuned to 60 Hz or lower will do the trick. Distort the sine wave slightly for added harmonic character or reach for a more complex wave (e.g., sawtooth). The resulting kick packs the sort of wallop that, if misused, can damage loudspeakers. Done well, it draws a crowd onto the dance floor.

A noise source (pink noise, white noise, or electric guitar) can be gated on and off with each snare hit, adding significant spectral complexity to the naturally noisy buzz and rattle of any snare drum.

Less-obvious pairings can be made. Use a vocal track as the key input to an expander with guitar feedback. Let a hi-hat open and close on a piano. Mixing rewards imagination.

TREMOLO

Rather than attenuate between notes and phrases, one can attenuate *during* the note with a more regular rhythm. The result is tremolo, the familiar wobble of slow amplitude modulation. Many guitar amps and keyboard patches offer this feature. It helps draw attention to the instrument and set the vibe for the performance. This signature sound often has the convenient side effect that because of the ear-grabbing amplitude motion of the tremolo, it is simply easier to hear.

A throbbing, twangy guitar approach is common, but in the studio tremolo effects are applied freely to any signal. Engineers don't need a tremolo knob on the instrument or the amp to get this effect. Effects units can do this instead.

A multieffects unit may very well have a patch labeled "tremolo." Done. But tremolo is hidden in another effect: the autopanner. The autopanner is found on many multieffects devices. A single input is fed to a pan pot that can be programmed to move in some desired patterns: left to right, right to left, back and forth repeatedly, and so on. The speed, depth, and shape of the panning are often controllable as well (much like the modulation section of a delay; see Chapter 7).

Patch up an autopanner set to pan continuously, but use only one output. The volume will increase as it pans toward the active output. It will decrease as it pans toward the unused autopanner output. Tremolo results.

The Leslie cabinet (see Chapter 8) is an acoustic signal processor that, among other things, causes a form of tremolo as the sounds spin toward and then away from us.

The most sophisticated tremolo effect comes through clever use of an expander (Figure 6.6). Insert the expander into the signal to be amplitude-modulated. Feed the key input of the expander with a musically relevant, consistent performance.

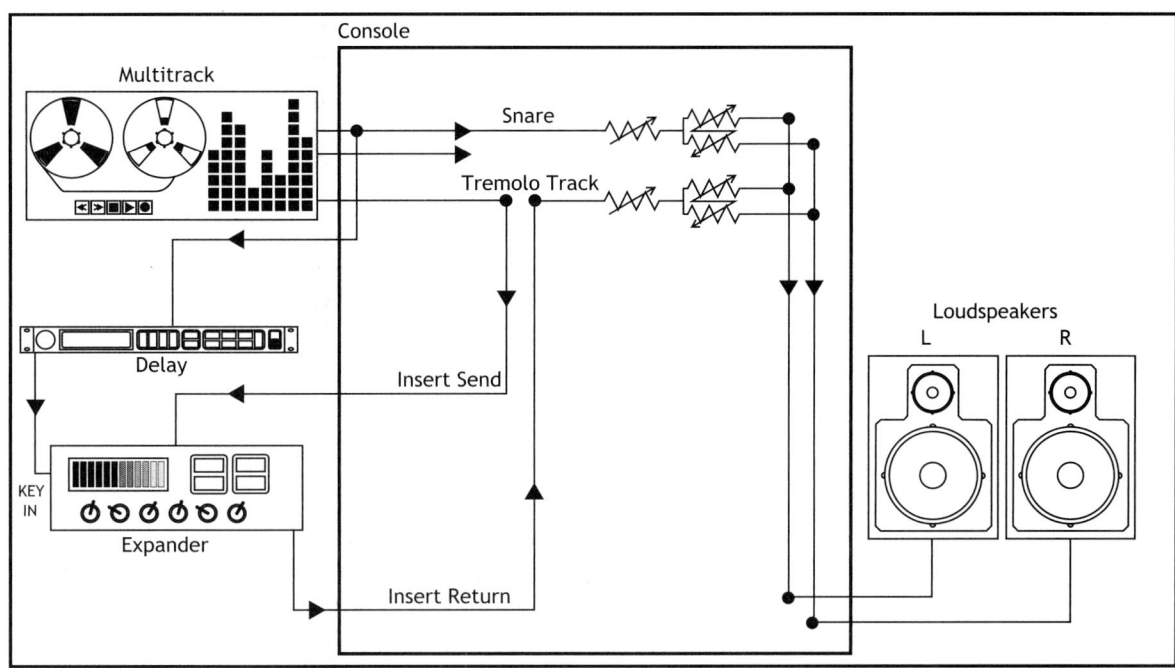

FIGURE 6.6
Creating tremolo through keyed expansion.

The backbeat of a snare drum might make an ideal candidate. Feed a bit of snare to a delay. Set the delay to an eighth-note time interval. Raise the delay feedback so that it repeats several times. Feed the delay output to the side chain input of the expander. A little (okay, more than a little on the first few tries) massaging of parameters and the result is tremolo, locked into the tempo of the song. Each snare hit and each delayed repetition of the snare hit enters the level detector of the expander and triggers some attenuation. The expander returns to unity smoothly in between these snare triggers. The amplitude of the signal is continuously and regularly turned down and back up again.

Adjust the delay time for slower or faster tremolo. Select a different side chain track for possibly more interesting, less consistent tremolo (e.g., hi-hat). Access the click track or tempo-mapped MIDI to create any pattern of tremolo you wish.

▶ An expander offers the opportunity to add tremolo to *any* track. Add tremolo to a piano. Obviously, no real piano can do this at symphony hall. But any piano recording can do this in your control room. Try tremolo on a reverberant decay. Add tremolo to a *compressed* reverberant decay. Apply tremolo to an extremely long reverb tail, turning a 30-second decay into a pad of reverberant energy that seems to play along like a musical instrument. Clearly, the expander/ gate is a source of a broad range of musically valuable effects that might feature any track or any other effect in your mix.

6.4 SUMMARY

As you master compression and limiting, you are also gaining insight into the use of expansion and gating. Probably the biggest challenge is developing the intuition to know what turning things down can do for the sound that remains.

You've got to practice a bit with expansion, as the parameters are every bit as obtuse and counterintuitive as compression parameters. Master the logic before you try to make music.

Once you are comfortable with the idea, start fixing noisy tracks with gentle expansion and noise gating. Explore the limits of naturalness and the cool side effects of unnaturalness as you take it to extreme settings and even edit out portions of the signal.

Keyed gating—particularly using the close-microphone snare track to key open a gate across pretty much any other track in the universe—is a rich area to explore, leading to effects from the subliminal to the unmissable. More apparent forms of gating let us, as mix engineers, add textures, pulses, and musical hooks to a track in need of a little studio push.

The most underappreciated effect: stop wasting time, without delay.

MIX SMART QUICK START: Delay
GOALS

- Different effects families live in specific delay-time windows: long, medium, and short.
- Master the concept that time effects can have spectral implications.
- Fit tracks into a musical groove and offer key tracks some delay-based support.
- Feature a track through attention-getting slap echo, flanging, double-tracking, and chorus.

GEAR

- Sources of delay include tape, analog circuits, and—mostly—digital.
- Exploit the full production potential through strategic use of regeneration, a low-pass filter, polarity reverse, modulation rate, modulation depth, and modulation shape.

The delay line is the single most important signal processor used for manipulating the time axis of an audio signal. In concept, it could not be simpler: delay all signals fed to the device. Audio goes in. It waits the designated amount of time. Audio comes out. Simple on the surface, this *process of holding back a sound for some defined amount of time* becomes the building block for a vast range of effects and is a required processor in all recording studio environments.

This chapter begins with some theory, summarizing delay technologies frequently used in recording studios and describing the basic signal flow through a delay processor. This summary is followed by a detailed look at the many creative applications of delay common in professional music production.

7.1 PARAMETERS

The delay in time may be generated in a number of ways—in software or hardware, analog or digital. The digital delay is perhaps the most intuitive source of delay. Once audio is made digital, it is a string of numbers as willing to be stored in random access memory as any word-processing document or web page. All that is needed is a bit of memory management and a user interface. Any computer should be able to do this. Not surprisingly, digital audio workstations are very good at this. Dedicated digital devices—standalone units that offer digital delay outside of the computer—are also available.

Purely analog delay lines also exist, though they become increasingly rare as digital audio becomes ever more available, affordable, and capable. Inductors and capacitors are the basic analog circuit components used to introduce some time delay to an analog signal.

An analog tape machine—perhaps sitting underutilized somewhere in your studio—is also a clever source of delay (see Figure 7.1) and is the basis for a whole class of delay effects, discussed in more detail in the following sections.

7.1.1 Basic Controls

! Typically, an aux send is used to get our signal into the delay unit. What's going on at the delay processor itself? Most delay devices have available the controls shown in Figure 7.2: input and/or output level, delay time, and regeneration

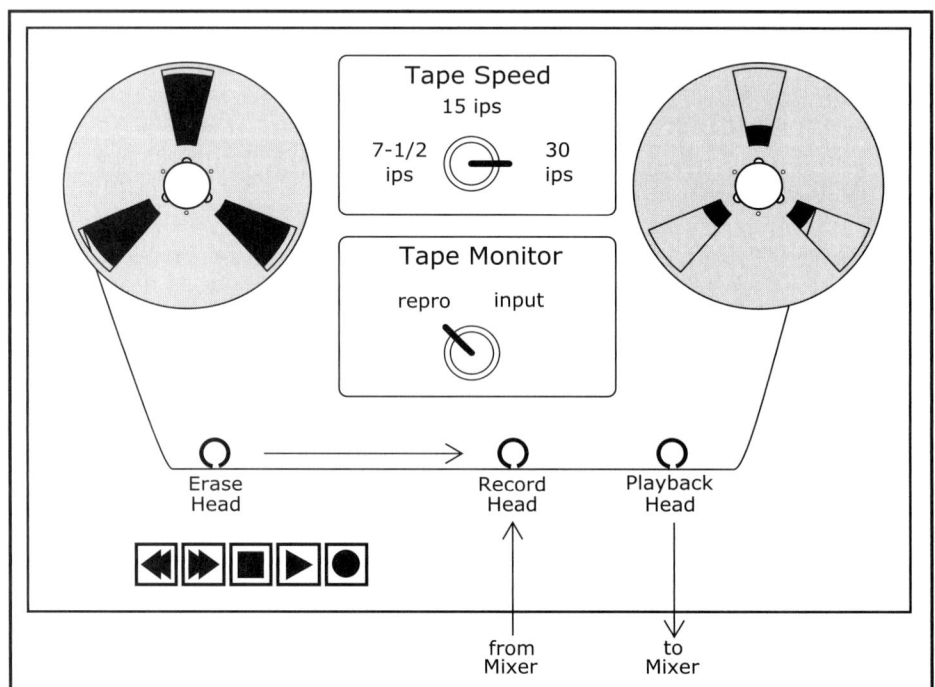

FIGURE 7.1
Tape delay.

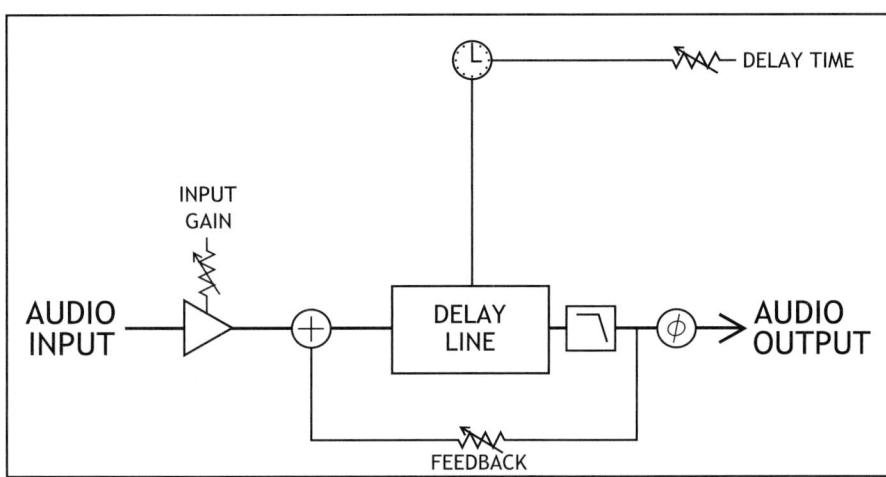

FIGURE 7.2
Typical signal flow within the delay unit.

control. Input/output levels are self-explanatory. Levels are typically set so that there is unity gain through the unit. Watch the input level to make sure the device isn't driven into distortion (unless, of course, distortion is part of the intended effect; see Chapter 4).

DELAY TIME

Of course, we need a means for adjusting the delay, so the parameter is provided. On a tape delay, your only means of changing the delay time is to change the tape speed. On a digital delay, it seems almost trivial. Once upon a time, not so long ago, the delay-time setting was a bit cruder, offering no numeric read-out. On analog delay lines and most early digital delays, you turned the delay-time setting to the left to shorten the delay, and to the right to lengthen the delay. No numeric readout confirmed it. This bit of anachronistic charm is mentioned to emphasize an essential point not lost on any experienced engineer: the delay time can be set by ear, not by eye. I'll go further. The delay time *should* be set more by ear than by eye. Listen, please listen, to the sonic implications of your delay-time setting. What might make mathematical sense when you set up the device may not make much musical sense when you listen to it.

We have a delay-time control. Set it to the right *sounding* value.

REGENERATION

The *regeneration control*, sometimes called the *feedback control*, sends some of the output of the delay right back to the input of the delay. In this way, a delayed signal can be further delayed by running it through the delay again. This process is how a comb filter is made more resonant and an echo is made to repeat more than once.

LOW-PASS FILTER

The ability to conveniently introduce a low-pass filter (see Chapter 3) directly into the signal flow of the delay proves quite useful. Many delay effects benefit from having some of the high-frequency content of the signal removed. Discussed later in this chapter and in the detailed discussion of distance in Chapter 9, high-frequency attenuation can be an effective way to communicate distance. In addition, attenuating the presence range of the delayed signal can help prevent some delay effects from cluttering the mix and distracting the listener. The low-pass filter is a common part of many delay effects and is therefore a clever stock feature in any good delay unit.

POLARITY REVERSE

The need to reverse the polarity of the delay effect is common enough that it, too, is a frequent delay line feature. It can take many delay-based effects and make them a bit stranger, a little wilder. It shifts the spectral qualities of some delay effects, giving the mix engineer greater flexibility at the touch of a button.

7.1.2 Modulation

The simple controls of Figure 7.2 empower the delay to become a fantastically diverse signal processor, spawning the many effects detailed throughout this chapter. The effect becomes more interesting still when the delay time is allowed to *change*. Many types of signal-processing effects are built on varying delays. Figure 7.3 adds a modulation section to the standard controls on a digital delay device. These controls are an elegant addition to the signal-processing capability of a delay.

Figure 7.4a describes a fixed delay time of 100 ms, which creates a slap echo, as discussed in detail shortly. The delay unit takes whatever signal it is sent, holds it for the delay time specified (100 ms), and then releases it. That's it. Throughout

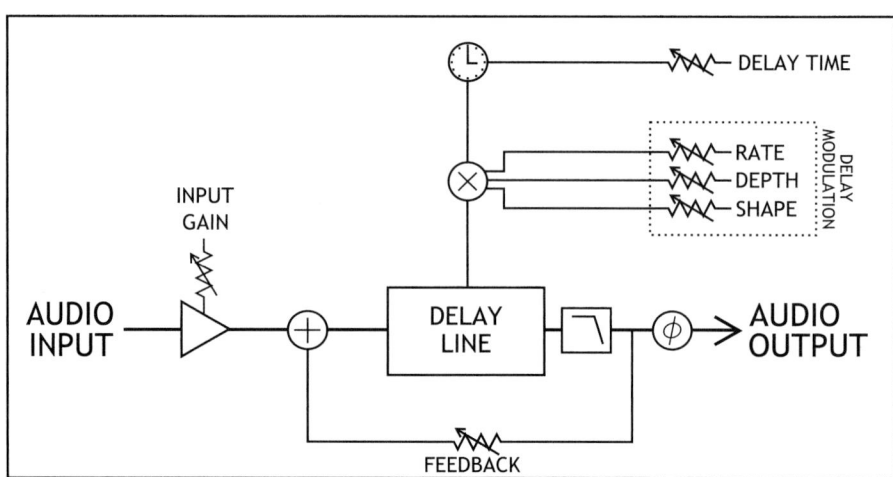

FIGURE 7.3
Signal flow for a delay line with modulation section.

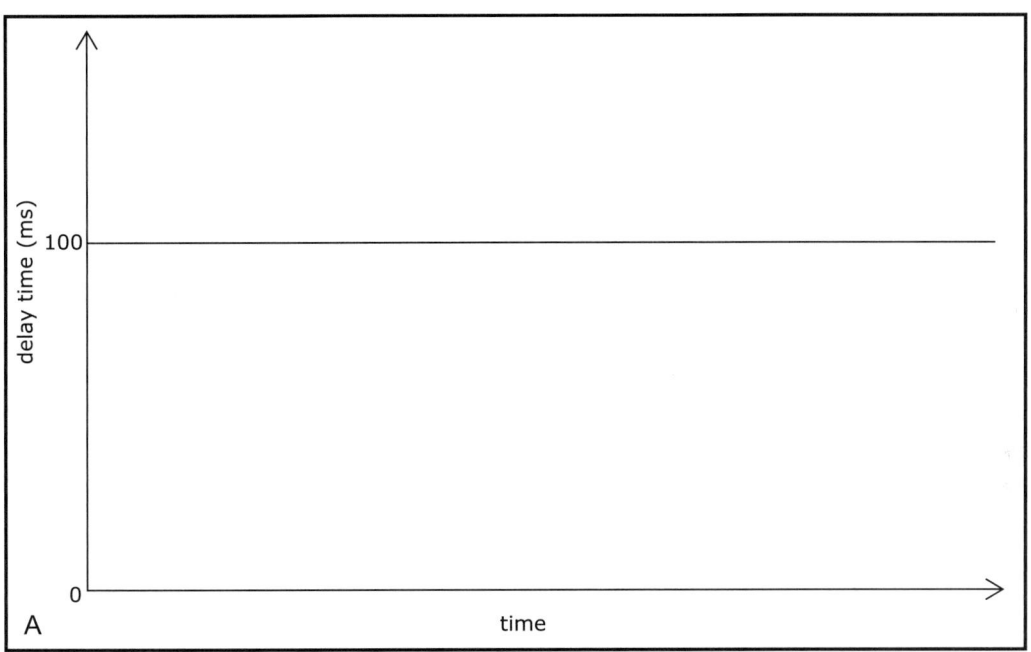

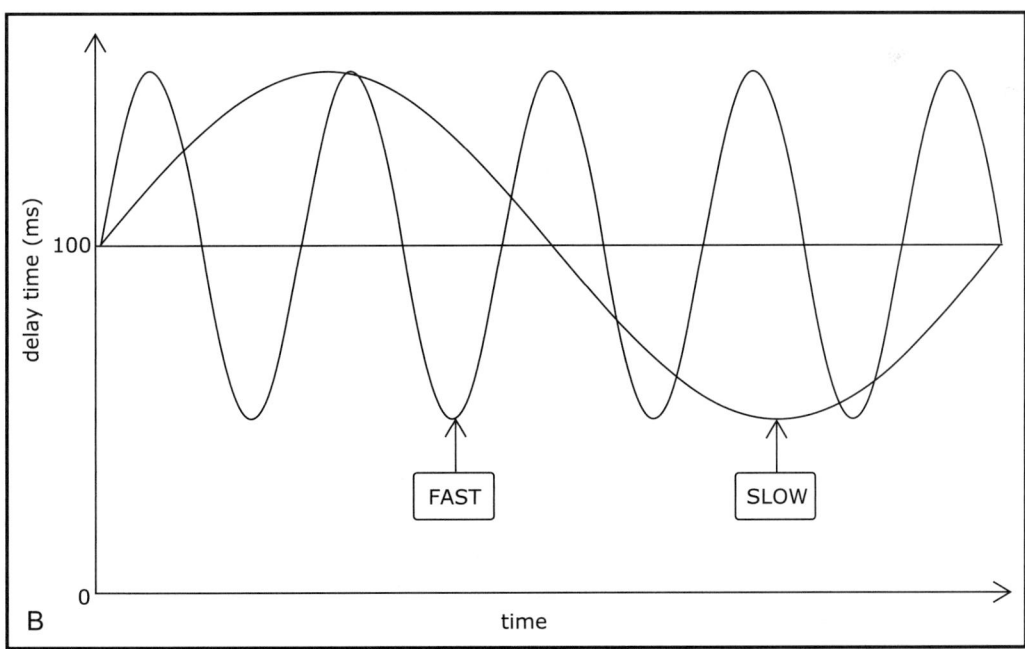

FIGURE 7.4
(a) Fixed delay; (b) changing delay: rate;

Continued

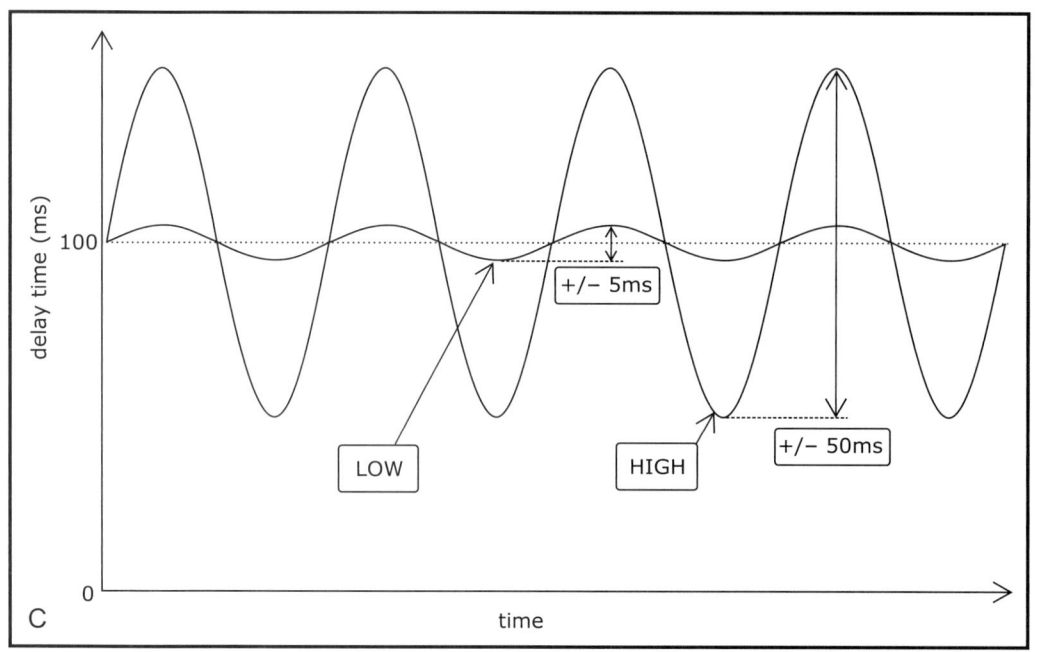

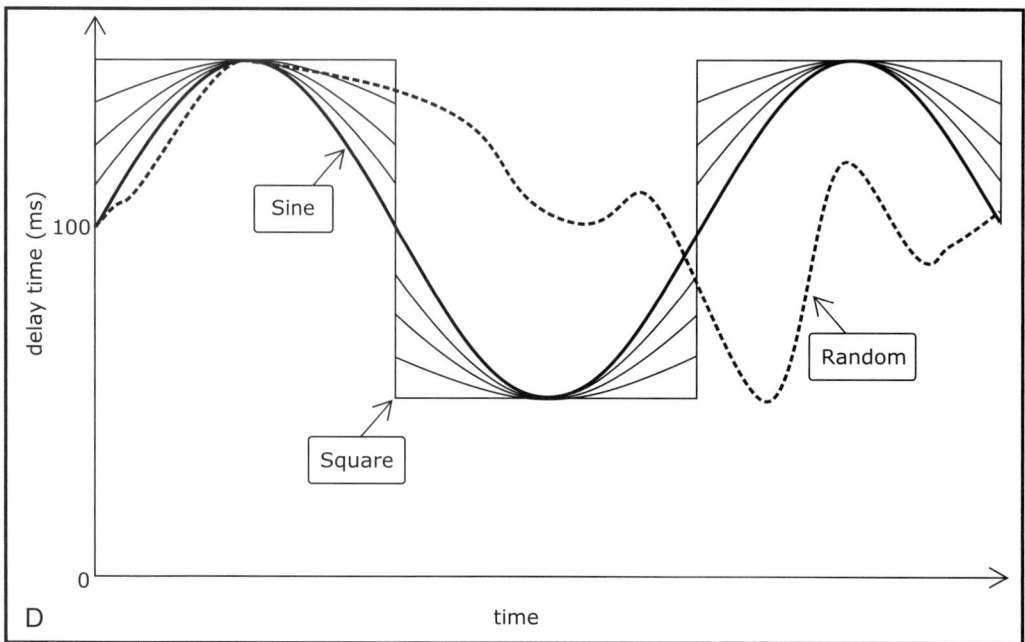

FIGURE 7.4—CONT'D
(c) changing delay: depth; and (d) changing delay: shape.

the song, all session long, the delay time remains exactly 100 ms; all signals sent to it—guitar, vocal, or ukulele—experience the exact same amount of delay. That's a delay without modulation.

The graphs of Figure 7.4 may seem peculiar at first, plotting time on the vertical axis versus time on the horizontal axis. Seemingly similar, the two time axes are wholly independent variables. The vertical axis represents the delay-time parameter within the delay unit. The horizontal axis represents the typical human experience of time, flying by, if we're having fun. Figures 7.4a through 7.4d, therefore, show the delay-time parameter as time goes by.

Some beautiful and powerful effects require the use of a changing delay, using the modulation controls: rate, depth, and shape. *Rate* controls how quickly the delay-time parameter with the effects device is changed. Figure 7.4b gives a graphic representation of what happens when this control is changed. Engineers find cases in which they want to sweep the delay time imperceptibly slowly; at other times, they need a fast, very audible rate.

Depth controls how much the delay is modulated. It bounds the delay time at the extreme, defining the shortest and the longest delay times allowed. Figure 7.4c graphically contrasts two different settings. The original, fixed delay time might be increased and decreased by 5 ms, 50 ms, or more.

Shape describes the path taken by the device as it changes the delay time within the bounds set by the depth control at the speed determined by the rate control. As Figure 7.4d shows, it can sweep in a perfect, sinusoidal shape back and forth between the upper limit and the lower limit specified (those upper and lower delay limits were set with the depth control described previously). Alternatively, there may be a need for a square-wave trajectory between delay times, in which the delay time snaps, instead of sweeps, from one delay value to the other. Figure 7.4d highlights an additional common feature of the shape control: it lets the engineer use a shape that is some mixture of the two—part sinusoid, part square.

Beyond these sine and square wave modulation shapes, the delay may have a sawtooth, triangle, or random setting under which the delay time moves less orderly between the two delay extremes.

Finally, some delay units let you use a combination of all of these controls, for example, varying the delay time in a slightly random, mostly sinusoidal general pattern. The shape control may make it possible to mix these shape options and set a unique contour for the delay's motion between its highest and lowest settings.

These three modulation parameters give us much-needed control over the delay so that we can play it like a musical instrument. They set how fast the delay moves (rate), they set the limits on the range of delay times allowed (depth), and they determine how the delay moves from its shortest to its longest time (shape).

💡 If you are familiar with the use of a low-frequency oscillator (LFO) to modulate signals in some synthesizers, for example, you will recognize that the modulation section of a delay unit relies on a simple LFO. Rate is the frequency of the LFO. Depth is the amplitude of the LFO. Shape, of course, is the type of LFO signal. Instead of modulating the amplitude of a signal as might be done in a synthesizer, this LFO modulates the delay-time parameter within the signal processor.

7.1.3 Delay Time

Not all delays are created equal. In order to understand the creative sonic possibilities for a delay effect, it helps first to separate them into different categories based on the length of the delay (Figure 7.5). Delays are classified into three broad categories, cleverly called long (greater than about 50 ms), medium (between about 20–50 ms), and short (less than about 20 ms).

LONG DELAY

Consider the long delay. A delay is classified as "long" when the delay time is no less than about 50 ms and as long as . . . well, as long as you are willing to wait around for the music—a fortnight, maybe! The idea of the long delay is that the output happens late enough after the source sound that we hear it as a separate event. It very much depends on the spectral qualities and amplitude envelope of the signal being auditioned, but it is fair to say that delay times of less than 50 to 60 ms are hard to hear as separate events; they are a different class of delay-based effect, not perceived as an echo.

An echo is a classic example of a long delay. A sound happens, and we hear a separate repetition of that sound. Transient and percussive sounds like drums reveal an echo at shorter delay times, closer to 50 ms. Slower, more sustained and low-frequency-dominated signals like a whole note on an upright bass played with a bow make it harder to hear the separate echo—60 or more milliseconds are needed. Philosophically, we consider a delay time to be long when it accesses the family of effects built on echoes and the musical use of an audible repetition.

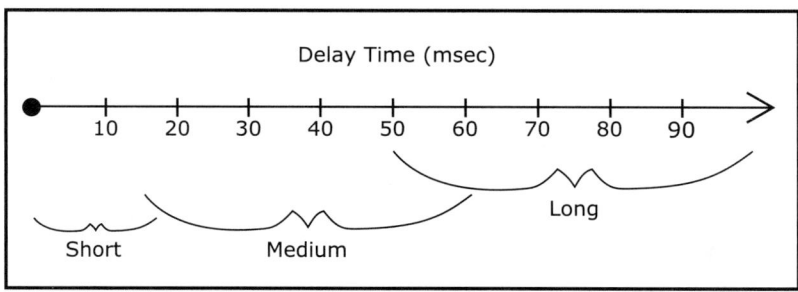

FIGURE 7.5
Delay effects divide into three classes based on the delay time.

SHORT DELAY

Delays times can be so short that they aren't perceived as echoes. The delayed sound happens too quickly for the brain to notice it as an event separate from the original sound. When delays are short enough, they are perceptually fused with the undelayed sound. The mix of signal plus delayed signal becomes a single entity with a different sound quality.

🔊 So as the delay time falls below about 50 ms, the sound of the delay is no longer an echo. It becomes, well, something else. It is not that the delay is impossible to hear, just that it has a different perceptual impact when the delay time is short. In fact, very short delays can have an important *spectral* effect on the sound and lead to a whole class of special effects. Here's how.

Constructive and destructive interference

Let's do some math, visually more than numerically. Doing so will give you great insight into the beautiful production potential of short delays. Sine waves—with their familiar, faithfully repeating wave shape—are helpful in illustrating the frequency-dependent implications of the short delay. Mixing together—at the same volume and pan position—the original signal with a delayed version of itself might have results like the two special cases shown in Figure 7.6.

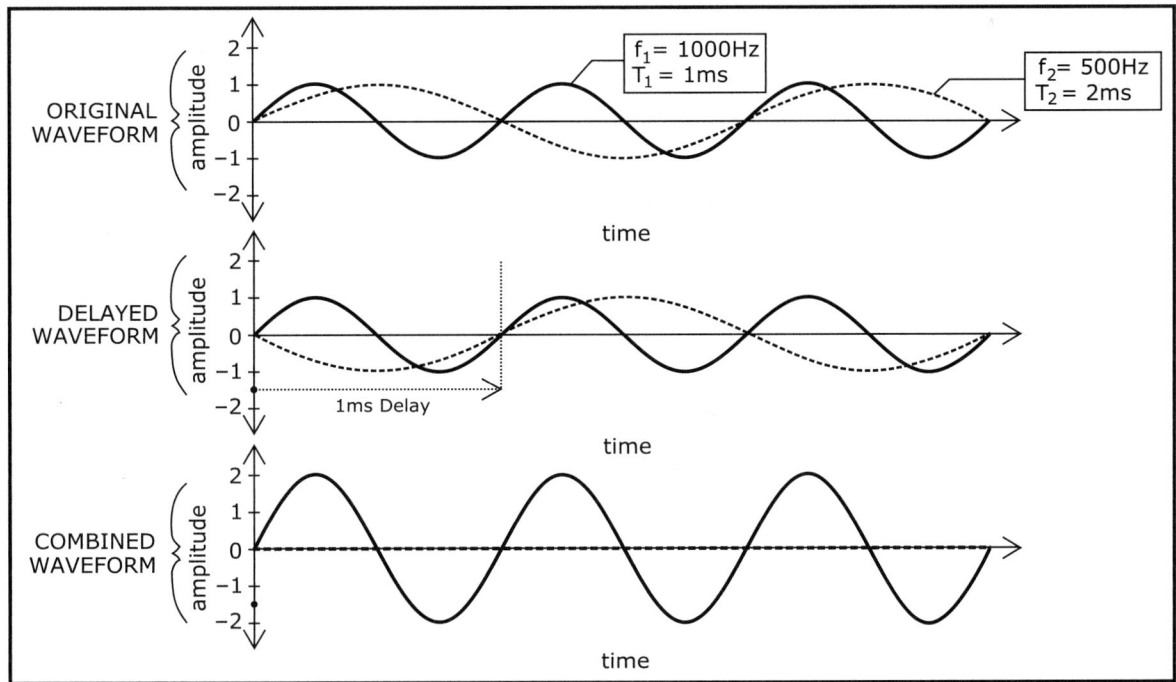

FIGURE 7.6
Constructive/destructive interference.

If the delay time happens to be exactly the same as the period of the sine wave, constructive interference like that shown with the solid line in Figure 7.6 results. That is, if the delay time of the processor is equal to the time it takes the sinusoid to complete exactly one cycle, then the two signals will combine cooperatively. The net result is a signal of the same frequency, but with twice the amplitude.

The situation shown by the dashed line in Figure 7.6 represents another special case. If the delay time is set to equal exactly half of a period (half the time it takes the sine wave to complete exactly one cycle), then the original sound and the delayed sound move in opposition to each other. For a sound wave in the air, pressure increases meet pressure decreases for a combined result of no net pressure change. For a voltage signal on a wire, positive voltages meet equal magnitude but negative voltages and sum to zero volts. In a digital audio workstation, positive numbers cancel out negative numbers. The combination results in zero amplitude—silence.

With access to a sine wave oscillator (either as test equipment or within a synthesizer), you can give it a try; 500 Hz is a good starting point. This frequency isn't quite as piercing as the standard test tone of 1,000 Hz, yet the math remains easy. The time it takes a pure 500-Hz tone to complete one cycle is 2 ms:

$$\text{Period} = 1 / \text{Frequency}$$
$$= \frac{1}{500}$$
$$= 0.002 \text{ seconds}$$
$$= 2 \text{ ms}$$

Mixing together equal amounts of the original 500-Hz sine wave and a 2-ms delayed version will create perfectly constructive interference very much like the solid line in Figure 7.6. Lower the delay time to 1 ms—creating the dashed-line situation of Figure 7.6—and the 500-Hz sine wave is essentially cancelled.

As the bold waveforms on Figures 7.7 and 7.8 show, these doublings and cancellations happen at certain other higher frequencies as well. For any given delay time, certain frequencies line up just right for perfect constructive or destructive interference.

The math works out as follows. For a given delay time (t expressed in seconds, not milliseconds), the frequencies that double are described by an infinite series: $1/t$, $2/t$, $3/t$, and so on. These frequencies all possess the feature that when the delay time t is reached, they are all back exactly where they started, in this case beginning another repetition of their cycle. The frequencies that double in amplitude have cycled exactly once, twice, or some other integer multiple at the instant time t occurs (Figure 7.7).

The frequencies that cancel are $1/2t$, $3/2t$, $5/2t$, and so on. These are the frequencies that are one-half-cycle shifted when the delay time is reached. They've cycled $1\frac{1}{2}$ times, $2\frac{1}{2}$ times, or any integer-and-a-half times (Figure 7.8).

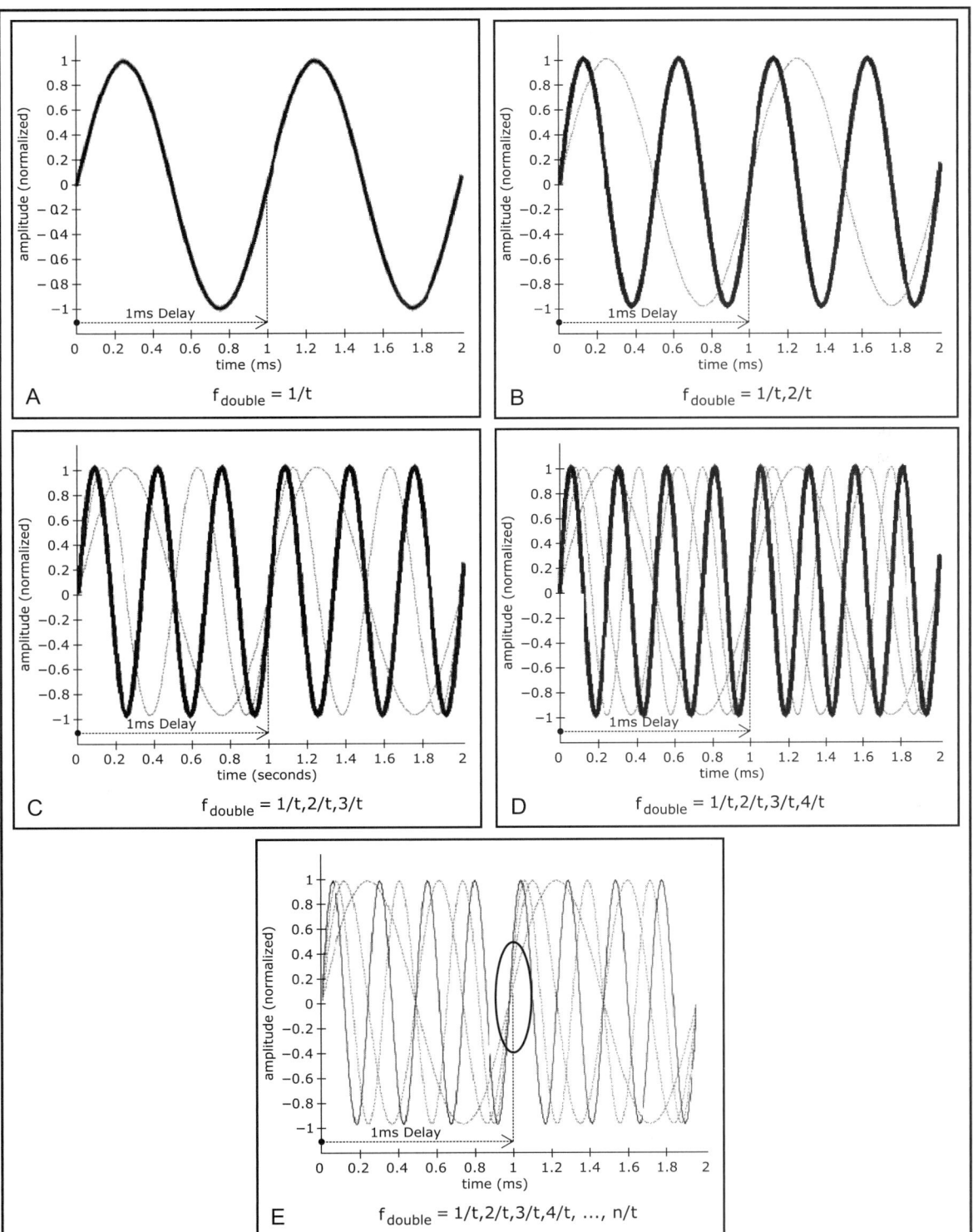

FIGURE 7.7
Amplitude doubling: 1 ms delay.

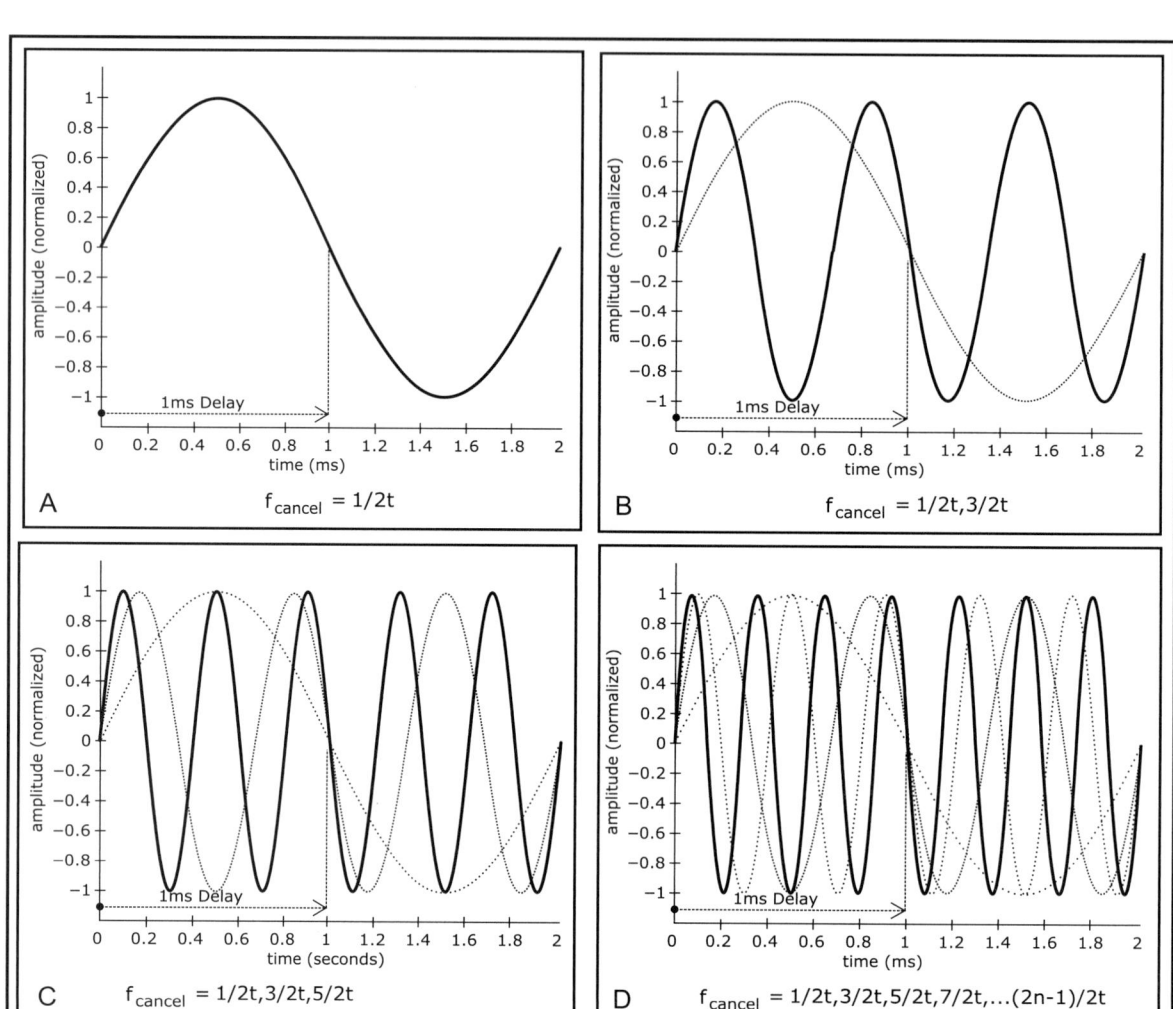

FIGURE 7.8
Amplitude canceling: 1 ms delay.

Using these equations, one can confirm that a signal combined with an equal amplitude 1-ms delay (t = 0.001 seconds) of the same signal has spectral peaks at 1,000 Hz, 2,000 Hz, 3,000 Hz, and so on with nulls exactly in between at 500 Hz, 1,500 Hz, 2,500 Hz, and so on. This result is consistent with the earlier observation in Figure 7.6 that combining a signal and its 1-ms delay can cancel a 500-Hz sine wave. A 2-ms delay has amplitude peaks at 500 Hz, 1,000 Hz, 1,500 Hz, and so on and nulls at 250 Hz, 750 Hz, 1,250 Hz, and so on. The math reveals that the peaks and dips happen at several frequencies, not just one. Although this pattern theoretically occurs for all frequencies without limit, audio engineers focus on those peaks and valleys that fall within the audible spectrum from about 20–20,000 Hz.

🔊   Another way to explore this further would be to set up the mixer so that it combines a sine wave with a delayed version of itself set to the same amplitude and panned to the same location. Sweep the sine wave frequency higher and lower, watch the meters, and listen carefully. With the delay fixed to 1 ms, for example, sweep the frequency of the sine wave up slowly beginning with about 250 Hz. You should hear the mixed combination of the delayed and undelayed waves disappear into silence at 500 Hz, reach a volume peak at 1,000 Hz, fall silent again at 1,500 Hz, reach a loud peak again at 2,000 Hz, and so on. A *delay*, not an equalizer, changes the amplitude of the signal as a function of frequency. A *delay*, not a fader or a compressor, changes the loudness of the mix. The connection between time- and frequency-dependent amplitude is an audio surprise to be savored.

Comb filter

The constructive and destructive interference associated with short delays clearly leads to a radical change in the amplitude of an audio signal that varies with frequency. Is this a silly parlor trick or a valuable music production tool?

▶   To answer this question, one has to get rid of the pure tone (which pretty much never happens in pop music) and hook up an electric guitar (which pretty much always happens in pop music). Run a guitar signal—live from a sampler, or off the multitrack—through the same setup listed earlier. With the delayed and undelayed signals set to the same amplitude, listen to what happens.

Is it possible to find a delay-time setting that will enable the complete cancellation of the guitar sound? Nope. The guitar sound isn't a pure tone. It is a complex signal, rich with sound energy at a range of frequencies. No single delay time can cancel out all the frequencies at once. But mixing together the undelayed guitar track with a 1-ms delayed version of the same guitar track definitely has an audible affect on the sound quality.

It was already observed that a 1-ms delay can cancel entirely a 500-Hz sine wave. In fact, it will do the same thing with guitar (or piano, or ukulele, etc.). Musical instruments containing a 500-Hz component within their overall sound will be affected by the short, 1-ms delay: the 500-Hz portion of their sound can in fact be cancelled. What remains is the tone of the instrument without any sound at 500 Hz.

But wait: there's more. It was also shown in Figures 7.6 and 7.7 that 1 kHz would double in amplitude when this 1-ms delay was added to the signal. So for the guitar, the 1,000-Hz portion of the signal gets louder.

Taking a complex sound like guitar, which has sound energy at a range of different frequencies, and mixing in a delayed version of itself at the same amplitude, will cut certain frequencies and boost others. This is called *comb filtering* (Figure 7.9) because the alteration in the frequency content of the signal looks like teeth on a comb.

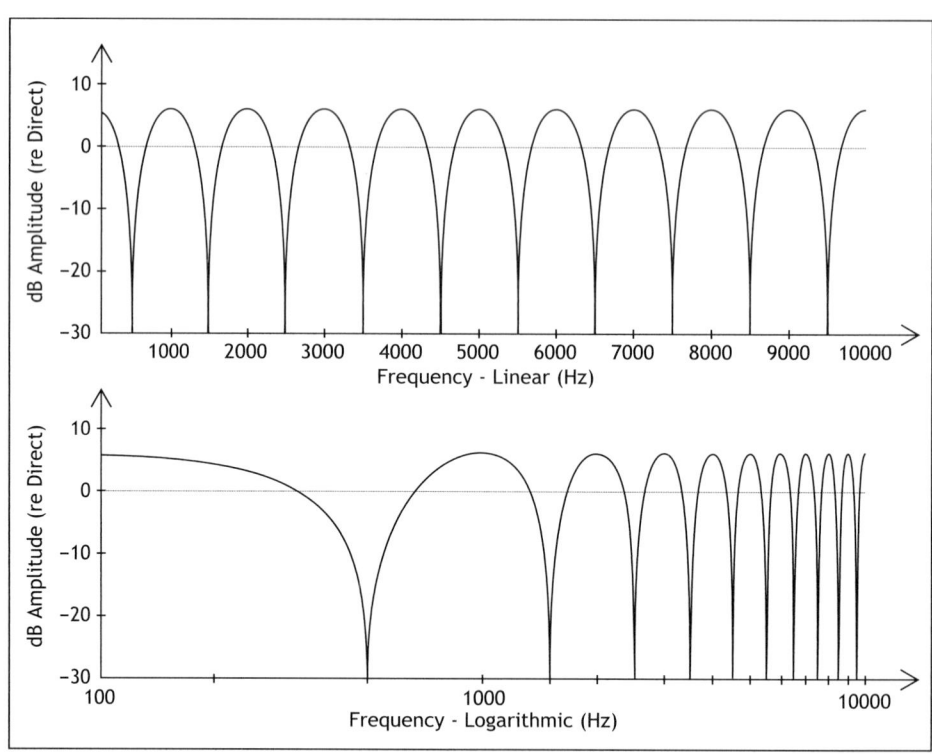

! Combining a musical waveform with a delayed version of itself radically modifies the frequency content of the signal. Some frequencies are cancelled, and others are doubled. The intermediate frequencies experience something in between outright cancellation and full-on doubling. Flip the polarity of the delay and you swap the peaks with the notches, and the notches with the peaks. Radical frequency sculpting is at your fingertips. The changes in frequency content associated with combining a signal with a short delay of itself suggest that short delays are less like echoes and more like equalizers (see Chapter 3).

Short delays are too short to be perceived as echoes. In fact, they are so short that they start to offer a patterned interaction with the original, nondelayed sound, adding some degree of constructive (i.e., additive) or destructive (i.e., subtractive) interference to different frequencies within the overall sound. Figure 7.6 demonstrates this for a sine wave. Figure 7.9 summarizes what happens in the case of a complex wave—a more typical audio track like guitar, piano, saxophone, vocal, and so forth.

▶ Equalization (EQ) is not too far off the mark as a way to think about comb filtering, but it would be nearly impossible to actually do the same with a practical equalization device: a boost here, a cut there, another boost here, another cut there, and so on. In theory, one could simulate comb filtering with an equalizer, carefully dialing in the appropriate boosts and cuts. That's the theory. In fact, such an EQ move is unlikely. To fully imitate the comb filter effect that a 1-ms delay creates, the studio would need an equalizer with about 40 bands of parametric EQ (20 cuts and 20 boosts within the audible spectrum), each with its own, unique setting of bandwidth. Such an equalizer would be prohibitively expensive to build, would likely have other side effects (e.g., noise and phase distortion) in terms of quality, and would take a long time to set up. In fact, part of the point of using short delays in this way in music production is to create sounds that can't easily be achieved with an equalizer. A single short delay creates an unbelievably complex EQ-like frequency contour.

Short delays offer a very interesting extra detail: they create mathematical—not necessarily musical—changes to the sound. Study Figure 7.9, comparing the upper curve to the lower curve. Both plots show the same information. But the lower graph presents the information with a logarithmic frequency axis—the typical way of viewing music. The keyboard of a piano and the fingerboard of most stringed instruments lay things out logarithmically, not linearly and equally spaced (see Appendix B). If one looks at comb filtering with a linear (and nonmusical) frequency axis, as in the upper part of Figure 7.9, one finds that the peaks and dips in the filter are spaced perfectly evenly. In fact, it is only by looking at the spectral result in the linear domain that the name "comb filter" becomes clear. The logarithmic plot on the lower portion of Figure 7.9 would not make for a very useful comb, but it's the more musically and perceptually informative visual representation.

Pure tones, cycling with infinite reliability, would experience the comb-filtering effect at any delay time, not just short ones. A delay time of 1 ms has been shown to cause constructive and destructive interference at a precise set of frequencies. The logic still holds when the delay time is 1,000 ms, 1,000 days, or more. In music production, however, where the target of our signal processing is the complex music signals created by resonant mechanical systems (i.e., guitar, piano, tuba) rather than pure tone test signals, the effect is—practically speaking—limited to very short delays.

Two related phenomena are at work. First, as discussed in the earlier section "Long Delay," delays greater than about 50 ms become perceptually separate events. Our hearing system identifies the nondelayed and the long-delayed signals as two separate events, each with their own properties. Comb filtering relies on the fusing together of the nondelayed and the delayed signals into a single perceived event.

Second, music signals do not repeat with the rigid regularity over time of sine waves. Musical signals change constantly. The singer moves on to the next syllable, the next word, the next line. The guitarist moves on to the next note, the next chord. The pitch of even a fixed, sustained note will drift up and down, at least slightly. Long-delayed signals appear too late to interact directly with the nondelayed signal. Unlike a sine wave, the musical signal has progressed on to other things—new notes, new pitches, new phase relationships with the delayed signal that has finally arrived.

Even if the musical signal is relatively unchanging—the whole note of a bass guitar, or the double whole note of a sustained piano chord—it still falls well short of sine wave stability. A bass note held over time still wobbles a bit in pitch, which might be a means of expression as the performer gently raises and lowers the pitch for musical effect. It might also be a practical inevitability; new strings on a stable instrument hold a pitch more steadily than old strings on a weak instrument. If the pitch has changed, the undelayed and long-delayed signals are too different from each other for the mathematically predicted comb filtering to occur. In sound recording, the distinct spectral signature of comb filtering has a practical limit then. It audibly exists only for very short delays of 15 to 20 ms, or less.

MEDIUM DELAY

In between long and short, we naturally have medium. When the delay time is between the short and long thresholds, greater than about 15–20 ms and less than about 50–60 ms, it opens up a unique set of production possibilities worthy of their own study.

7.2 MIX STRATEGIES: DELAY

Delay represents one of the richest sources of mix ideas, although you might not expect this to be the case at first. Delays can be front and center, impossible to miss for all who hear your mix. Delays also enhance a mix at a subliminal level, so they might be in more mixes than you've realized thus far. We discuss the full production potential of the effect next.

7.2.1 The Fix

As with all the core effects discussed in these pages, we begin first by fixing any problems known to drag down any of our tracks.

TO COMB FILTER, OR NOT TO COMB FILTER, THAT IS THE MIX MOVE

▶ We can learn much from a simple electric guitar overdub session. With the guitar amplifier in the middle of the recording space on the typical concrete or hardwood floor, the tracking engineer places the chosen microphone maybe a few feet away and tries to capture the natural sound of the amp in the room. This is a good approach, represented in Figure 7.10. The potential problem—one that might be illustrative of adding a short delay in your mix—is that the sound reflected off the floor and into the microphone will arrive a split second later than the sound that went straight from the amplifier to the microphone. The path is longer via the reflected path, introducing a bit of short delay. The result is some amount of comb filtering. Recording a sound and a single reflection of that sound is a lot like mixing a track or sample with a delayed version of itself, as in the discussions previously. With any loud reflection representing a delayed version of a sound, it rapidly becomes apparent that comb filtering is an everyday part of recording.

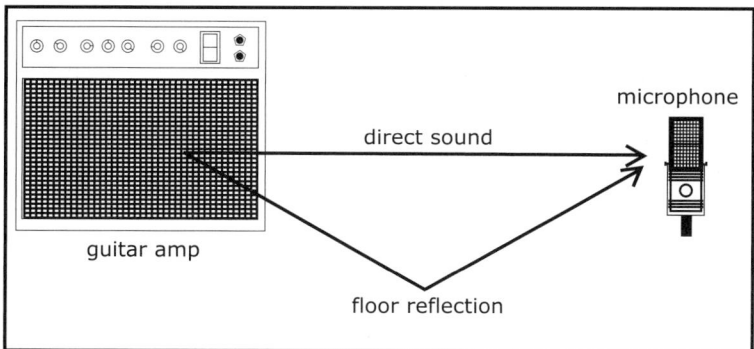

FIGURE 7.10
Delay from room reflections.

Fortunately, the sound reflected off the floor will also be a little quieter, reducing the comb filter effect: less-pronounced peaks and shallower notches. If the floor is carpeted, the comb filtering is a little less pronounced at high frequencies. Place a thicker pad of sound-absorbing material at the point of the reflection and the comb filtering is likely to be even less audible. An important part of the recording craft is learning to minimize the audible magnitude of these reflections by taking advantage of room acoustics when placing musical instruments in the studio and strategically placing sound-absorptive materials around the musical source. This approach is used by the tracking engineer to capture a nice sound at the microphone.

Better yet, learn to exploit these reflections and the comb filtering they introduce on purpose. For example, raising the microphone makes the difference in distance between the reflected path and the direct path even longer. Raising the microphone therefore lengthens the acoustic delay-time difference between the direct sound and the reflected sound, thereby lowering slightly the spectral locations of the peaks and valleys of the comb filter effect.

Of course, other factors are in play. Raising the microphone also pushes the microphone further off-axis from the amp, changing the captured timbre of the electric guitar tone as viewed by the microphone. Those frequencies not captured in both the direct sound and the reflected sound will not be subjected to the constructive and destructive interaction that leads to the classic comb filter pattern (Figure 7.9). A more complicated, less consistent alteration to timbre is created. It is the job of the tracking engineer to hear this out and ensure that the alteration is flattering to the guitar tone.

Another strategy is to raise the amp up off the floor, perhaps setting it on a piano bench or at the edge of a strong, rattle-free table. Alternatively, it is also common to tilt the amp back so that it faces up toward a raised microphone and away from the sound-reflective floor. These other approaches are designed to acoustically adjust the amplitude of the short delayed signal coming from the floor and combining with the direct signal at the microphone.

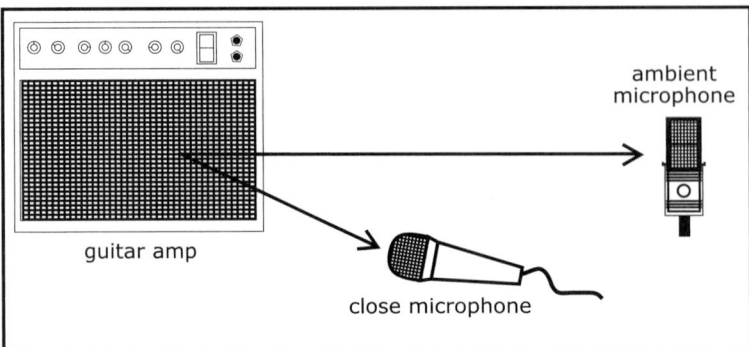

FIGURE 7.11
Delay from multiple microphones.

▶ As long as the guitar amp is set up, let's use it to illustrate another common short delay phenomenon. A common approach to recording a guitar amp—and many other instruments, for that matter—is to use a combination of two or more microphones to create the sound, even as it is recorded onto a single track. Consider the session shown in Figure 7.11: two microphones, one track. Here, a close microphone (probably a dynamic) grabs the in-your-face gritty tone of the amp and a distant microphone captures some of the liveness and ambience of the room. An engineer might label the channel fader controlling the close microphone something like "close" and the channel fader governing the more distant microphone "room." The tracking engineer adjusts the two faders to get an attractive combination of close and room sounds and prints that to a single track of the multitrack.

That's only half the story. As the recording engineer adjusts the faders controlling these two microphones, not only is the close/ambient mix changed, but the amount of comb filtering introduced into the guitar tone is also affected. These two microphones pick up very similar signals but at different times. Mixed together, they act very much like the signal-plus-delay scenario just discussed. Moving the distant microphone to a slightly different location is just like changing the time setting on the delay unit. It effectively selects different key frequencies for cutting and boosting using the exact same principles explored in Figures 7.6, 7.7, 7.8, and 7.9. Sound travels a little farther than one foot per millisecond. To lengthen the delay time difference by about a millisecond, move the distant microphone back about a foot. To get a 10-ms delay increase, move the distant microphone back about 10 feet. It's that simple.

Naturally, there's too much to keep track of. Each of these microphones receives reflected sounds from the floor and the ceiling and all the other room boundaries, all in addition to the obvious direct sound from the amp. The engineer tries to orchestrate the complex interaction of the many components of guitar sound radiating out of the amp. The direct sounds into multiple microphones arrive at different times, leading to some amount of comb filtering. The reflections from the various room boundaries into each microphone arrive at a later time than the direct sound, adding additional comb filtering, with peaks and dips falling at different spectral locations. There are an infinite number of variables in recording. Understanding comb filtering is part of how tracking engineers master the vast recording process.

! A wise mix engineer must be aware of this issue. At mixdown, we listen for comb filter artifacts in any tracks we are offered, paying particular attention to tracks recorded with more than one microphone and tracks which might have been tracked near a prevalent acoustic reflector (like the floor). We embrace any deliberate comb filtering we inherit from tracking, recognizing it as a spectral imprint that the band and the producer likely sought out while recording the parts, however many months ago. We try to make the unique tone work in the mix.

But we also watch out for unintended comb filtering. If it doesn't sound good to you, and if you have some doubts about the quality of the tracking experience, make note of the comb filtering and plan to reshape it with EQ and some strategic additions of more short delays to obscure the unpleasantness with new comb-filtering contours based on new delay times.

▶ We don't have to rely on the tracking engineer for good versions of these effects. For any track in need of some character, consider fabricating some spectral fun through the addition of some short delays. Perhaps you seek a tough, heavy, larger-than-large guitar tone. I know I often do. Maybe a comb filter–derived frequency response bump at 80 Hz is the secret. Or should it be 60 Hz? That's your mix decision. Explore this spectral issue by adjusting the delay time to taste.

Electric guitar responds well to comb filtering. With energy across a range of frequencies, the peaks and dips of comb filtering offer a distinct, audible sound property well worth exploring. Other instruments reward this kind of experimenting. Try adding one, two, three, or more different short delays to a dense horn section, a simple synth line, or the lead ukulele. Experiment with the comb filter–derived signal processing to get a sound that is natural but beautiful, or one that is unnaturally beautiful.

One loud reflection can make or break a track. As mix engineers, we must forever be on the lookout. A similar heads-up must be issued whenever you mix a multitrack project dedicating two (or more) tracks to the same performance of the same instrument. The reason is clear: if the tracks were made using microphones that were the same distance from the instrument, the tracks can be combined freely, letting you mix for the best combination of any unique qualities in the tracks without fear of comb-filtering evils.

🔊 If, however, the microphones were different distances from the instrument—and this is the much more likely scenario—you have time-of-arrival differences in the tracks that can lead to audible comb filtering. To combine the tracks is to introduce comb filtering.

This issue isn't all bad. As we've discussed, well-placed frequency deviations might reshape the sound in desirable ways. But don't leave it all to chance. If you don't like the spectral characteristics of the comb filtering that occurs when you combine the two or more tracks, simply tune it. That is, introduce a short delay to one of the tracks and adjust it to taste. The unique quality of the delayed track alone doesn't change. If it has a gorgeous tone courtesy of a gorgeous microphone, it retains its own tone. If it has a tasty bit of reverberant ear candy because it was tracked in a great-sounding room (with walls made of acoustical candy, no doubt), you retain that room sound. Introducing the delay affects only the spectral locations of the peaks and valleys in the comb filtering that comes from combining the tracks.

Adjust the delay time until those peaks and valleys please you. Adjust the levels of the two tracks to both combine the individual attributes of each track and to control the amount of comb filtering. Keep in mind that the comb filtering is most pronounced when the two similar tracks are at the same level. Whenever one track—either track—is set louder than the other, the peaks recede, the notches fill in, and the comb-filtering contour becomes less extreme.

If no comb filtering is desired, time-align the two different tracks, nudging the more distant mike forward and/or the closer mike backward. In this way, we implement a sort of time travel miracle. We get the benefit of multiple microphone placements, but we reunite them with little to no spectral damage by sliding them into time alignment.

Two tracks on one instrument gives you many mix options. Don't just mix and match the tracks to taste; tune the amount and quality of any comb filtering caused by their interaction. Work the time and the amplitude.

7.2.2 The Fit
Delay offers clever, compositional approaches to fitting together the many multitrack elements that lead to an effective, mixed whole.

GROOVE

Well-timed long delays or *echoes* are an excellent way to fill in part of the rhythm track of a song. Reggae is famous for its cliché echo. Drum programmers have been known to put in an eighth- or quarter-note delay across the entire groove in some dance music. Guitarists use delay too. U2's the Edge has made delay a permanent part of his guitar rig and prevalent part of his guitar sound. A classic example is apparent from the introduction of U2's "Bad" on *The Unforgettable Fire*. The quarter-note triplet delay isn't just an effect; it's part of the riff. The Edge has composed the delay element into the song. Used in this way, echo becomes part of the groove—a driving musical force for the tune.

When you feel like the excitement of the rhythm section as performed sinks lower when the composition needs it to stretch higher, consider adding excitement with a bit of echo-based groove enhancement. Care and musicality are essential, though. The style needs to tolerate this rather obvious effect. Tucking the driving rhythmic enhancement in at a low level might offer a subtle push. If you lack the compositional experience for this kind of addition to the production, it is a good idea to consult the songwriter, the arranger, and the drummer. This sort of mix move is well within the domain of the mix engineer, but its impact on the intentions of the band and the composer cannot be overstated.

SUPPORT

The addition of a pulsing, rhythmic echo to a track, by way of a long delay, has surprising potential even when it is not a prevalent part of the groove of the song. We add sonic support to some tracks through the addition of a constant echo, placed almost subliminally in the mix and nearly hidden by the other sounds in the mix. A soft echo underneath the lead vocal can give it added richness and support—a bed of energy to help the singer through a fragile, difficult performance. This approach can strengthen the sound of the singer, especially when the melody heads into falsetto territory. Pulsing, subliminal echoes feeding a long reverb can create a soft and delicate sonic foundation under the vocal of a ballad (a reverb effect discussed more in Chapter 9).

Then there's the vulnerable rock-and-roll singer in front of his mate's Marshall stack. After the last chorus, the singer naturally wishes to scream "Yeaaaaaaaaah!" and hold it for a couple of bars. It isn't easy to overcome the guitarist's wall of sound. Help the singer out by pumping some in-tempo delays into the scream.

The best "Yeaaaaaaaaah!" ever recorded in the history of rock and roll (according to your author, based on no data whatsoever) is Roger Daltrey's in "Won't Get Fooled Again" on *Who's Next* by the Who. The scream occurs right after the reintroduction, when those cool keyboards come back in, and right before the line, "Meet the new boss, the same as the old boss." This scream is a real rock-and-roll icon.

Listen carefully (especially at the end of the scream) and you'll hear a set of delayed screams underneath. It's Roger Daltrey, only more so. It's half a dozen Roger Daltreys. It makes quite a statement. Anyone can do this. All they need is a long delay with some regeneration—and young Roger.

7.2.3 The Feature

Highlight tracks in your mix with sounds only a delay can create.

ECHO

Music (available in the mind's ear or in the studio) will illustrate the musical potential of the long delay quite nicely. Hum or sing along with Pink Floyd's tune "Comfortably Numb" from *The Wall*. There is that all-important first line: "Hello . . . hello . . . hello. Is there anybody in there?" That repeating of "hello" is a classic use of a long delay. The dreamy, disturbed, out-of-mind state of our walled-in friend and hero, Pink, is enhanced by (the entire, brilliant rock-and-roll arrangement, including) this repeating, gently fading echo.

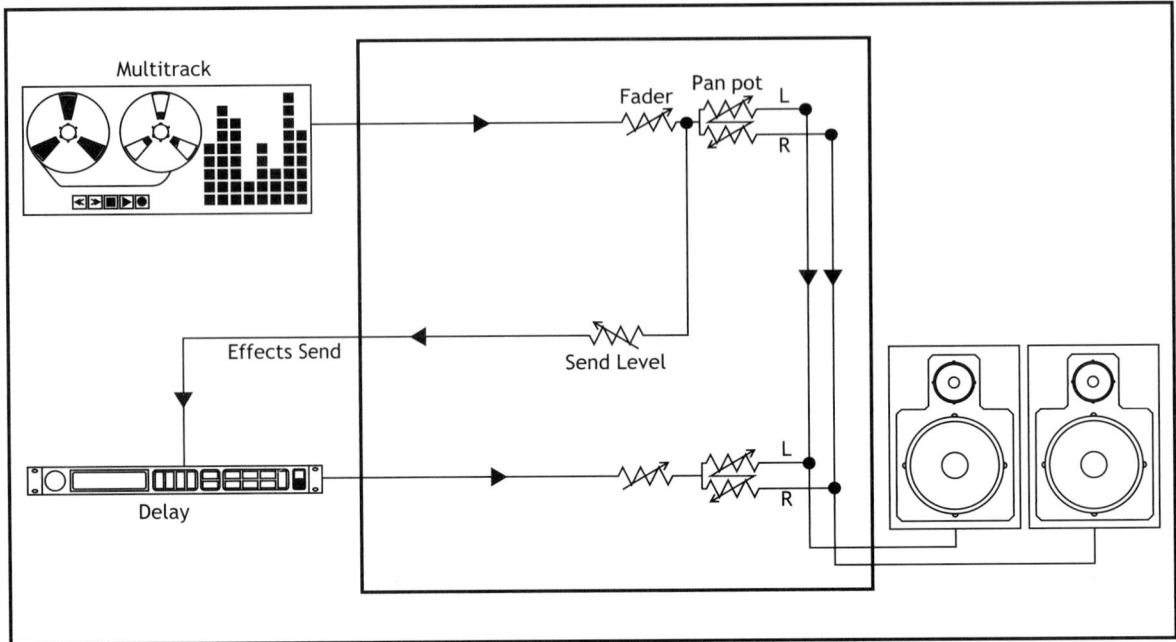

FIGURE 7.12
Signal flow for delay.

▶ How's it done? Perhaps the simplest way is to use a postfader aux send from the vocal to the delay, which is returned on a separate fader (as shown in Figure 7.12). The voice is sent to the delay processor, which when set to a long delay, creates the echo.

But there is a headache here. Hooking it up as shown in Figure 7.12 adds an echo to the entire vocal performance, not just the word "hello." Each and every word repeats. This approach leads to a tiresome, distinctly nonmusical bit of vocal chaos. It's hard to find the melody. The lyrics are difficult to understand, obscured by the relentless output from the delay as each and every word happens two, three, or more times. Echo applied to the entire vocal track in this way is rarely useful.

The solution on an analog console (in which the send control itself isn't automatable) is the creation of an automated send as shown in Figure 7.13. Now the engineer has a fader and cut button dedicated to the control of the *send* into the delay, not just the return. The echo send remains cut most of the song. Open it up briefly for the single word "hello" and, presto, that single word starts to repeat and fade. The rest of the vocal line, not sent to the delay, fits nicely around this discrete, one-word echo. No distraction. No trouble understanding the words. The melody is easy to follow and the rest of the musical arrangement is easy to hear.

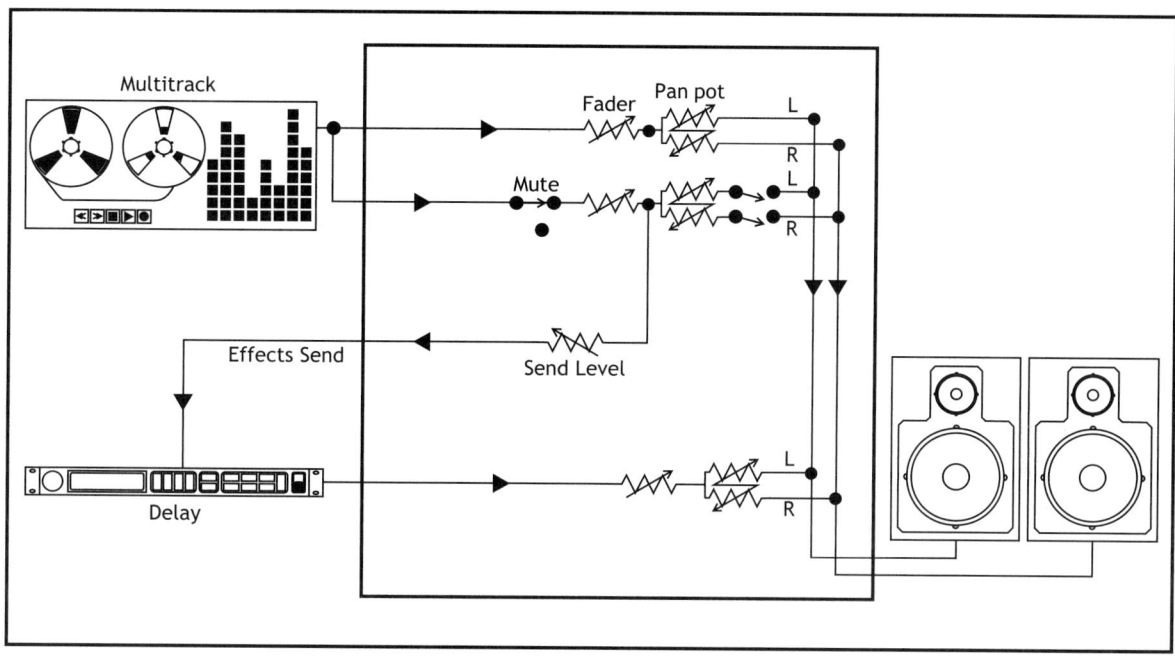

FIGURE 7.13
The "automated" effects
send.

On a digital audio workstation, two approaches are valid. Automate the send on
and off as needed. When the aux send is automatable, as in a DAW, there is no
need to create the extra strip dedicated just to creating the echo effect.

It may take some practice getting the feed into the delay unit just right. You must
unmute the effects send just before the word "hello" and mute it again just after.
You may find it helpful to imagine yourself singing the line. With a musician's
ear for every nuance of a performance, learn the exact phrasing of this track.
Know when the singer breathes, learn how long the singer sustains each vowel,
and master the exact rhythm of each syllable.

Getting into the musical performance in this way, you find that brief spaces
between words open up and become easier to find. With practice, this trick
becomes an easy, intuitive bit of hand/ear coordination. Store this unmute/
mute gesture into the mixer automation system, and a perfectly tailored delay
effect occurs each and every time this part of the tune is played.

Another DAW alternative is to make a copy of the entire vocal track and strip out
all of the words *except* those you wish to enhance with an echo. In this way, you
have a new track of words for echo-emphasis. Unassign this new track from the
mix bus and instead send it only to the delay. It sits silent most of the time, playing only the words you wish to embellish with echo.

💡 Approximately 99.9 percent of the time (and I'm just guessing here; it might be 99.99 percent), echoes like these should be set to a time that makes musical sense. You don't simply pick a random delay time and mix away. A *musical* delay time is carefully dialed in. Should it repeat with a quarter-note rhythm, an eighth note, a triplet? It's worth thinking this essential step through carefully.

Tuning by ear

🔊 One proven way to find the right length of delay relies on your sense of rhythm. Try using the snare to "tune" a delay—to set a delay time that makes musical sense. Even if the plan is to add delay to the vocal, piano, or guitar, it is usually easiest to use the snare for setting the delay time both because it is a rhythm instrument and also because it hits so often. So much of pop music has a backbeat—the snare falling regularly on beat two and beat four. Send the snare to the delay and listen to the echo. Starting with a long delay time of about 500 ms, adjust the delay time until it falls onto a musically relevant beat. This method can be extremely confusing at first. It may help initially to pan the snare track off to one side and the delay return to the other. It's pretty jarring to hear a delay fall at a nonmusical time interval. But when it is adjusted into the time of the music, you will instantly feel it. It is perhaps easiest to find a quarter-note delay, but with practice and concentration, finding triplet and dotted rhythms becomes perfectly intuitive.

After the delay-time value appropriate to the tempo of the song is found by ear, don't forget to pull the snare out of the effects send, then send the vocal (or whichever track is to be treated with echo) to the delay device instead. That use of snare was a device for tuning the delay, not a mix move. Once the correct delay time has been found, undo the snare patching and return the snare to its rightful place in the mix. The delay now sits, ready to add the perfect echo to any track you feed it.

Tuning through math

Sometimes, in our search of a specific musical time value, we *calculate* a delay time instead. How is this calculated? It is time for a useful bit of algebra. If the tempo of the song is known in beats per minute (bpm) and the length of a quarter-note delay in milliseconds (Q) is desired, perform the following straightforward calculation:

- Convert beats per minute into minutes per beat by taking the reciprocal: bpm beats per minute becomes 1/bpm minutes per beat.
- Convert from minutes to milliseconds: 1/bpm minutes per beat $\times$ 60 seconds per minute $\times$ 1,000 milliseconds per second.
- Putting it all together, the length of time of a quarter note in milliseconds per beat is:

$$Q = (60 \times 1,000)/(\text{bpm})$$
$$Q = 60,000/\text{bpm} \qquad (7.1)$$

For example, a song with a tempo of 60 beats per minute ticks like a watch, with a quarter note occurring exactly once per second. Try using Equation 7.1:

$$bpm = 60$$
$$Q = 60,000/bpm$$
$$Q = \frac{60,000}{60}$$
$$Q = 1,000 \text{ ms per quarter note, or}$$
$$Q = 1 \text{ second per quarter note}$$

Double the tempo to 120 bpm:

$$bpm = 120$$
$$Q = 60,000/bpm$$
$$Q = \frac{60,000}{120}$$
$$Q = 500 \text{ ms per quarter note, or}$$
$$Q = \text{half a second per quarter note}$$

Milliseconds are used here for two reasons. First, the millisecond is the magnitude that most delay units expect. Second, it typically leads to comfortable numbers in musical applications. For the frequencies we can hear, and for the effects we are likely to use, units in milliseconds generate numbers of manageable size—not too many decimal places, not too many digits. Using seconds, minutes, years, or fortnights would still work, theoretically. These are all units of time. But the millisecond is the more convenient order of magnitude. One could measure one's age in nanoseconds but using years turns out to be more convenient, leading happily to a smaller number.

Musical delay time

Calculating first the quarter-note delay makes it easy to then determine the time value of an eighth note, a sixteenth note, dotted or triplet values, and so on. Table 7.1 can be a useful tool. Use the snare drum approach or the bpm conversion to find the time equivalent of a quarter note. Then use this table and some musical judgment to dial in the right type of echo-based effect.

In Pink Floyd's "Comfortably Numb" example, a dotted eighth-note delay is cleverly used. This is a deliberate production decision, not a happy accident. It is worth transcribing it for some production insight.

The tune is dreamy and lazy in tempo, moving at about 64 bpm. The two syllables of "hello" are sung as sixteenth notes. To count quarter notes, just count, "one, two, three ...." To count eighth notes, insert the syllable "and" in between the beats: "one and two and three and ...." To count the more complicated sixteenth-note rhythm, stick in two additional syllables to identify all parts of the sixteenth-note pattern: "e," which rhymes with "free," and "a," like the a in "vocal." Sixteenth-note time thus becomes, "one e and a two e and a three e and a ...."

Table 7.1 Converting Tempo into Time

bpm	quarter	eighth	sixteenth	dotted quarter	dotted eighth	dotted sixteenth	triplet quarter	triplet eighth	triplet sixteenth
40	1500	750	375	2250	1125	562.5	1000	500	250
41	1463	732	366	2195	1098	549	976	488	244
42	1429	714	357	2143	1071	536	952	476	238
43	1395	698	349	2093	1047	523	930	465	233
44	1364	682	341	2045	1023	511	909	455	227
45	1333	667	333	2000	1000	500	889	444	222
46	1304	652	326	1957	978	489	870	435	217
47	1277	638	319	1915	957	479	851	426	213
48	1250	625	313	1875	938	469	833	417	208
49	1224	612	306	1837	918	459	816	408	204
50	1200	600	300	1800	900	450	800	400	200
51	1176	588	294	1765	882	441	784	392	196
52	1154	577	288	1731	865	433	769	385	192
53	1132	566	283	1698	849	425	755	377	189
54	1111	556	278	1667	833	417	741	370	185
55	1091	545	273	1636	818	409	727	364	182
56	1071	536	268	1607	804	402	714	357	179
57	1053	526	263	1579	789	395	702	351	175
58	1034	517	259	1552	776	388	690	345	172
59	1017	508	254	1525	763	381	678	339	169
60	1000	500	250	1500	750	375	667	333	167
61	984	492	246	1475	738	369	656	328	164
62	968	484	242	1452	726	363	645	323	161
63	952	476	238	1429	714	357	635	317	159
64	938	469	234	1406	703	352	625	313	156
65	923	462	231	1385	692	346	615	308	154
66	909	455	227	1364	682	341	606	303	152
67	896	448	224	1343	672	336	597	299	149
68	882	441	221	1324	662	331	588	294	147
69	870	435	217	1304	652	326	580	290	145
70	857	429	214	1286	643	321	571	286	143
71	845	423	211	1268	634	317	563	282	141
72	833	417	208	1250	625	313	556	278	139
73	822	411	205	1233	616	308	548	274	137
74	811	405	203	1216	608	304	541	270	135
75	800	400	200	1200	600	300	533	267	133
76	789	395	197	1184	592	296	526	263	132
77	779	390	195	1169	584	292	519	260	130
78	769	385	192	1154	577	288	513	256	128
79	759	380	190	1139	570	285	506	253	127
80	750	375	188	1125	563	281	500	250	125
81	741	370	185	1111	556	278	494	247	123
82	732	366	183	1098	549	274	488	244	122

Table 7.1	Converting Tempo into Time—cont'd								

				Milliseconds per note value					
				dotted			triplet		
bpm	quarter	eighth	sixteenth	quarter	eighth	sixteenth	quarter	eighth	sixteenth
83	723	361	181	1084	542	271	482	241	120
84	714	357	179	1071	536	268	476	238	119
85	706	353	176	1059	529	265	471	235	118
86	698	349	174	1047	523	262	465	233	116
87	690	345	172	1034	517	259	460	230	115
88	682	341	170	1023	511	256	455	227	114
89	674	337	169	1011	506	253	449	225	112
90	667	333	167	1000	500	250	444	222	111
91	659	330	165	989	495	247	440	220	110
92	652	326	163	978	489	245	435	217	109
93	645	323	161	968	484	242	430	215	108
94	638	319	160	957	479	239	426	213	106
95	632	316	158	947	474	237	421	211	105
96	625	313	156	938	469	234	417	208	104
97	619	309	155	928	464	232	412	206	103
98	612	306	153	918	459	230	408	204	102
99	606	303	152	909	455	227	404	202	101
100	600	300	150	900	450	225	400	200	100
101	594	297	149	891	446	223	396	198	99
102	588	294	147	882	441	221	392	196	98
103	583	291	146	874	437	218	388	194	97
104	577	288	144	865	433	216	385	192	96
105	571	286	143	857	429	214	381	190	95
106	566	283	142	849	425	212	377	189	94
107	561	280	140	841	421	210	374	187	93
108	556	278	139	833	417	208	370	185	93
109	550	275	138	826	413	206	367	183	92
110	545	273	136	818	409	205	364	182	91
111	541	270	135	811	405	203	360	180	90
112	536	268	134	804	402	201	357	179	89
113	531	265	133	796	398	199	354	177	88
114	526	263	132	789	395	197	351	175	88
115	522	261	130	783	391	196	348	174	87
116	517	259	129	776	388	194	345	172	86
117	513	256	128	769	385	192	342	171	85
118	508	254	127	763	381	191	339	169	85
119	504	252	126	756	378	189	336	168	84
120	500	250	125	750	375	188	333	167	83
121	496	248	124	744	372	186	331	165	83
122	492	246	123	738	369	184	328	164	82
123	488	244	122	732	366	183	325	163	81

Continued

Table 7.1   Converting Tempo into Time—cont'd

				Milliseconds per note value					
				dotted			triplet		
bpm	quarter	eighth	sixteenth	quarter	eighth	sixteenth	quarter	eighth	sixteenth
124	484	242	121	726	363	181	323	161	81
125	480	240	120	720	360	180	320	160	80
126	476	238	119	714	357	179	317	159	79
127	472	236	118	709	354	177	315	157	79
128	469	234	117	703	352	176	313	156	78
129	465	233	116	698	349	174	310	155	78
130	462	231	115	692	346	173	308	154	77
131	458	229	115	687	344	172	305	153	76
132	455	227	114	682	341	170	303	152	76
133	451	226	113	677	338	169	301	150	75
134	448	224	112	672	336	168	299	149	75
135	444	222	111	667	333	167	296	148	74
136	441	221	110	662	331	165	294	147	74
137	438	219	109	657	328	164	292	146	73
138	435	217	109	652	326	163	290	145	72
139	432	216	108	647	324	162	288	144	72
140	429	214	107	643	321	161	286	143	71
141	426	213	106	638	319	160	284	142	71
142	423	211	106	634	317	158	282	141	70
143	420	210	105	629	315	157	280	140	70
144	417	208	104	625	313	156	278	139	69
145	414	207	103	621	310	155	276	138	69
146	411	205	103	616	308	154	274	137	68
147	408	204	102	612	306	153	272	136	68
148	405	203	101	608	304	152	270	135	68
149	403	201	101	604	302	151	268	134	67
150	400	200	100	600	300	150	267	133	67
151	397	199	99	596	298	149	265	132	66
152	395	197	99	592	296	148	263	132	66
153	392	196	98	588	294	147	261	131	65
154	390	195	97	584	292	146	260	130	65
155	387	194	97	581	290	145	258	129	65
156	385	192	96	577	288	144	256	128	64
157	382	191	96	573	287	143	255	127	64
158	380	190	95	570	285	142	253	127	63
159	377	189	94	566	283	142	252	126	63
160	375	188	94	563	281	141	250	125	63
161	373	186	93	559	280	140	248	124	62
162	370	185	93	556	278	139	247	123	62
163	368	184	92	552	276	138	245	123	61
164	366	183	91	549	274	137	244	122	61
165	364	182	91	545	273	136	242	121	61
166	361	181	90	542	271	136	241	120	60

| **Table 7.1** | Converting Tempo into Time—cont'd | | | | | | | | |

| | | | | Milliseconds per note value | | | | | |
| | | | | dotted | | | triplet | | |
bpm	quarter	eighth	sixteenth	quarter	eighth	sixteenth	quarter	eighth	sixteenth
167	359	180	90	539	269	135	240	120	60
168	357	179	89	536	268	134	238	119	60
169	355	178	89	533	266	133	237	118	59
170	353	176	88	529	265	132	235	118	59
171	351	175	88	526	263	132	234	117	58
172	349	174	87	523	262	131	233	116	58
173	347	173	87	520	260	130	231	116	58
174	345	172	86	517	259	129	230	115	57
175	343	171	86	514	257	129	229	114	57
176	341	170	85	511	256	128	227	114	57
177	339	169	85	508	254	127	226	113	56
178	337	169	84	506	253	126	225	112	56
179	335	168	84	503	251	126	223	112	56
180	333	167	83	500	250	125	222	111	56

To appreciate the perfection in Pink Floyd's dotted eighth-note delay time, let's consider two other perhaps more obvious choices: a quarter-note delay or an eighth-note delay (see Table 7.2).

In Table 7.2, the quarter-note delay strongly emphasizes the time of the song; it's orderly and persistent:

Hello x x hello x x hello . . .

Each repeat of the word falls squarely on the beat, making it seem like Pink is being nagged or pushed around, and the very orderliness of the quarter-note repetition takes away from the soporific state intended by this composition.

| **Table 7.2** | Evaluating the Musical Timing of Delays | | | | | | | | | | | |

1	e	&	a	2	e	&	a	3	e	&	a	the beat
Hel-	lo											sung word
x	x	x	x	hel-	lo							**quarter-note delay**
x	x	hel-	lo									**eighth-note delay**
x	x	x	hel-	lo								dotted eighth-note delay
x	x	x	hel-	lo		x	hel-	lo	x	hel-	lo	**with regeneration**
Hel-	**lo**	x	hel-	lo		x	hel-	lo	x	hel-	lo	**net effect**

An eighth-note delay, on the other hand, forces the words to fall immediately and persistently one after the other, with no rest in between:

Hello hello hello hello hello

which is simply annoying, and stressful, which is not the desired musical effect.

The delay time chosen in the released recording has the effect of inserting a sixteenth-note rest in between each repeat of the word. "Hello" is sung on the downbeat. The echo never again occurs on a downbeat. First, it anticipates beat two by a sixteenth note, then it falls on the middle of beat two (called the "and" of two, because of the way eighth notes are counted). It next lands a sixteenth after beat three. Finally, it disappears as the next line is sung.

This timing scheme determines that "hello" won't fall squarely on a beat again until beat four, by which time the next line has begun and "hello" is no longer audibly repeating. It's really a pattern of three in a song built on four. This scheme guarantees it a dreamy, disorienting feeling. It remains true to the overall "numb" feeling for the song's atmosphere, giving an uncertain, disconnected feeling. The result is a premeditated creation of the desired emotional effect. And it's a catchy hook—a real Pink Floyd signature.

Consider using the low-pass filter to add some ponderous mystery and distance to the echo. Consider flipping the polarity to make the delay slightly, impossibly, unnaturally different. Consider modulating the delay time slightly to make the echo drift around a bit in time and in pitch. Are you getting the sense that creative options abound?

It's a curious idea: adding an echo to a singer, or a piano, or a guitar, or whatever. It doesn't seem to have any motivation based on reality. The only way to hear an echo on the vocal of a song without the help of studio signal processing is to go to a terrible-sounding venue (like an ice hockey rink or the Grand Canyon) and listen to music. The sound of an echo across the entire mix that occurs in these places—places not designed for music listening—is quite an unpleasant experience. It is almost always sonically messy and distinctly nonmusical. The echoes found in pop music tend to be used with more restraint. In some cases, the echo is added to a single track, not the whole mix. To keep things from becoming too confusing, the output of the delay is often mixed in at a low level, so as to be almost inaudible. As Pink Floyd so ably demonstrates, another valid approach is to apply echo only to key words, phrases, or licks. It's counterintuitive, but echo offers rich production potential for the creative mix engineer.

SLAP

🔊 A staple of 1950s rock is sometimes part of a contemporary mix: slapback echo (or slap echo). Music fans pretty much never heard Elvis without it. Solo work by John Lennon, therefore, often had it. Guitarists playing the blues tend to like it. Add a single audible echo somewhere between about 80 ms and as much as 200 ms, and each and every note bounces and pulses a bit, courtesy of the single, quick echo. On a vocal, a slap echo adds a distinct, retro feeling to the sound. Elvis and his contemporaries reached for this effect so often that it has become a cliché evocative of the period. Pop-music listeners today have learned to associate this effect with those happy days of the 1950s.

On guitar, slap echo makes a performance sound more live, putting the listener in the smoky bar with the band. Most music fans have experienced the music club, with that short echo of sound bouncing of the back wall of the venue. Slap echo can conjure up that specific experience for many listeners.

Before the days of digital audio, a common approach to creating this sort of effect was to use a spare analog tape machine as a generator of delay. During mixdown, the machine is constantly rolling, in record. The signal is sent from the console to the input of the tape machine in exactly the same way one would send a signal to any other effects unit—using an echo send or spare track bus. That signal is recorded at the tape machine and milliseconds later is played back. That is, although the tape machine is recording, it remains in *repro* mode so that the output of the tape machine is what it sees at the reproduce or playback head. As Figure 7.1 shows, the signal goes in, it gets printed onto tape, the tape makes its way from the record head to the playback head (taking time to do so), and finally the signal is played back off the tape and returned to the console. The result is a tape delay.

The signal is delayed by the amount of time it takes the tape to travel from the record head to the repro head. The actual delay time then is a function of the speed of the tape and the particular model of tape machine in use (which determines the physical distance between the two heads).

Want to lengthen the delay time? Slow the tape machine down. There might be two, maybe three, choices of tape speed: 7½, 15, or 30 inches per second. Need it longer still? Use both tracks of the stereo tape deck. Audio goes in to the left track and gets played back from the left side of the repro head. Patch that output to the input of the right side of the tape machine. It is now delayed again as it makes its way to the right output. Double the delay time by sending your signal sequentially through each side of the tape machine. None of these delay times seem exactly right? Maybe the tape machine has vari-speed, which lets the engineer achieve tape speeds slightly faster or slower than the standard speeds listed previously.

Can't make these delay times fit into the rhythm of the song? Now you are faced with the rather expensive desire to acquire another analog two-track machine—one with a different head arrangement so that the delay time will be different.

A single-tape machine, which might cost several thousand dollars, is capable of just a few different delay settings. A three-speed tape machine used this way is like a really rather expensive effects device capable of only a handful of delay time settings. Tape delay was originally used because it was one of the only choices at the time.

To help out, manufacturers made tape delays. These were tape machines with a loop of tape inside. The spacing between the record and playback heads was adjustable to give the engineer more flexibility in timing the delay. Now, in the twenty-first century, studios have more options. Life is good. Today, we can buy a digital delay that is easily adjustable, wonderfully flexible, cheaper than a tape machine, and either fits in one or two rack spaces or exists conveniently in a pull-down menu on our digital audio workstation.

▶   But those who have a spare open reel tape machine that has perhaps been sitting unused ever since they made the investment in a CD burner or started printing their mixes back into the DAW have the opportunity to create tape slap. It can even be a cassette deck if it has a tape/monitor switch to allow monitoring of the playback head while recording.

Why bother? Tape delay is more trouble and more expensive than many digital delays. But there is no denying it: some great old recordings made effective use of it. That is reason enough for some engineers. Retro for retro's sake.

Ah, don't fall for the gimmicks. You should go to the trouble to use a tape delay when you want that "sound." An analog tape machine introduces its own subtle color to the sound. Mainly, it tends to add a low-frequency hump to the frequency content of the signal. The exact frequency and gain of this low-frequency emphasis depends on the tape machine, the tape speed, the tape gauge, and how the machine is calibrated.

If you push the level to the tape delay into the red, that signature analog tape compression is introduced. At hotter levels still, analog tape saturation distortion results (see Chapter 4).

Tape delay becomes a more complex, very rich effect now. It isn't just a delay. It is a delay plus equalizer plus compressor plus distortion device. It can be darn difficult to simulate digitally. It is sometimes the perfect bit of nuance to make a track special within the mix.

If the slap is a bit too distracting, engage a low-pass filter and strategically dull the sound a bit. The slap is now audible without causing a rhythmic hiccup.

EMPHASIS

🔊   Adding a long delay to a key word, as in the Pink Floyd example, is a way to emphasize a particular word. It can be obvious, like the "hello" that begins the song. Simulating a call-and-response type of lyric, the delay is often a musical hook. The echo invites others to sing along. Alternatively, it can be more subtle. A set of emphasizing delays hits key words throughout "Synchronicity II" on the Police's final album, *Synchronicity*. The first line of every chorus ends with the word "away," which gets a little delay-based boost. Listen also to key end words in the verses: "face," "race," and, um, "crotch." These are enhanced with a quick dose of several echoes, courtesy of the regeneration control. The Wallflowers' "One Headlight" on *Bringing Down the Horse* offers a great example of emphasis through delay. Listen carefully to the third verse. The words "turn" and "burn" each get a single, subliminal dotted quarter-note delay. The rest of the vocal track receives no such effect.

Similar, sparing use of echo occurs on the phrases of slide guitar lines in the middle solo section. The guitar is panned left. The first echo falls dead center, and the second echo is panned to the right. It's a nice detail in a seemingly straightforward arrangement.

It's not unusual to apply a low-pass filter to these sorts of delays. Attenuating some of the high-frequency content from each repeat of the sound makes it sink deeper into the mix. When sound travels some distance in a room, the high-frequency portion of the signal is attenuated more quickly than the low-frequency part. High-frequency attenuation due to air absorption is an inevitable part of sound propagating through the air. As a result, a well-placed low-pass filter can be used to make any element of the multitrack project seem more distant. Our brain seems to infer the air absorption and perceptually pushes the sound away from us.

! Good delay units provide engineers with this filter as an option (as shown in Figures 7.2 and 7.3). Moreover, there is often the ability to double the delay time on outboard digital delays by pressing a button labeled "X2," meaning "times two," which cuts the sampling rate in half. With half as many samples to keep track of, the amount of time stored in a fixed amount of memory effectively doubles—hence the "times two" label. Halving the sample rate also lowers the upper-frequency capability of the digital device. Engineers today know this from their pursuit of higher sampling rates in their productions: 44.1 kHz, 48 kHz, 88.2 kHz, 96 kHz, 176.4 kHz, 192 kHz, and beyond. A key benefit of an increased sampling rate is improved high-frequency response. Even as sampling rates creep up on many studio tools (especially DAWs and multitrack recorders), an engineer may purposefully *lower* the sampling rate on an outboard digital delay device. Doing so lengthens the delay time of the device and low-pass-filters the signal—an often-desirable mix move when the subtle emphasis of a mix detail through delay is your goal.

FLANGER

Dialing in a very short delay time and modulating it via the three delay modulation controls leads to an effect known as *flanging*. The only rule is that the delay time needs to be in that range short enough to lead to audible comb filtering. That range suggests a starting delay setting of less than about 10 ms, though the effect may be more obvious at delay times closer to 5 ms. This ensures that audible comb filtering will occur. Set the delay modulation controls to taste.

That ringing, whooshing, ear-tickling sound that is created by a flanger comes from the simple comb-filter effect enhanced by these modulation controls. Consider two different short delay settings. One delay time causes the peaks and valleys in the frequency content of the combined signal to occur at one set of frequencies. A different delay setting results in a different set of peaks and

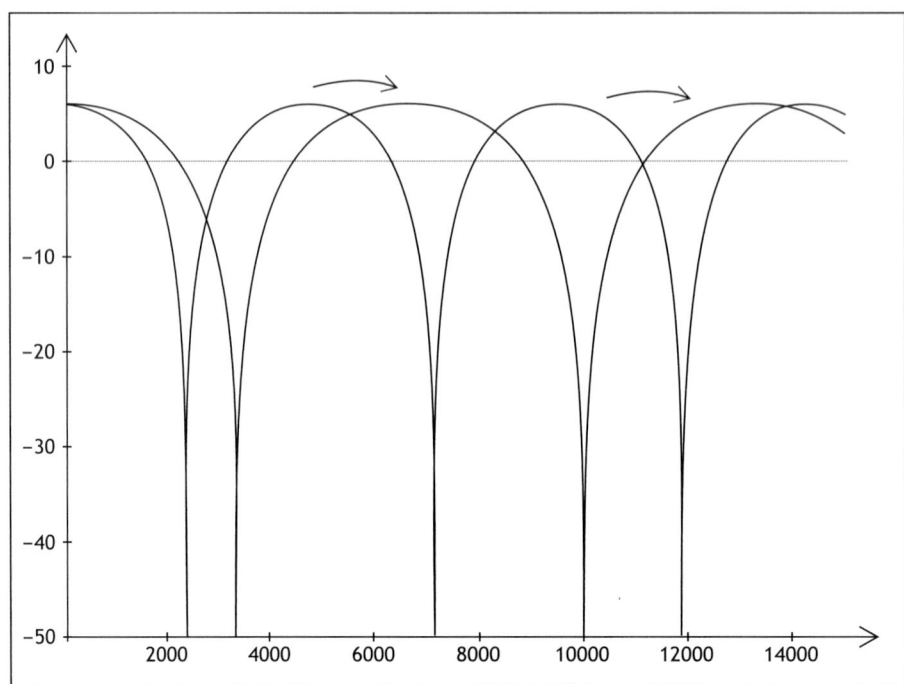

FIGURE 7.14
Swept comb filter.

valleys (Figure 7.14). This result suggests that when a continuously modulated delay value sweeps from one delay time to another, the comb filter bumps and notches sweep also. Figure 7.14 shows the result, lovingly known by the curious moniker of flanging.

◀ Sometimes there's no holding back—the entire mix gets flanged. "Life in the Fast Lane" on *Hotel California* by the Eagles presents a classic example at the breakdown near the end of the tune.

In other applications, the effect is applied to a single instrument—just the drum kit or just the rhythm guitar. You might flange just a single track. Or you might limit the effect to just one section of that track (e.g., only on the bridge). Audio experience—from your time as a mixer and as an avid listener of other mixes—gives you the maturity and judgment needed to know when, how, and how much of effects such as this to use. It becomes a matter of personal opinion and musical taste.

◀ Pop music is full of examples of flanging. "Then She Appeared" from XTC's last effort as a band on the album *Nonsuch* offers a good case study of a gently sweeping flange. Notable from the first time the words "then she appeared" occur, a bit of your traditional flange begins, courtesy of a set of short vocal delays being slowly modulated. In this example, the flange comes and goes throughout the song, offering us a good chance to hear the vocals with and without the delay treatment. A more subtle approach occurs on Michael Penn's

"Cover Up" from the album *Resigned*. A bit of flanging appears on the vocal for the single word "guests" near the end of the second verse. That's it. No more flanging on the vocal for the rest of the tune. It's just a pop-mix detail to make the arrangement that much more interesting. The flange effect actually softens the rather hard-sounding, sibilant, and difficult to sing "sts" at the end of the word "gue*sts*."

The simple effect that comes from mixing in a short, modulated delay offers a broad range of audio effects. Flanging invites your creative exploration.

DOUBLE-TRACKING

In the previous discussion on long delays (those greater than about 50 ms), you saw how they are used for a broad range of echo-based effects. Short delays of about 20 ms or less create the radical comb-filtered effect that—especially when modulated—we call flanging. What goes on between 20 and 50 ms?

This middle range of delay times does not cause echoing or flanging. The delay is too short to be perceived as an echo. It happens too fast. But the delay is too long to lead to audible comb filtering for most musical signals.

▶ Try a 30-ms delay on a vocal track for a good clue about what is going on. A medium delay sounds a little bit like a double track—like two tracks of the same singer singing the same part.

It is a common multitrack production technique to have the singer double a track. The engineer, producer, and musician work hard to capture that elusive "perfect" take. It can require several tries. It might even require several hours. It can get a little grim, but it might even take several different sessions over several weeks or months before everyone is happy with the vocal performance. Once this track is finally accomplished, have the singer record the part again, on a separate track. The vocalist sings a new track while listening to the just-recorded "perfect" track. The resulting sound is stronger and richer. It even shimmers a little.

🔊 If you are unfamiliar with the sound of doubled vocal tracks, a clean example can be found at the beginning of "You Never Give Me Your Money" on the Beatles' *Abbey Road*. Verse one begins with a single vocal. On the words "funny paper," the doubling begins. The vocal remains doubled for the next line and then the harmonies commence. Naturally, each harmony part is doubled, tripled, or more.

🔊 Roy Thomas Baker is famous for, among other things, pushing doubling to the hilt. Check out the harmonies, doubled and tripled (and beyond), throughout the Cars' first album, *The Cars*. For example, listen to the first harmonies on the first song, "Good Times Roll," when they sing the hook, "good times roll." It sounds deep and immense; the vocals take on a slick, hyped sound.

🔊 The deep layering of multiple vocal tracks by the same vocalist reached its zenith back in 1951 when Les Paul and Mary Ford covered "How High the Moon."

Mary Ford sings lead vocal. Mary Ford sings multipart harmonies. And most ear-tingling of all, Mary Ford doubles and triples any and all of her vocal parts. Pop music has always, from the beginning, found occasion to be more than a little over the top.

This layering of tracks borrows from the tradition of forming instrumental sections in orchestras and choirs. Consider the sound of one violin. Then imagine the exact same piece of music with twelve violins playing the same line, in perfect unison.

The value of having multiple instruments play the same musical part is almost indescribable. Adding more players doesn't just create more volume. The combined sound is rich and ethereal. It transports the listener.

🔊 A crystal-clear application of doubling can be found on Macy Gray's "I Try" from the album *On How Life Is*. Typically, double tracks support the vocal, adding their inexplicable extra bit of polish. They are generally mixed in a little lower in level than the lead vocal, reinforcing the principle track from the center or panned off to each side. The Macy Gray tune turns this on its head. At the chorus, where pop production tradition calls for a good strong vocal, the vocal track panned dead center does something quite brave: it all but disappears.

The chorus is sung by double tracks panned hard left and right. It's brilliantly done. Rather than support the vocal, they *become* the vocal. The chorus doesn't lose strength. The tune doesn't sag or lose energy at all. The doubled tracks—panned hard but mixed aggressively forward—offer a contagious hook that invites the listeners to sing along, to fill in the middle.

Although pop vocals, especially background vocals, are often doubled, any other instrument is fair game. A common track to double is rhythm electric guitar. The same part is recorded on two different tracks. On mixdown, they might appear panned to opposite sides of the stereo field.

The two parts are nearly identical. Change just one variable in the recording setup: switch to a different guitar, a different amp, or a different microphone, or slightly change the tone of the doubling track in some other way. Maybe the only difference between the tracks is the performance. As no two performances are ever identical, the resulting pair of guitar tracks will vary slightly in timing. The "chugga chugga" of the left guitar track is slightly early in one bar and slightly late in the next. The result is that, through the interaction of the two guitar tracks, the ears seem to pick up on and savor these subtle delay changes. At times, the two tracks are so similar that they fuse into one meta-instrument. Then one track pulls ahead and gets noticed on its own. Then the other track pulls ahead in time and temporarily draws the listener's attention. Then, for a brief instant, they lock in together again. And so it goes. Doubled guitars are part audio illusion and part audio roller coaster; they are an audio treat.

As mix engineer, you can simulate the layering and doubling tracks through the use of a medium delay. If it isn't convenient, affordable, or physically possible to have the singer or guitarist double the track, reach for a medium delay. Some amount of delay modulation is introduced so that the synthesized double track moves in time a little relative to the source track. This effect helps it sound more organic instead of like a cloned copy of the original track. The addition of a bit of regeneration creates a few additional doubling layers of the track underneath the primary one.

Some delay units have the ability to simultaneously offer several delay times at once (called *multitap delays*), each modulated at its own rate. Use several slightly different delay times in the 20- to 50-ms area and synthesize the richness of many layered vocals. Using pan pots, spread the various medium-delay elements out left to right and front to back (in surround applications) for a wide wall of sound. As with all studio effects, make sure that the sonic result is appropriate to the song. The solo folk singer doesn't usually benefit from this treatment. Neither does the jazz trumpet solo. But many pop tunes welcome this as a special effect on lead vocals, backing vocals, keys, strings, pads, and—of course—ukulele.

CHORUS

An alternative name for the heavy use of modulated medium delays to simulate more than one player is *chorus*. The idea is that through the use of several different delays in the 20-to-50-ms range, one could upgrade a single vocal track into a simulation of the sound of an entire choir. Thus the term "chorus."

! Naturally, stacking up 39 medium delays around a single vocalist will not convincingly sound like a choir of 40 different people. Think of it instead as a special electronic effect, not an acoustic simulation. And it isn't just for vocals. John Scofield's trademark guitar tone includes a strong dose of chorus (and distortion, a sweet guitar, and brilliant playing among other things). It is not uncommon to add a bit of chorus to the electric bass in a pop ballad. This medium-delay concoction is a powerful tool in the creation of musical textures.

To see how "out there" the effect can be made, reach for the album *Throwing Copper* by Live and listen to the beginning of the tune "Lightning Crashes." It's difficult to impossible to know exactly what kind of signal-processing craziness is going on just by listening. The guitar sound includes short and medium delays, among the panning, distortion, and phase shifting effects going on. A fair summary is that the delay is being modulated between a short, flanging sort of sound (around 10 ms) and a longer, chorus sort of delay time (around maybe 40 ms). Note especially the sound of the guitar in the second verse, when the effect and the relatively clean sound of the guitar are mixed together at similar volumes. This mix presents a good taste of chorus. Such amazing projects are a great inspiration to the rest of us to do more with the effect.

7.3 SUMMARY

A simple delay unit offers a broad range of audio opportunity, representing a nearly infinite number of sound qualities. Short delays create that family of effects called flanging. Medium delays lead to doubling and chorusing. Finally, when the delay is long enough, it separates from the original signal and becomes its own perceptible event: an echo.

🔊 Take a quick tour of all of these effects with a single album: *Kick* by INXS. To hear a terrific use of flange, listen to "Mediate." This is a textbook bit of flanging. For a true doubling, listen to "Need You Tonight" and those hard-panned questions: "How do you feel? What do you think? Whatcha gonna do?" There's a groove-essential echo on that descending set of electric guitar chords that first appear after the triple hit of the opening chord in the intro. It pans off to the left and fades away. With these echoes, the feel of the tune is elevated to a much more intriguing, interwoven state of complexity. Finally, the same album demonstrates a classic application of chorus to an electric guitar. Check out the rhythm guitar in "New Sensation" and the steely cool tone the chorusing adds.

Flanging, doubling, chorus, and echo are three very different kinds of effects that come from a single kind of effects device: the delay does it all.

CHAPTER 8
Pitch Shift

The most irresistible and, all too often, unbearable effect:
use your power for good, not evil.

MIX SMART QUICK START: Pitch Shift

GOALS

- Fix out-of-tune tracks that can't be re-recorded correctly.
- Fit bits of a crowded mix together with the help of a pitch-based spreader, thickener, or layering effect.
- Feature tracks with obvious new pitch qualities, composing new musical parts, and imparting a new shape to any melody or chord sequence.

GEAR

- Extraordinary signal processing sophistication is accompanied by complex and rarely intuitive user interfaces.
- Musical use of pitch correction tools requires mastery of the parameters and practice with the processor.
- Listen carefully. Pitch perfection is not always the best answer. Learn to expect unexpected artifacts.

! Manipulating the pitch of the notes we record is one of the most important new capabilities to enter the recording studio in a decade. It offers extraordinary tactical advantage when a sour note undermines an otherwise stunning performance. It presents rich aesthetic possibilities as the full creative potential is far from fully discovered. You are among the first generation of innovators who get to set trends with this effect that may influence all generations of mix engineers. But monkeying around with pitch runs the serious risk of being overused, big time.

Pitch shifting is the deliberate offset of the performed pitch up or down by a specified amount. Use a playback sample rate that is higher than the original sample rate used when the audio was recorded, and the pitch goes up. Play an audiotape at a slower speed than intended and the pitch of the recording goes down. Adjust for time so that the new pitch follows the original phrasing, and you've got the chance to place the pitch-shifted track right back into your mix. Pitch-shifting alone offers interesting production possibilities.

Add to this the ability to analyze the pitch of a signal—to compare the performed pitch to an ideal, in-tune pitch—and pitch correction becomes possible. *Pitch correction is the coordinated use of pitch detection plus pitch shifting, in which a performed pitch is adjusted by whatever amount necessary to make the new pitch acceptably close to a desired pitch.*

8.1 PARAMETERS

🔊 Before any discussion of mix techniques that deliberately use pitch shifting, it is worth noting that pitch shifting is a natural part of other effects. Recall the chorus effect that comes from adding a slowly modulated delay of about 20–50 ms (see "Medium Delay" in Chapter 7). Chorus is difficult to describe in words. To study it, one must listen to it. A careful, critical listen to the richness that the chorus effect adds to a vocal or guitar reveals a subtle amount of pitch shifting. Beyond that blurring of things in time through the use of perhaps several medium delays, pitch shifting is a fundamental component of that effect known as chorus. Because a chorus pedal relies on a *changing* delay, it introduces a small amount of pitch shifting. As the delay time sweeps up, the pitch is slightly lowered. As the delay time is then swept down, the pitch is then raised ever so slightly. One can't have chorus without at least a little pitch shifting.

For deliberate pitch shifting, specify the amount of pitch shift musically, in semitones, or mathematically, in cents. A semitone is $1/12$ of an octave, with each semitone shift representing an amount equivalent to moving to an adjacent key on the piano. To zoom in to a finer scale, it can be useful to specify the shift in cents; there are 100 cents per semitone.

! With the exception of the perfect octave, no fixed shift in pitch will always be in the key of the song. Any note in a scale, shifted up or down by any number of octaves, lands on a note in the scale. Shift every note in a scale by any other musical interval—a perfect fifth, a minor third, a tritone—and you'll find the destination, pitch shifted note may or may not be within the scale. It is not always necessary to be in the key of the piece, but when you wish to raise or lower a note by a musical amount or when you wish the modified pitch to be in the key of the piece, many pitch-shifting algorithms allow you to specify the key signature or the acceptable scale to target, which allows the pitch shifter to adjust up or down as needed to a new pitch that is in tune with the harmony of the song.

A real, human, expressive performance is rarely a practice in perfect pitch precision. Musicians slide into some pitches and frequently embellish with breaks, bends, and vibrato. Pitch-shifting tools anticipate this variance by letting you tailor the reaction time of the process. In this way, expressive bends and wobbles in pitch aren't whittled away immediately: they can be allowed through, pitch imperfections and all. The goal might be to adjust the overall tonal center of the performed note to a closer, more in-tune pitch, but allow the expressive bends to deviate from this new center. The pitch shifting might sound more expressive if the early parts of the notes are allowed a wider window of tolerance, but sustained notes are then forced to acceptable pitch. When it works, you've got better tuning without loss of emotion and musicality.

Pitch *correction* builds on the pitch, scale, and reaction time parameters of straight-ahead pitch shifting. Pitch correction is pitch detection plus pitch shift. The performed pitch is determined, the target pitch is specified, and the pitch-correction algorithm applies the necessary, ever-changing amount of pitch correction to the track, note by note. Automatic pitch correction frees the software to adjust the pitch continuously based on its own best guess of the amount needed to pull the note to the desired pitch, without being distracted by expressive bends, vibrato, and other musical pitch liberties. Manual pitch correction lets you review that best guess pitch modification from the algorithm and personally revise it in an effort to protect vibrato, savor blue notes, and otherwise preserve musicality.

One last parameter needs careful consideration: formant. Extreme or crude pitch shifts can make a vocal sound very unnatural—think baby groundhogs and giant sci-fi creatures. Using a pitch-shifting processor to raise or lower the pitch of a note is not the same as singing or playing the higher or lower note. There are resonant spectral properties of an instrument that aren't much changed when the performed pitch is changed. When you sing a note, you have your own characteristic spectral peaks indicative of your voice. Your vocal sound is, in part, defined by the size and resonant qualities of your chest cavity, your throat length, your nasal passage, and so on. Sing the same syllable, the same note, at slightly different pitches—a little higher and a little lower—and your identifying spectral signature stays very nearly the same. Ditto for instruments. Each saxophone, tympani, and ukulele has its own characteristic spectral qualities that it brings to every note it plays. The formant is that relatively unchanging spectral fingerprint that each instrument brings to a note, regardless of the pitch.

You are probably familiar with one way to meaningfully change your own personal singing formant—use helium. (It's a party trick to be done with care. Consult a doctor. Don't do it for very long. The time you spend singing on helium is like time spent holding your breath. You aren't getting much oxygen, and oxygen is pretty essential to your ability to survive and sing the next line in the song.) The lower density of the gas (helium versus air) shifts the frequencies of the various resonant bits of your voice up. When you sing on helium, you sing the same frequencies through a resonant system whose characteristic formant resonances are raised. You sound unnatural, inhuman, and—let's be honest— *unbearable* on repeated listening.

🔋 For truly natural-sounding pitch shifting, we need a processor that separates the formant from the musical pitch. On the most capable pitch shifters, you'll see parameters dedicated to this. Formant-savvy pitch shifters allow the notes to be tuned into the desired pitch without making the track take on those distracting, unrealistic chipmunk or robot qualities. They keep the helium effect out of the pitch-shifted track.

On the other hand, shifting the formant (never mind the pitch) is its own effect. Raising the formant can make a voice more feminine and can raise a sax up to the next higher instrument in the family (e.g., from tenor to alto, or even in between). Lower the formant to make a voice more masculine or to step down to a lower form of the instrument (e.g., from tenor to baritone).

8.2 MIX STRATEGIES: PITCH SHIFTING

Pitch-shifting effects are common in multitrack production—sometimes subtle, other times obvious; sometimes accidental, other times deliberate.

8.2.1 The Fix

You'll see shortly that pitch shifting has a variety of production possibilities; the most common effect, by far, is pitch correction. When parts of a performance are out of tune, we might be able to fix it through pitch correction.

Fixing through pitch shifting is typically hooked up using the *insert* or applied through *offline processing* (see Appendix A). We change the pitch because we don't like it. We replace the out-of-tune track with the newly tuned track.

PITCH CORRECTION

The best solution to an out-of-tune note is to record a new performance: an *in-tune* performance. Allow me to reiterate: the best way to get a track in tune is to record a performer who is in tune. Alas, recording sessions don't always go smoothly. Performers might connect with their fans through features other than pitch accuracy. Production schedules and funding can conspire to put us in a bind, pitch-wise. And so we find ourselves needing a pitch shifter to fix any annoyingly out-of-tune notes that can't be retracked.

Many a session goes like this: It's 4:19 a.m. It's the seventeenth take of the song. It's a great take. Then on the last repeat of the last chorus, the singer—understandably tired from singing for so long—drifts flat on a key word. And now they are too tired to hit it. What to do?

First, we should listen back to old takes to see if we can edit in an in-tune replacement for that phrase or that word that is still musically compelling and consistent with the rest of the performance. Otherwise, we pitch-shift the problem.

In the old days of multitrack production (and you are encouraged to try this manual approach), the sour note was sampled. Using a pitch shifter, it was then manually tuned by ear, based on your musical judgment. It was raised or lowered to taste. Then, the sampled and pitch-shifted note was re-recorded and edited back onto the multitrack. With the problematic note shifted to pitch perfection, no one was the wiser.

! Alternatively, and more frequently, we reach for pitch correction hardware or software. When it detects the sharp or a flat note, it shifts the pitch automatically by the amount necessary to restore tuning. We must listen acutely, watching the whole pitch-correction process and listening for anomalies that make the performance sound unnatural or disappointing.

First, specify the pitch center of the recording. If the whole band is tuned to A444, but the pitch-shifting processor is pulling everything to A440, you'll never get the track in tune. In this way, we determine what counts as in tune.

Second, determine the key and scale that identifies the acceptable notes for the performance.

Third, spend some time adjusting the reaction time of the pitch-correction tool. This is essential if you want the pitch-corrected performance to sound convincing, natural, believable. We discuss deliberately unnatural pitch-shifting strategies later in this chapter. For pitch correction that casual listeners won't notice as pitch-corrected, you need to allow a musical amount of missed, bent, or otherwise imprecise pitches through. It takes a light touch, sharp listening skills, and sound musical judgment to do it well. Too slow, and there will be too much that is out of tune in your tune. Too fast, and each note of the performance snaps unnaturally to the corrected pitch, making the performer sound crude or clumsy and creating a melodic contour that doesn't support the emotional intent of the melody.

Finally, pay attention to the formant. Different pitch-correction tools offer different approaches, but you generally have controls that identify the type of track to be pitch-shifted and that specify the amount of formant manipulation to be tolerated.

◀)) Pitch correction is simple in concept, but you must listen carefully. Practice with each make and model of pitch shifter you use. Some are better than others, generally. But some are better than others for certain situations. Some are designed for vocals, and others might work well on electric bass. It takes time to learn how to coax a convincing performance out of the algorithm. You'll need to practice with each plug-in.

As you develop facility with the parameters and figure out how to make it sound good, remain aware of pitch shifting's limitations.

First, do not forget that it is the job of the mix engineer to choose when the pitch-shifted track offers any improvement to the recording versus when the original performed pitch is to remain. There is a strong temptation to "perfect" each and every track in a multitrack project. And if you—the mix engineer—aren't the one tempted to do this, the performers will be. Don't be surprised to get a tap on your shoulder from some of the other players while you are tuning the vocal; they are going to want a little bit of that on their sax solo, their synth line, their bass part, and so on. Seeing pitch shifting in action plays on the confidence of a performer. If a meter displays pitch errors, a player is going to want to fix it and make it error-free—make it "perfect."

But "perfect" here is highly subjective. A track deemed perfect by pitch correction software may lack musical expressiveness and might be riddled with artifacts of the pitch-shifting process. Listen closely and stay critical. And don't be afraid to reject the pitch-corrected version. There is everything from slightly, vaguely mistuned, to very clearly out-of-tune performances throughout the history of music. Tuning a track doesn't always serve the art. Perfection is its own aesthetic, not suitable for all productions. When you don't necessarily need perfection, don't chase it.

‼ As recording engineers, we are used to making the trade-off between aesthetic beauty versus objective correctness when they are in conflict. As mix engineers, we know the most musical use of our studio isn't always the most technically correct. The members of the band don't have this experience or this context. You must help them hear the track, without looking at the pitch-correction display, and decide when and—more importantly—when *not* to use the correction effect.

Then there are the deliberate manipulations of pitch by the performer. Vibrato is an obvious example of the musical detuning of an instrument on purpose. And, if every note of a blues guitar solo were pitch-shifted into perfection, string bends unbent so that the pitch never strayed, how blue would those blues be? There is a lot to be said for a musical amount of "out-of-tune-ness." Remove all the bends and misses, and we risk removing a lot of emotion from the performance.

Finally, there are limits. Producers should not expect to create an opera singer out of a folk singer, or a pitch-perfect pop star out of a wobbly, wandering wannabe. There is no replacement for actual musical ability. If the bass player can't play a fretless in tune, don't let them play a fretless in your tune. If the violin player can't control his or her intonation, don't let him or her play on your session. Don't expect to rescue poor musicianship with automatic pitch correction. Use it to add to a stellar performance, not to create one. Musical sense and good judgment must motivate everything that is done in the recording studio. People generally want to hear the music, not the effects rack.

8.2.2 The Fit

We push back against the chaos and crowding that is a perpetual part of multi-track production through some strategic pitch shifting. The goal is to fit together the many pieces without allowing some parts to drown out others. We work to empower every track to fulfill its intended musical role. In these particular examples, we want to hear the pitch-shifted track *in addition* to the source track. An *aux send* can do this (see Appendix A), or we can make a copy of the appropriate track and process it *offline* with pitch shifting. Then both the processed and unprocessed tracks are mixed to taste.

SPREADER

In Chapter 2, use is made of a common effect built, in part, on pitch shifting: the spreader. Here is a quick review of the effect: the spreader is a "patch" that enables us to take a mono signal and make it a little more stereo-like. Contrasting some mono tracks, with other wider tracks, we are able to fit together more pieces into a more elaborate but stable mix.

▶ A single track is "spread" out by sending it through two delays and two pitch shifters. The delays are kept short, each set to different values somewhere between about 15–50 ms. If they are too short, the effect becomes a flange/comb filter; if they are too long, the delays stick out as distinct audible echoes. One delay might be 17 ms and the other 22 ms.

When using a spreader, the return of one delay output is panned left and the other is panned right. The idea is that these quick delays add a dose of support to the original monophonic track. In effect, these two short delays simulate some early sound reflections that one would hear if the sound were performed in a real room. The "spreader" takes a single mono sound and sends it to two slightly different, short delays to simulate reflections coming from the left and right.

That's only half the story. The effect is taken to the next level courtesy of some pitch shifting. Shift each of the delayed signals ever so slightly, and the mono source material becomes a much more interesting loudspeaker creation. Detune each delay a nearly imperceptible amount—maybe 5–15 cents. This is not a significant pitch change. An octave is divided into 12 half steps, representing adjacent keys on a piano or adjacent frets on a guitar. Each half step is further divided into 100 equal pitch increments called cents. The pitch shifting called for in the spreader, then, is just 5 to 15 percent of a half step—all but imperceptible except to the most trained listeners. The goal of the spreader is to create a stereo sort of effect. As a result, one seeks to make the signal processing on the left and right sides ever so slightly different from each other. Just as unique delay times are selected for each side of this effect, choose different pitch shift amounts left and right as well—maybe the left side is shifted down 8 cents while the right side is shifted up 8 cents.

Like so much of what is done in recording and mixing pop music, the effect has no basis in reality. When adding delay and pitch shifting, you aren't just simulating early reflections from room surfaces anymore. The spreader makes use of common studio signal-processing equipment (delay and pitch shifting) to create a wide stereo sound that only exists in loudspeaker music. This sort of thing doesn't happen in symphony halls, opera houses, stadiums, or pubs. It's a studio creation, plain and simple.

THICKENER

! Take this effect further and the sonic result might be thought of as more of a "thickener." There is no reason to limit the patch to two delays and two pitch changes. Ample signal-processing horsepower in most digital audio workstations makes it trivial to chain together eight or more delays and pitch shifts. Strategic selection of unique delay times, pitch-shift increments, and pan locations control the fullness, width, and coloration of the effect. It is likely to sound unnatural when used heavily, but a light touch of this effect on vocals, guitars, or keyboard parts can help those tracks sound larger, fuller, and more exciting. Slowly modulate each of those delays like a chorus, and more complex pitch shifting is introduced. Added in small, careful doses, this densely packed signal of supportive, slightly out-of-tune delays will strengthen and widen the loud-speaker illusion of the track.

LAYERS

Recall our discussions on double tracking and the chorus effect using delay in Chapter 7. Pitch shifting offers an enhancement to the effect, a variation on the theme that lets us add additional, more convincing layers to a track.

🔊 Even when a track has no objectionable pitch problems, you might find it productive to send that track through a pitch-correcting processor and *add* it the mix—don't *replace* it. Think of the pitch-altered version as a double track for the original, un-pitch-corrected track. It's a faux double track; the performer never had to play it twice. The pitch shifter synthesizes a new version of the track, one with slightly different pitch, note by note. The processing latency associated with the plug-in further distinguishes it from the original track. Introduce the modulating medium delays of the double tracking and chorus effects (see Chapter 7), running the outputs of those delays through their own pitch-correction processors, and you've created triple, quadruple, and more performances. In this way, a single element of your multitrack arrangement—a boring, single, monophonic, hard to fit into a crowded mix track—becomes a richly layered, densely textured, lush fabric to lay into your mix. Contrast these with other razor-sharp, narrow mix elements of mono tracks that have no spreading, thickening, or layering applied to them, and you can fit an elaborate mix together.

8.2.3 The Feature

When we want to feature a sound, we don't hesitate to *insert* a pitch shifter or add new pitch-shifted tracks to the mix and present the effect front and center.

LESLIE

Pitch-shifting processing made some clever advances well before things went digital, when audio was all analog, and gain stages were all tube. Hammond B3 organs, many blues guitars, and even vocals are sometimes sent through a rather unusual device: the Leslie cabinet. The Leslie sound is a hybrid effect built on pitch shifting, volume fluctuation, and often a good dose of tube overdrive distortion. The Leslie cabinet can be thought of as a guitar or keyboard amp in which the speakers sound as if they rotate. A two-way system, the high-frequency and low-frequency parts work in slightly different ways.

The high-frequency driver of a Leslie is horn-loaded. The driver is fixed, but it fires into a rotating horn. The perceived location of sound is near the end of the horn, which moves toward and then away from the listener as the horn spins. The spinning horn is like a two-prong propeller. One prong is the horn, and we hear sound whiz by. The other prong is a "mute" horn, emitting no sound—there only to mechanically balance the spinning apparatus. So the high-frequency part of this two-way speaker rotates through two extremes, from facing us directly to turning away from us entirely.

It would be very difficult to spin the large, low-frequency driver to continue the effect at low frequencies. Instead, the static woofer is enclosed inside a kind of cylinder. The cylinder has a few large holes in it. While the woofer conveniently remains fixed, the drum rotates, opening and closing the woofer sound. The result is a low-frequency approximation of what the Leslie is doing with the horn at higher frequencies.

In addition, the spinning system has three speeds. The rotating horn and drum may be toggled between off (the amp stays on, but the rotating mechanisms stop), slow, and fast during the performance.

The sound of the Leslie is fantastic. With the cylinder and horn rotating, the loudness of the music increases and decreases, creating a kind of throbbing tremolo or amplitude modulation (see Chapter 6). With the high-frequency horn spinning, a Doppler effect is created: the pitch increases as the horn comes toward the listener/microphone and then decreases as the horn travels away.

The typical example used in the study of the Doppler effect is a train going by, horn shouting its tone. The sound of the pitch dropping as the train passes by is caused by this phenomenon. Sound sources approaching a listener with any appreciable velocity increase the perceived pitch of the sound. As the sound source departs, the pitch similarly decreases.

The high-frequency portion of the Leslie sound is heard through a horn. The perceived location of that sound source is the bell of the horn. So while the driver sits fixed within the Leslie cabinet, the high-frequency sound source is moving. The Doppler effect results.

The low-frequency driver, which is housed within a spinning drum, does not experience a pronounced pitch shift. As the holes within the drum rotate by, the low-frequency signal gets slightly louder. Continued rotation of the drum causes the sound of the woofer to be again attenuated, when the holes haven't opened up the woofer to the listener/microphone. The apparent location of the low-frequency driver doesn't move, however, so there is no Doppler shift. Amplitude modulation without pitch bending is the signature low-frequency sound of a Leslie.

The net result of the Leslie system then is a unique fluttery and wobbly sound built on a rather surprising bit of pitch shifting on the high band only. To feature any track in your mix, you are free to use intricate applications of pitch alteration in any and all ways you can dream up. Let this time-proven contraption from the last century inspire you to assemble, in your studio, your own variation on the theme.

BIG SHIFT

The straightforward pitch shifting that is the basis for the spreader and the thickener can be used in a more forward, don't-try-to-hide-it way. The hazard with an obvious pitch shift is that it can be hard to get away with musically. Special effects—in movies and on some records—in which a vocal is shifted up or down by an octave or more can create the perfect new monster sound. Misused, however, it provides a comedic effect. If it is too low, the pitch-shifted vocal conjures up images of big-throated alien androids invading the mix to intimidate weaker beings. Too high, and the singer becomes a hedgehog on helium.

🔊   In the hands of talented musicians, aggressive pitch shifting really works. Prince famously lowers the pitch of the lead vocal track and takes on an entirely new persona in the song "Bob George" from *The Black Album*. The effect is obvious, and an incredible story results.

🔊   No effort was made to hide the deeper than typically found in nature bass line of "Sledgehammer" on Peter Gabriel's classic *So*. The entire bass track seems to include the bass, plus the bass dropped an entire octave. The octave-down bass line is mixed right up there with the original bass. There is nothing subtle about it.

Getting lower lows is particularly popular in most styles of music destined for clubs and dance floors. As mix engineer, feel free to chase deep lows with the help of aggressive pitch shift, but be sure to check your mix on average loudspeakers. The tune needs to work with and without subwoofers. Mix in some regular lows with your super-low lows so that even small speaker systems get a strong sense of your thunderous mix intentions.

HARMONY, COUNTERPOINT, AND DISSONANCE

The pitch-shifting effect can be used to add two-, three-, or four-part harmony if the engineer is so inclined. Feature a key melody by opening it up to multi-part harmony.

! Get out the arranging book, though, because the pitch shifter makes it easy to inadvertently add a dissonant interval or two. If each pitch-shifted note is a fixed interval above whatever note occurs in the source track, some notes won't be in the key of the tune. Only the octave shift always stays in tune with the harmony of a song. The perfect fourth and perfect fifth, tempting as they are, lead to a couple of nondiatonic notes when applied as a fixed interval above all the notes of the scale. The major third applied rigidly to every note in a scale leads to dissonance and confusion; many notes will be out of tune with the key of the song.

You can create new lines without unwanted dissonance using those pitch shifters that allow you to manually set the pitch of each and every note, or specify the key and the desired scale for the song. Now the pitch change is adjusted to the appropriate note for the harmonic structure of the song. Use this feature to create a harmony line from the lead vocal, to create a countermelody to the guitar solo, and so on.

The pitch shifting can be tied to Musical Instrument Digital Interface (MIDI) note commands enabling you—or, better yet, someone in the band—to dictate the harmonies from a MIDI controller. The pitch shifter raises or lowers the pitch of the recorded track according to the notes played on the keyboard or other MIDI controller. The result is a harmony or countermelody line fabricated from the source track, adding all the harmony and dissonance desired.

This production tool can reach beyond single harmonies. You can use pitch shifting to turn a single note into an entire chord. String patches can sometimes be made to sound more orchestral with the judicious addition of some perfect octave and perfect fifth pitch shifting (above and/or below) to the patch.

🔊 It doesn't stop with simple intervals. Chords loaded with tensions are okay too when used well. Progressive rock band Yes put it front and center on the angular guitar solo in "Owner of a Lonely Heart" on the album *90125*. Single-note guitar lines are transformed into something more magical and less guitar-like using pitch shifters to create the other notes.

STOP TAPE

🔊 An obvious pitch-shifting effect worth mentioning is the *stop tape effect*. As analog tape risks extinction, this effect may one day be lost on future generations of recording musicians. When an analog multitrack tape is stopped, it doesn't stop instantly; it takes a brief instant to decelerate. Large reels of tape, like two-inch 24 track, are pretty darn heavy. It takes time to stop these large reels from spinning. If one monitors the tape while it tries to stop (and many fancy machines resist this, automatically muting to avoid the distraction this causes during a session), one will hear the tape slow to a stop. Schlump. The pitch dives down as the tape stops. Sometimes this sound is a musical effect. It's not just for analog tape, as Garbage demonstrates via a digital audio workstation effect between the bridge and the third chorus of "I Think I'm Paranoid" on their second album, *Version 2.0*.

START TAPE

The stop tape effect can be turned around, at least in the analog domain. Have the performer start playing before the tape recorder is rolling. Go into record immediately as the tape machine gets up to speed. The result, on playback at full speed, is a high-pitched descent into proper pitch. While the tape machine was coming up to speed, the signal was being recorded at an improperly slow speed. Playback at proper speed raises the pitch of that portion of the signal, and a unique pitch shift results.

🔊 The intro to "Synchronicity II" on the album *Synchronicity* by the Police demonstrates this effect quite clearly. The squealing electric guitar in feedback pops into the mix with a sharp pitch bend courtesy of some coordinated start tape effects.

THE PERFECTION AESTHETIC

Earlier in this chapter, we discussed the use of pitch correction as a way to adjust the pitch of a mispitched track, seeking to fix the pitch seamlessly without anyone being the wiser. We can outdo ourselves here.

You might, as a production goal, seek out impossibly perfect pitch, across all tracks, without trying to hide the effect. Grunge music of the 1990s celebrated its rawness, its imperfections. Bands went to a lot of trouble to sound as if they weren't trying. The sound of the garage was perfected. With obsessive, compulsive pitch-correcting passion, you might rebel against those rebels. Tune everything with fault-intolerant precision and a new flavor results. Polished, careful music isn't for everyone, but it is a valid goal for the mix engineer if it suits the goals of the songwriters and the performers. As you synthesize new sounds, invent new textures, and otherwise push the envelope as a mix engineer, you have available to you the perfection aesthetic: everything is in tune, in time, in its place. It takes time. It takes patience. But it represents a production paradigm unique to our time. The digital audio workstation, equipped as it is with tools that didn't exist in the analog domain (e.g., pitch correction, among others), offers the signal processing potential for a kind of theoretical or mathematical perfection. If you try to find an artistic potential in that—and be careful, because perfection can sound lifeless and cold—surely pitch correction will be a part of it. It's not that the rigid tuning of stylized tracks is always the perfect answer. But it could be; for certain artists in certain genres, perfectly tuned tracks are a mix engineer's answer to the search for a new style.

QUANTIZED PITCH

Finally—and cautiously—we must consider *quantized pitch*, whereby each and every single note performed is snapped to a pitch grid, deliberately trying to sound processed. Artists are perpetually motivated to sound different. Their muse pushes them to innovate. Their fans demand that they entertain them with the new hit without imitating their last. The history of music is full of technological innovations offering up a new sound, and artists exploiting it a far as they can: that slap echo we associate with Elvis's voice; the overdrive distortion we associate with rock-and-roll guitar; the heavy use of synth textures that says it's the 1980s; the turntable gymnastics that reveal the dexterity of a DJ repurposing the sound of vinyl. We must add to that list the sound of quantized pitch shifting. Listeners are pretty sure that a human sang or played the part, but we changed the sound so much when we mixed it that it sounds like a machine. There is some amount of fascination, for now, among artists and music fans alike for that half-human, half-machine sound. Our brain is seduced a bit by a "how could that be?" phenomenon. Quantized pitch effects also make new hooks possible. Vocal lines, guitar riffs, and horn parts can take melodic contours that no human would ever be capable of playing.

▶ For quantized pitch, we simply engage pitch correction more aggressively, without regard for believability. We snap it to a scale not sung. We flatten out all pitch performance variations. We make it follow a melody of our own invention, different from that which was sung. We might deliberately have the singer sing a simpler or different part. We then pitch shift it to the desired melody, and we celebrate the unbelievable precision and associated artifacts. It's a technological hook in a pop tune that makes its living hooking in listeners.

! This approach may make it possible for a musician of mediocre or worse talent to sound interesting and exciting, at first. The infatuation with this effect will soon wane, and we'll return to seeking out musicianship in recorded art. Vapid ideas decorated in gimmicky sounds won't hold a candle to the work of prodigious talents whose musicianship also includes stretching the concept of even quantized pitch shifting into new artistic directions. This effect requires coordinated musical synergy among the performer, the writer, and you—the mixer.

8.3 SUMMARY

Pitch shifting—and its more popular cousin, pitch correction—do what they say. Change the pitch by any amount, or change the pitch by just the right amount to bring it in line. The challenge with pitch shift is in the anomalies and artifacts that too often result. It's as intuitive to use as an equalizer (see Chapter 3). EQ accompanies its intuitive ease of use with relatively few side effects. With pitch effects, you quite likely know what you need to achieve, but will have trouble getting the processor to do it just right.

The signal processing of pitch shift is very complicated. In fact, pitch-shifting hardware and software represent some of the most sophisticated signal processing in the entire studio. Complicated processes can lead to messy results. A great mix engineer must have many hundreds if not thousands of hours of experience and practice before he or she can make musical use of such a sophisticated signal processor.

Glitches in timing and flinches in timbre lead to hitches in believability. Allow yourself enough time in the session to listen carefully, not just to the new pitch, but also for any related sonic side effects. When significant application of pitch shift is called for, expect to zoom in and chase some problems note by note. With the patience of a saint, you can alter pitch for correction, spreading, thickening, and layering. If you don't hold back, you can create the Leslie effect; introduce ear-grabbing obvious shifts; fabricate harmony, counterpoint, and dissonance; create special effects; pursue the perfection aesthetic; and explore quantized pitch in musical ways.

CHAPTER 9
Reverb

Easy to hear, hard to control, critical to success.

MIX SMART QUICK START: Reverb
GOALS

- Fix tracks with too much of that close-microphone sound and replace noisy endings.
- Fit tracks together using reverb to create spatial contrasts, distance, width, duration, and a united sense of ensemble.
- Feature mix elements through added liveness, simulated space, fabricated space, altered timbre, and reshaped texture.
- Invoke a scene change or decorate a track with a gated burst of energy, an ear-grabbing reverse effect, or an ear-tingling regenerative or dynamic wash of energy.

GEAR

- Key parameters of reverb time, bass ratio, and predelay are starting points for designing and refining a reverberant sound.
- Internalize the sound quality signatures of all reverberant sources: rooms, chambers, springs, plates, and digital.

Reverb is the sound of our sounds still sounding after we stop the sound. Turn out the lights, and the room snaps to darkness. Stop the music, and the sound lingers on a bit. When we make music, it rockets outward, at exactly the speed of sound—over 750 miles per hour in air on a typical day. The sound ricochets off the various surfaces in the room—walls, floor, ceiling, rack gear, computer screens, pizza boxes, tequila bottles, and so on. Gradually, the sound energy spreads out and is absorbed by the air and the surface materials in the room. But until that energy is absorbed we are going to hear it—it's going to reverberate.

Reverberation is the resonant sound of a space: the brief wash of energy that follows the direct sonic event as the sound waves propagate within at least a partially confined space, bouncing about, as the energy dissipates and the sound decays.

In the creation of recorded music, we leverage reverb from two sources: recording rooms and signal-processing devices. Although reverb is a concept born from room acoustics, the reverb of sound recordings serves many purposes. In its most literal application, reverberation can help evoke the sound of an architectural space, placing elements of our mix in a symphony hall, an opera house, a cathedral, a cave, an empty oil tanker, a tea kettle, or elsewhere. But reverberation is also used by mix engineers for more abstract applications: widening the perceived size of a sonic image, pushing mix elements away from the listener, extending the duration of a subtle mix detail, altering the timbre of a sound, invoking a scene change to help the song form tell a story, synthesizing entirely new sounds, and more.

This chapter tours the technologies used to create the reverb used in sound recording and illustrates the broad range of studio effects it generates.

9.1 PATCHING AND PARAMETERS

Words and numbers fail us when we try to fully describe something as wonderfully complicated as reverb. How do we meaningfully communicate the sonic difference between one symphony hall versus another to someone who has heard neither hall? Nevertheless, we mix engineers focus on a short list of quantities to summarize the coarse properties of reverb: reverb time, bass ratio, and predelay. We hook it up and dial it in.

9.1.1 Patching and Plugging In

Reverb effects, almost without exception, are parallel processes using *effects sends*, not serial processes using inserts (see Appendix A). Patch up aux sends into the reverbs so that a single effect can be accessed by any and all parts of the multitrack production. The reverb outputs are patched to effects returns or spare monitor paths feeding the mix bus.

We add reverb to a track while mixing to augment the sound. We don't replace the track with a reverberation-modified version. We add as much reverb as we want, of whatever kind of reverb we need, to any audio track that needs it. Therefore, aux sends are the preferred approach. Inserting a reverb is almost never done.

It is common to want to share a single reverb across many tracks. Running one reverb process and feeding any and all tracks desired via aux sends is the most logical approach. Great reverbs are expensive—in dollars and in computing resources. Buying multiple outboard reverbs and setting them to the same algorithm is a waste of money. Inserting multiple identical reverbs is a waste of computing power. Share the wealth by using aux sends for reverbs.

▶ This recommendation doesn't mean that you should have only one reverb going at a time. Far from it! It is common to have half a dozen different reverbs in a mix—or more! But it is foolish to have any of those reverbs running the same program. Aux Send 1 might feed a large, warm, lush hall while Aux Send 2 feeds a bright hall. Aux Send 3 might feed a medium room, Aux Send 4 a rich chamber, Aux Send 5 a plate, Aux Send 6 a spring, and so on. Multiple aux sends feeding multiple different reverbs lets you conveniently add any reverb of any type to any track. We discuss in this chapter the many motivations for reverb; running six or more reverbs won't seem unusual or intimidating by the time you finish this chapter.

9.1.2 Reverb Time (RT$_{60}$)

The perceived liveness of a hall is measured objectively by *reverb time*, or RT_{60}. Perhaps the most noticeable quality of a room's acoustics, reverb time describes the duration of the reverberant wash of energy. More specifically, it is the length of time it takes the sound to decay by exactly 60 decibels.

Allow sound to play in the hall. It could be punk music or pink noise. Abruptly cut off this sound. The hall does not instantly fall silent. It takes a finite amount of time for the hall to return to silence again. RT_{60} is the standard measure of this length of time. In a highly sound-reflective, large enclosed space—think cathedral or symphony hall—it takes two or more seconds for the sound to decay by 60 dB. Outdoors, in a flat desert (don't worry, we've packed plenty of water and we have plenty of sun protection) with only the ground to offer a single reflection, the sound ends all but instantly. The reverb time approaches zero. In the studio, we have available to us everything from zero to two to ten or more seconds of decay time, if we can find a use for it.

9.1.3 Bass Ratio (BR)

Reverb time is better understood across a range of frequencies. The resonant quality of a space is rarely consistent across all frequencies. High frequencies might decay more quickly than low frequencies. Some specific frequency ranges might have particularly long decay times; others might die off abruptly.

Rather than relying on a single number, RT_{60}, to describe reverberation, it might prove more helpful to the mix engineer to think of reverb in terms of a low-frequency, mid-frequency, and high-frequency reverb time. Or measuring reverb time in octave bands might be a better approach: $RT_{1,000}$ describes the length of time it takes sound energy in the one-octave band centered on 1,000 Hz to decay by 60 decibels. RT_{500} describes the 60-dB decay time one octave below, centered on 500 Hz. As our range of hearing spans some 10 octaves, this more refined method of calculation suggests that reverb is better described by 10 numbers. In truth, even 10 frequency-dependent decay times don't come close to fully describing reverb—it is too wonderfully complex to be so narrowly defined. On the other hand, a set of 10 numbers is too much information to keep up with in a mix session when we are responsible for so many other aspects of the recording. Out of necessity, we further simplify.

Bass ratio, sometimes called *bass multiply*, offers a useful and common simplification. It is a single number comparison of lower octave reverb times to middle frequency reverb times. Specifically:

$$BR = (RT_{125} + RT_{250})/(RT_{500} + RT_{1000}) \qquad (9.1)$$

where BR = bass ratio, and RT_X = 60-dB decay time in the octave band centered on the frequency X.

If the lower-octave reverb times are longer than the middle-frequency reverb times, the bass ratio will be greater than 1. The perceived overall warmth and low-frequency richness of a performance space is very much influenced by its bass ratio, and ratios slightly greater than unity are often the design goal for a hall expecting to host romantic classical music. In recorded music, we may honor this symphony hall tradition—or deliberately head the other way, pushing bass ratios much less than 1 for some tracks, and much greater than 1 for others.

Due to the importance of reverb time as a function of frequency, RT_{60} stated alone is generally understood to be a middle-frequency reverb time (usually 1,000 Hz; maybe 500 Hz). Bass ratio adds required extra context, describing the general reverberant trends below this middle-frequency reverb time.

9.1.4 Predelay

In addition to the duration of the reverberant decay and the relative duration along the frequency axis, low versus mid, every recording engineer must understand a third reverb parameter: *predelay*. In an actual performance hall, a gap in time exists between the arrival of the direct sound straight from the sound source to the listener and the arrival of any sound reflections or the reverberant wash of energy that follows. Predelay is the difference in time of arrival between the direct sound and the subsequent first associated reflection.

The size and shape of the performance space is the key determinant of predelay time. In orchestra halls, it is often the side walls that create the first reflection. Reflections off of the ceiling or the rear walls arrive much later. Therefore, a narrow hall is likely to have a shorter predelay time than a wide hall. In smaller spaces, the ceiling may be the closest reflecting room partition. With reverb-generating signal processors, of course, predelay is simply an adjustable parameter almost without limits.

Navigate through the presets of any digital reverb and you'll find many more parameters, but these three are the most informative starting points for defining the overall sound of a reverb. With these set at target values, dig deeper into the other parameters as needed to fine-tune your sound.

9.1.5 Reference Values

The very idea of reverb for music comes from real spaces, such as symphony halls and houses of worship. The reverb used in recording studios is typically generated by signal-processing devices. These user-adjustable pieces of equipment are wonderfully—and sometimes frustratingly—independent of the physics of sound constrained by architecture. The total freedom to synthesize any kind of reverberant sound is at times paralyzing for the novice engineer and has been known to bog down even veteran engineers. It is useful to bracket the range of studio reverb parameters based on the architectural acoustics of classical performance venues. An engineer is welcome to venture beyond physically realizable reverb properties, but clever engineers know when they have done so.

The symphony halls most adored by conductors, orchestras, critics, and enthusiastic music fans represent perhaps the highest form of achievement in reverb for romantic orchestral music. Three halls are consistently rated among the best halls in existence today and serve as our reference points in the recording studio:

1. Boston Symphony Hall, Boston, Massachusetts, United States
2. Concertgebouw, Amsterdam, The Netherlands
3. Musikvereinssaal, Vienna, Austria

Detailed analysis of these halls and other halls approaching their quality leads to a useful set of representative reverb values (Table 9.1). Setting up a studio reverb so that its parameters fall within these preferred values does not, of course, guarantee success. This quality of reverb may not sound appropriate for the music production at hand. These values have proven themselves appealing for the live performance of romantic orchestral music, not all forms of music. Also, a hall that falls within the desirable ranges for these three values might fail on other fronts.

As mentioned earlier, something as complex as the reverb within a hall isn't fully defined by so few numbers. These particular values could be met, yet for reasons not captured in these quantities, the hall remains disappointing. Even within the great halls, more contemporary classical music may not be flattered by this reverb. Opera, jazz, and rock each sound best with at least slightly different reverberant qualities. There is no universally right, best, perfect reverb. Use these values as a reference. Step out of these defined ranges deliberately, armed with an understanding of why the current multitrack production needs a shorter reverb and a lower bass ratio, yet a longer predelay than what might sound good for a Mahler symphony.

Table 9.1	Preferred Ranges for Three Key Reverb Parameters, in Halls for Romantic Orchestral Music (from Beranek, 1996)
Parameter	**Value**
Early decay time	2.0 to 2.3 s
Bass ratio	1.1 to 1.45
Initial time delay gap or predelay	20 ms or less

9.2 DEVICES

To add reverb, we have two choices: record natural reverberation separately (i.e., onto separate tracks of the multitrack recorder, often called *room tracks*) at the time of the original music performance or employ reverberation devices that create the desired spatial, ambient, or other qualities at mixdown via signal-processing effects.

9.2.1 Room Tracks

▶ It helps for the mix engineer to know what happened during the recording phase of the project. During the recording of overdubs with close-microphone techniques, the tracking engineer might also have placed distant microphones in the studio in order to capture the room's natural reverberance. These ambient signals are typically recorded as separate audio tracks so that they later appear as fully adjustable elements of the multitrack arrangement when we mix. Though a single mono room track has value, stereo productions often record room tracks in stereo pairs. Surround productions will record room tracks in four- or five-track (or more) sets. During mixdown, these room tracks can themselves be adjusted and processed in any way we choose. Some high-end recording studios are prized for the sound of their live rooms that have proven particularly effective in recorded music. Architecturally much smaller than opera houses and symphony halls, these live rooms typically offer supportive early reflections with a dose of relatively short reverberant decay.

The history of recorded music documents the value of recording even popular music in reverberant spaces. Many of the most important works of recorded art were recorded in studios much larger than is common today. Studios existed in converted churches and giant loft spaces. The talented engineer was able to capture recorded ambience at the tracking stage that worked for the production when finally mixed and enjoyed over loudspeakers. Few engineers today get the chance to record in such large, live spaces.

Although it does not happen very often, multitrack music productions sometimes go to the trouble of recording elements in very reverberant spaces. The session leaves the small recording studio and tracks the drums in a church or a string section in a hall and background vocals in a solid concrete basement. Room tracks are a critical part of such sessions.

9.2.2 Chambers

Room tracks can be added after the recording session if we just play back some of our tracks in a great-sounding room. The room becomes our reverb unit. Clearly, it is impractical to utilize an orchestra hall as an effects device. Such buildings are rented at great expense and cannot be proximate to every recording studio. As a result, the natural reverberation of a large hall is rarely part of a pop or rock multitrack music production. We do seek out other, more convenient, more affordable spaces and use them to provide a bit of natural reverb to our mix—these are called *reverb chambers*.

Reverb time is directly proportional to a room's cubic volume and inversely proportional to a room's total sound absorption. So we seek out large, sound reflective spaces—as best we can. As the rooms nearest studios are rarely as large as the 3,000-seat performance hall, we focus more on having a highly sound-reflective, sound-diffusive space.

A reverb chamber is equipped with loud-speaker(s) and microphones (Figure 9.1). To achieve a long reverb time, what the space lacks in cubic volume (Figure 9.2) is overcome through high sound reflectivity. Plaster, concrete, stone, and tile are effective materials for such a space.

There is no such thing as a "typical" reverb chamber. Recording studios install them wherever they can find affordable space. Custom-built rooms just for reverb are an expensive endeavor for recording studios, which are often located in urban areas (e.g., New York, Los Angeles, London, Tokyo) where real estate costs in dollars per second of reverb time will turn your stomach.

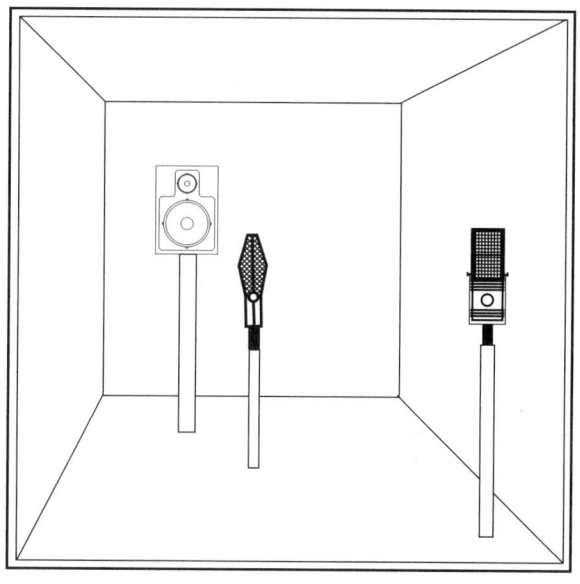

FIGURE 9.1
Chamber: reverb from a small, highly sound-reflective room.

Practicality has motivated recording studios to award the lofty title of "reverb chamber" to such unlikely spaces as multistory stairwells (Columbia on 7th Avenue, Avatar/Power Station); attics (Motown, Capital Studios on Melrose Avenue); bathrooms (project studios around the world); and basements (Columbia 30th Street, Capital Studios on Hollywood and Vine, and Abbey Road).

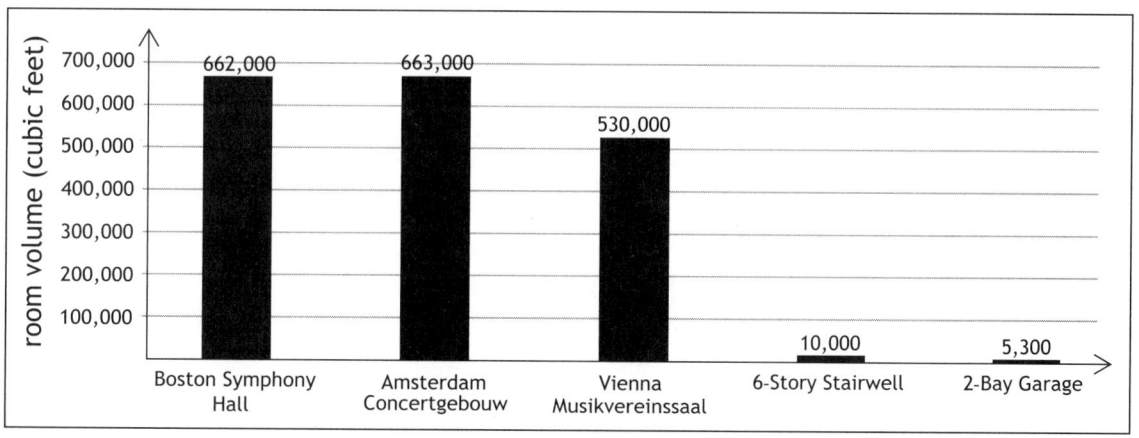

FIGURE 9.2
Chambers must generate reverberation within a significantly smaller room volume that symphony halls.

💡 Some trends in the architecture of the proven chambers guide us when we set up our own. Highly sound-reflective materials finish these spaces. Concrete, stone, brick, and tile are common. Frequently, thick layers of paint cover the naturally occurring pores in the concrete and masonry for minimum sound absorption. For reasons of tradition and superstition as much as science, irregular shapes and sound-diffusing elements are frequently sought out. Stairs, columns, pipes, highly articulated surfaces, and nonparallel walls are the norm.

Selection and placement of the loudspeakers and microphones within are critical to the sound of the chamber. The loudspeakers are often selected for their dynamic range capabilities as much as their sound quality. To energize the chamber and overcome any extraneous noises within, the ability to create very loud sounds is desired. Horn-loaded, sound-reinforcement speakers make a good choice, with efficient ability to handle high power. Practicality influences the loudspeaker selection, too. Last year's control room monitors are frequently repurposed as this year's chamber speakers. Reusing old monitors is financially more appealing than buying new ones for the chamber. As coloration is so prevalent in chamber reverbs—often deliberately sought out and emphasized in the mix (as you'll see later in this chapter)—loudspeakers with a flat frequency response aren't strictly required. Nonflat frequency response actually becomes a creative variable. Clever engineers turn the spectral imperfections of the loudspeakers in combination with the frequency anomalies of the space into a chamber reverb with a distinct and hopefully pleasing flavor.

Chamber microphones are placed very much in the same way room microphones are placed in a recording session. Engineers develop intuition about where in a room a microphone might sound best. Experimentation follows, and the microphone placement is revised as desired. Condenser microphones are most common, ranging from large diaphragm tube condensers to small diaphragm electret condensers. Though omnidirectional microphones may seem best for fully picking up the sound of the chamber, directional microphones are common, too, as they reward the engineer for even slight changes in location and orientation.

Generally, a line of sight between the loudspeakers and microphones is avoided. The loudspeakers are placed so as to energize the space generally. The microphones are placed to capture the subsequent reverberation. The direct sound from speaker to microphone amounts to an acoustic delay only, whose level is likely to be much higher than the reverberation that follows. Avoiding that direct sound maximizes the reverb that is captured by the chamber microphones and returned to the mixing console. Reverb is the intended effect, after all. Directional loudspeakers (horns) and microphones (cardioid and bidirectional), though not required, help the mix engineer achieve this.

🔊 Because reverb chambers are so small, a certain amount of modal coloration is unavoidable. It is hoped that the coloration this causes will sonically flatter any "dry" sounds sent to it. Equalization (see Chapter 3) is a signal processor purpose-built to introduce coloration to the sound. Engineers reach for reverb generated in a chamber when they think the chamber's coloration will lead to a desirable change in sound quality. Otherwise, the strong coloration must be avoided (no chamber reverb is added); ignored (other elements of the multitrack arrangement mask the coloration problem); or perhaps deemphasized with some EQ on the send and/or return.

💡 The typically small dimensions of a chamber lead to a predelay that is on the order of just a few milliseconds, well short of the 20 ms associated with large halls. Typically, a delay line (see Chapter 7) is inserted on the send to the chamber so that all reverb is appropriately delayed when it is combined with the other elements of the multitrack production.

9.2.3 Springs

A simple spring can be used to create a kind of reverberation. A torsional wave applied to a spring will travel the length of the spring. Upon reaching the end of the spring or encountering any change in impedance along its length, some of that twisting wave reflects back down the length of the spring from which it came. In this way, the wave bounces back and forth within the spring until the energy of the wave is converted into heat and dissipated through friction. Analogous to reflections between just two walls, a spring mechanically emulates the sound of a reverberant space—a *one-dimensional* space. Combine several springs of different lengths, thicknesses, and spring constants into a network of coupled springs and the pattern of reflections can be made more complicated.

The spring reverb has a unique sound but falls well short of simulating a real, physical, reverberant space. Its most important qualities in this regard are:

- The frequency response of the spring reverb system has limitations in the audio band and is far from flat.
- The number of reflections in even a multispring system provides only a finite imitation of the nearly infinite collection of reflections in a real room.
- The modal density of so simple a system is insufficient to prevent strong coloration.
- The buildup of reflection density does not grow exponentially with time as happens in a real room.

🔊 The spring reverb is simply not a room simulator. Yet it remains in use today. Affordable, analog, and portable, it is a common reverb in many electric guitar amps. Artists and engineers have locked onto it for its unique—if unreal—sound. In the right musical and sonic context, you should use spring reverb precisely because of this different sound quality.

9.2.4 Plates

The concept of the spring reverb is improved through the use of a metal plate instead of a spring, which upgrades the mechanical reverb to a two-dimensional design. A thin plate of metal has attached to it a driver that initiates a bending wave. This bending wave propagates through the plate, reflecting back at the edges, leading to an accumulation of reverb-like energy. A pickup transducer captures this reverberated signal. Reverberation on this plate behaves in a way very much analogous to a room with two pairs of opposing walls (but without a floor or ceiling).

The plate offers further advantages over the spring. By applying any of various means of damping (e.g., placing vibration-reducing materials against the plate), the reverb time is adjustable across a useful range, giving us some useful production flexibility.

The plate reverb, like the spring reverb, also fails to accurately simulate a real, physical, reverberant space. Its most important shortcomings versus acoustic reverberation are:

- The frequency response of the plate reverb system is far from qualifying as high fidelity. There is usually a lack of response in the lowest portion of the audio band and spectral content falls off quickly above the upper middle frequencies, leading to a distinctly metallic sound.
- The modal density of even a large plate is still insufficient to prevent strong coloration.
- Although the buildup of reflection density does grow exponentially with time, the two-dimensional, rectangular, mechanical reverberator does not grow at the same rate as a real, three-dimensional space with a complex shape.

🔊 As with the spring reverb, the plate reverb is simply not a room simulator. However, plates are extremely popular in multitrack production even today, largely because of their odd, unique sound qualities. Artists and engineers have found this sound to be useful for many applications, which we discuss shortly.

9.2.5 Digital Reverb

Although the synthesis of reverberation through math can require an intense amount of calculation horsepower, the resources are certainly available to do so. In fact, digital reverberators have been the most common tools for adding reverb in popular music mixing for more than a quarter of a century.

Countless algorithms exist for generating reverb digitally, but they generally fall into one of two categories. Infinite Impulse Response Filter networks are resonating digital signal processing algorithms based on recirculating delay systems, nesting together multiple comb filters and all pass filters and using elaborate modulation, feedback, and feed forward schemes.

Convolution reverbs are an alternative source of digital resonance. These reverbs use the impulse response of a system as a sort of blueprint for describing the timing and amplitude of all the individual reflections that make up its reverberation. Convolution is the mathematical process that applies that blue print to our digital audio waveform, imparting that system's spectral decay on any track we feed it. The impulse response that drives the convolution might be measurement data from an existing space or device, calculated data from computer modeling analysis, or wholly synthesized data fabricated by the creative sonic artist.

In the final analysis, all of these reverb technologies—chambers, springs, plates, and digital reverbs—are valid today. Each possesses unique advantages that the informed engineer knows can be strategically leveraged as needed on any production, depending on the production goal at hand.

9.3 MIX STRATEGIES: REVERB

This single device, reverb, is not a single effect. It is a vast range of effects. Reverb is a multitalented tool that offers the clever mix engineer countless ways to solve technical problems and pursue creative goals. It takes many years to master the wealth of reverb-based studio effects, but the work pays big dividends. Exquisite use of reverb sets apart the truly great productions from the merely average ones.

9.3.1 The Fix

Reverb can provide a clever solution to a technical problem or production frustration.

CLOSE MICROPHONE COMPENSATION

! Much of the audio in popular music is recorded using close-microphone techniques. Placing the microphone very near the musical instrument leads to tracks full of intimate detail. Done well, this can certainly be a pleasure to listen to; it is a proven technique for many important and successful recording artists. It is very much part of what makes the recorded music sound "better than real." However, a multitrack project consisting entirely of close-microphone tracks may be too intense, too intimate, too exaggerated, or too unnatural for some styles of music. On any given individual track of the multitrack arrangement, the close-microphone intimacy may not suit the creative goals of the art.

In this case, try a reverb processor offering a good dose of early reflections (generally a simple cluster of strategically chosen delays) and very short decay time (digital or plate reverb with a reverb time of 1.0 second or less) mixed in at a just-noticeable level. The goal is a slight diminishing of close-microphone immediacy. The sonic result, to the recording engineer, is that the recording sounds as if the microphones were further away from the musician at the time the recording was created. To the less-critical listener, the sonic result is a believable and pleasing reduction in the intimate detail of the particular track. It sounds more realistic, less exaggerated, offering welcome relief from too much up-close intensity.

This type of effect might use any of a number of reverb technologies, typically a plate or a digital reverb. You might even get away with using a "large hall" type of reverb setting at low level. But note that, in this application, the reverb isn't called on to offer that long, sonorous wash of heavenly decay. It is used only to diminish and blur the spectral detail of a track to better assemble a loudspeaker performance that is easier to understand. It is more about the dense early energy and less about the late reverberant decay.

THE CLEAN ENDING

! Sometimes the mix engineer must fabricate the illusion of some of the best-sounding performance spaces in the world. No pressure. Specific spaces—such as Boston Symphony Hall or Carnegie Hall—are sometimes the goal. We need to add reverberation to a recording, and we may wish to stay true to the original recording venue.

It is intuitive enough. A recording made in Carnegie Hall is often required to have the reverberant signature of Carnegie Hall. The end of a movement often reveals noise within the hall: people coughing, traffic just outside, air conditioning, and so on. The solution is often to edit out the end of the recording, replacing it with artificial reverb at that critical moment. The artificial reverb provides a clean, realistic musical ending free of extraneous noises. Any additional reverb added in the recording studio must, in this case, be perfectly consistent with the sound of—or our memories of the sound of—Carnegie Hall. Reverb devices are selected and the parameters are adjusted to achieve this.

A spring or plate will fail us here. Tailor a digital reverb by starting with the large hall preset that sounds closest, and then tweak the parameters into submission.

Convolution with an actual impulse response measurement at the recording session might make matching the hall decay much, much easier.

9.3.2 The Fit

With those two problems fixed, reverb is an essential tool for piecing multitrack elements together into a fulfilling loudspeaker playback experience. Making each track easily heard is a constant goal while mixing. As the multitrack gets filled with performance ideas, getting each track heard becomes a significant challenge for the mix engineer. We reach for reverb to help with this in a few ways. The use of reverb to set up some contrast among otherwise similar-sounding instruments is an essential method for preventing a big mix from becoming a big mess.

SPATIAL CONTRASTS

▶ Although the creation of an illusion of a single unifying space may be a goal for many pieces of recorded music, it is perfectly reasonable to pursue contrasting spatial qualities among various musical tracks in a multitrack production instead. The vocal might be made to sound as if it were in a warm symphony hall with one reverb; the drums may appear, sonically, to be in a much smaller and brighter room courtesy of another reverb; and the tambourine gets some high-frequency shimmer from yet another reverb. Applying different processing to different elements of a multitrack production can distinguish and set apart each sound or group of sounds treated with a unique reverb—it may make them easier to hear.

This technique enables a broad range of spatialities to coexist in a single, possibly crowded multitrack recording. In popular and rock music, this approach is the norm. Even simple productions commonly run three, four, or more different reverb units at once, each creating very different reverb qualities.

As so many pop productions have shown, the sound of a voice in a hall coexisting with the drums in a small room does not lead to any mental dissonance. Pop music fans are rarely troubled by the fact that such sounds could never happen naturally in a live performance. Listeners are motivated by what sounds beautiful, exciting, intense, and so on. Listeners are drawn to each of the different tracks embellished with a flattering reverb and other effects. And because of the contrasting effects, they are also better able sort out the dense collection of musical tracks within the multitrack production. The pieces and the whole become easier to hear, and easier to enjoy.

DISTANCE

Contrasts in distance are a productive lever to pull. Let's review some of what we learned many years ago in geometry class. A line possesses just one dimension. A plane is two-dimensional. A cube is three-dimensional. The pan pot, used with such care in assembling our stereo mixes, helps us work one—and only one—dimension: that line between the loudspeakers from left to right. Productions that simply pan things left to right are boring, one-dimensional works that miss so much opportunity. You and I want so much more, and so do the artists we work for.

Although home, car, and ear bud listening is a long way from providing fully three-dimensional sound recordings, even in surround, there is no excuse not to make all of our stereo recordings at least two-dimensional. To get beyond the narrow left-to-right dimension, we must learn to pull things forward and push things back, away from the listener.

As individual elements of your multitrack arrangement are pushed back into the soundstage, the overall stereophonic image of your work grows much more interesting. A crowded mix stretched taut in a line from left to right loosens up to make room for more tracks and more effects front to back.

Audio engineers ought to think hard about the sonic cues associated with distance. Gently roll off some highs to simulate air absorption and tell your listeners that a track is some distance away. Shift the wet/dry mix of your reverb effects more in favor of the reverb to push things back into a space and separate that sound source from your listener. Avoid overly present or bright reverbs; instead, use warm, natural programs that roll off high-frequency content in the reverb tail very much in the spirit of air absorption and distance. Use a light touch with compression so that some notes are deliberately, expressively obscured in the mix by other tracks, suggesting overlap and distance. Even pulling the fader down a smidge to evoke the lower level associated with distance can be a powerful effect.

These mix moves enable you to tiptoe a track backward into the depth dimension of the loudspeaker performance space. Be forewarned: all of these mix moves take practice, finesse, and a little bit of guts.

! Pulling out the highs to simulate air absorption is logically and perceptually valid. The trouble is that the first few times you do it, it's going to sound more than a little disappointing. The brighter track probably sounds better than the gently filtered track on its own. The reverb presents a similar dilemma. The drier track sounds clearer than the more reverberant track. And it's hard to pull a fader down. The louder track sounds instinctively better than the quieter track. Implementing these mix moves in isolation seems to indicate that things are only getting worse. Mature engineering judgment eventually teaches us otherwise. The key is to have a light touch and to evaluate these properties of the signal in the full context of the mix.

All of the effects for distance discussed thus far are low-magnitude adjustments. Distance is evoked with just a few decibels of high-frequency filtering and/or as little as a couple of decibels more reverb. A little massaging of the signal goes a long way (pun intended).

Consider a rock tune with strings. A midrange clutter of strings, doubled or tripled guitars, maybe some keys, and definitely some vocals presents itself. The strings—when soloed—might lose desirable texture when attenuated, filtered, and treated to extra reverb. Auditioned in the context of the entire mix, they drift back and away from the guitars and singers and find their own lush space behind the band. In the full context of the arrangement, the strings stop fighting the guitars and vocals and starting adding to the mix. The strings are now easier to hear, even though they are perceptually farther away. Probably more important, the guitars on the sides and the vocals in front all reveal themselves. With the strings deeper into the sound stage, the midrange competition is over.

♀ Engineers who have the strength of character to push select tracks back in the mix will find that they begin to fabricate bigger, deeper, more complicated images between two loudspeakers. Listeners are seduced right in. You may not need more effects to do this—just better coordinated use of the ones you have. It's just EQ, reverb, and faders, after all. A two-dimensional sound stage that spreads out left to right and front to back offers the band, their fans, and you a far richer listening experience.

WIDTH

Augmenting a sound with reverberant energy that is at least slightly different left versus right can create an engaging effect. A narrow, mono, boring track panned dead center in your mix offers the listener nearly identical signals at each of their ears, if they've set up their listening experience reasonably well. Interesting things happen when the signals at each ear starts to become unique. Any track—a vocal, a snare, a ukulele—can be processed with reverb to create subtle differences at the two ears. This approach can lead to a perceived widening of the apparent width of the sound source. Taken further, it can create a more immersive feeling for the listener. Surround sound offers us the chance to make this effect more pronounced and more robust.

Every reverb with two or more outputs is capable of widening a track. You chose the reverb you like best, based on experience, or the one that also simultaneously serves any other goals described elsewhere in this chapter. Through reverb, any single track of your multitrack production can occupy everything from a narrow point in space to as wide as the loudspeakers are placed—and perhaps even a little wider. Work the source width dimension to create contrast among your tracks—narrow solo instruments in front of a wide landscape of a rhythm section, for example. Use the widening effect of reverb to get a track noticed in a crowded mix without simply succumbing to the tempting urge to just push up the associated fader. Spatial width can fill in perceptual holes in the loudspeaker illusion we mix engineers create and is critical to making simple productions seem richer, deeper, and more complex.

DURATION

For short sounds (less than about 200 ms long), duration directly drives audibility. Really short sounds—think percussive sounds, from cowbells to snare drums—suffer a bit for being so short. All other things being equal, a snare sound that lasts a little longer will be easier to hear than an equivalent but shorter one. When we increase the length of a short sound a little, we raise its audibility in the mix without having to raise the fader—that's generally a good thing (see "The Case Credo" in Chapter 10).

🔊   Fast-releasing compression (see Chapter 5) and gated reverb (see Chapter 6) can do the trick, but simplest of all, you can also add a short dose of dense reverb. Plate reverb is an excellent choice. It offers strong spectral presence that our ears can't miss. Tune it to a short reverb time (under 1.0 second) and tuck it into the snare sound. The goal is to add a subliminal morsel of reverb to the short-lived snare hit. We aren't trying to push the snare to a distant location through reverberant obscurity. Nor do we wish to create the illusion that the snare drum is in an entirely other space. We seek only an increase in the duration of the sound. The casual listener may not even notice there to be any reverberation. The mix engineer adds the short explosion of reverberation simply to increase the net decay time of the snare hit, making it easier to hear and fitting it into a crowded mix.

ENSEMBLE

▶   We don't always work to separate tracks. Sometimes we deliberately unite different tracks into an integrated sound by making them more similar, through reverb. Taste and style might dictate that the entire multitrack production have the same reverb signal processing, putting the entire mix in a single space as if it were a single, live performance. In this case, the accumulated overdubs, instrument by instrument, are processed with the goal of placing each of them in the same single space spread out between and among the loudspeakers. The lead vocal, drums, piano, and hand percussion are all treated with reverb so as to create the illusion that all these players are in the same room together. That the players played separately, at different times, possibly in different studios, does not necessarily diminish this illusion of a single space. Signal processing is applied with the goal of uniting these instruments together in a single fabricated sonic space.

! If not the whole band, then any subset of the tracks might be fused together through reverb. A hazard of the close-microphone craft is that even sections of musicians (a string section, a horn section, a choir, etc.) might be tracked in relative sonic isolation. That is, each member of the section has an individual, independent track on the multitrack recorder and therefore its own, isolated sound. It is not blended with the other members of the section by the acoustics of a stage or a room. The close microphone picks up the individual member with very little acoustic energy from the other players in the section. Valid, practical motivations push pop-music recording in this direction. This motivates us to add reverb later, as desired, to reassociate the players into a single ensemble.

💡 Multitrack production gives artists the freedom to have potentially every single element of the multitrack arrangement fall into its own, unique ambient environment—put the singer in a hall, the drummer in a medium room, and the ukulele in a cathedral. This freedom is not always exercised. Using a single kind of reverb effect on many instruments helps to perceptually reattach these isolated tracks of audio back into a single section again. A collection of isolated point sources is merged into a single, broad section of players coming out of the loudspeakers with the unifying sonic signature of a single ensemble in a single space. Any reverb type is appropriate to this application, but digital reverbs set to "hall" or "medium room" are a good starting point.

9.3.3 The Feature

With problems fixed and the pieces coarsely fitted together, we now get to be creative, using reverb to feature aspects of the sound not easily accessed through any other effect. These advanced uses of reverb are essential for making a mix sound larger than life when played back through home speakers, ear buds, and laptops. That challenge cannot be underestimated.

💡 Reverb devices have their roots in room acoustics. The human experience in reverberant spaces motivated scientists and engineers to invent reverberant devices to modify tracks recorded in the highly absorptive rooms of recording studios so that they sound as if they were recorded in a larger, less absorptive spaces. Although there is much to master in that concept alone, this historical cause (room acoustics) and effect (reverb devices) does not mean that reverb is used only for evoking the size and sound of a space. We apply reverb signal processing throughout our mix for many other reasons. We might reach for reverb for the aesthetic change it brings to the production. A given reverb might be sought out for its own subtle, subjectively beautiful sonic character. Another reverb might be utilized for its ease of use or its playability. Engineers select and adjust reverbs to create timbres and textures that support the music, to influence the audibility of the various elements of the multitrack arrangement, and to synthesize completely new sounds; all of these results are not exactly spatial in nature. Reverb signal processing finds many other functions. In this way, reverb devices take on many of the attributes of musical instruments. Plan to work hard—to study and to practice—if you really want to make some reverb-enhanced music.

LIVENESS

We strategically deviate from reality when it suits our creative goals. One effective way to understand reverb is to identify two distinct elements within. First are the early reflections—the first- second- and third-order bounces of sound off nearby reflective surfaces. Second, is the reverberant decay—the brief waterfall of sound that follows later. Employing reverb signal processing to simulate a collection of early reflections only—without the wash of late energy that we associate with a large space—can perceptually diminish the close-microphone qualities of a track and contribute to the illusion that the audio track was performed in a real space by real humans. It makes the recorded performance feel alive.

Early reflections mixed in with a close microphone track suggests to listeners that the sound didn't just happen, but that it moved some air and it propagated across at least a small distance. A clinical, zoomed-in, close-microphone track is backed off slightly and made more natural. It reminds us that the performance is a human expression, not a machine-made concoction. Depending on the qualities of those reflections, it can help evoke in the listener's mind a real location such as a room or stage.

In addition, a direct sound followed by a volley of reflections indicative of a real-room geometry can make for a more robust stereophonic illusion of a location (left to right). What the brain figures out about a left-to-right phantom image can be reinforced by some added early reflections—more data to support an assessment of location.

We reach for digital reverbs to create a sense of liveness through early reflections. Springs and plates lack early reflections. These mechanical devices initiate dense reverberation too quickly. For sparse early energy, under our complete creative control, digital devices are the most effective choice. It's a concept born from sound in architecture. In a real space, the pattern of the early reflections—their relative amplitude and time of arrival—is determined by the room geometry and acoustic properties of the materials used to construct the space. The overall timing of these delays is determined by the scale of the space. Larger rooms will have these reflections spread out in time more than smaller rooms. Many digital reverbs emulate this property and give us some influence over them (Figure 9.3).

The addition of this kind of reverb to your mix adds a pattern of early reflections of a space, whether small or large, to a recorded signal with an independently adjustable amount of the associated late decay of a reverberant room. With digital reverb, it is possible to have early reflections evocative of a space with certain size, without any of the reverberant energy that such a space would physically

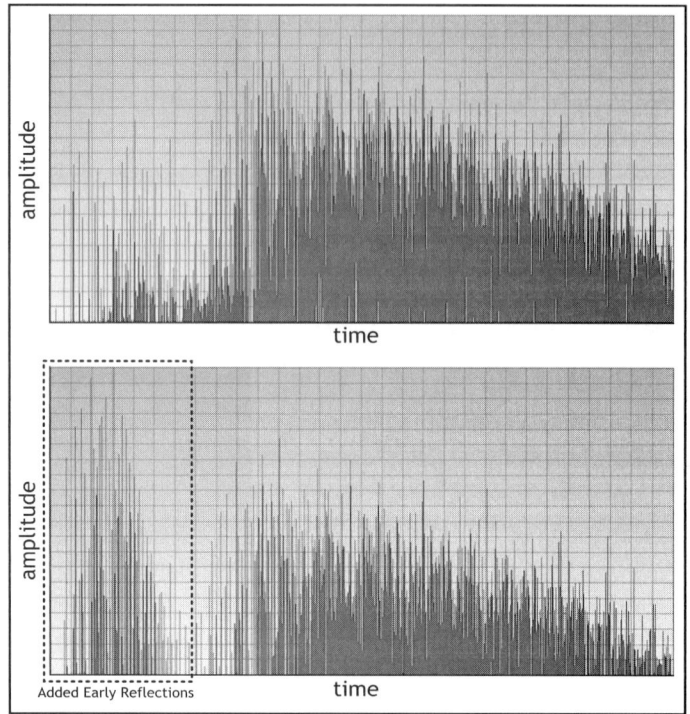

amplitude

time

amplitude

Added Early Reflections

time

FIGURE 9.3
A digital reverb with and without early reflections.

be required to create. The audio track achieves a more precise placement within a believable, real space without the reduction in clarity and intelligibility that the late reverberant energy would typically cause.

Note that the creation of the sonic illusion of a large or small space without the associated fog of confusion caused by reverb benefits not only the individual track being processed but also the many other elements of the mix seeking to be heard clearly. A long, reverberant wash on the snare drum can obscure the performance subtleties of other tracks—the intelligibility of the lead vocal, the voice leading of the rhythm guitar, the expressive phrasing of the ukulele solo. The decaying energy of a real space can interfere with the audibility of other important elements in the multitrack production, requiring us to pay careful attention while mixing. Or we just cheat, keeping the reflections only, and deleting or significantly attenuating the reverberant tail. A studio reverb adjusted for the presence of early reflections only, without the late wash of energy, evokes a kind of spatiality and believability without the associated clouding of the rest of the mix. Though room acoustics sometimes conspire to deny us, studio reverb flexibly empowers us. Where you want simple liveness in a crowded mix, seize this opportunity.

! Multitrack projects are often congested, with 48 or more tracks filling the arrangement simultaneously, each fighting to be heard. It requires great care on the part of the mix engineer and the arranger to prevent the mix from sounding cluttered and confused when played back over a humble pair of loudspeakers or headphones. Long reverb times, though sometimes irresistibly gorgeous and enveloping, can be the bane of clarity. Unburdened by the physics of room acoustics, we sometimes remove the late energy and use only the early energy, creating a live but intelligible sound. In this manner, recorded music is made to sound as if it were occurring in an actual space, while the perceptual detail of the tracks remains unnaturally, unbelievably vivid.

SIMULATING A SPACE

For many applications, evoking the sound of a space (and all the feelings, memories, and social importance invested in that space) is the goal of artificial reverberation. Three-dimensional space, as it exists in a church, concert hall, club, or canyon, can be a useful analogy for the mix engineer seeking to create a sonic landscape between or among the listener's loudspeakers. Among all of the reverb effects discussed in this chapter, simulating the sound of a space is the most common motivation for adding reverb to any track of a multitrack production.

Discrete elements of a multitrack production—or possibly the entire mix—might be processed to convert close-microphone studio recordings into a sonic illusion that the instruments were played in a larger space. Reverb devices available today easily create convincing patterns of early reflections and the associated reverberant wash evocative of spaces where music is heard. We can effectively transport our productions from whatever studio was used to any performance venue desired.

Although digital reverbs make it possible to simulate a specific, iconic hall or house of worship, we generally work at a more fictitious level. The need to simulate a realistic space doesn't necessarily require us to evoke a *specific* place. The goal of the reverb signal processing might simply be to create a believable sound of a likely, surprising, or otherwise appealing space. Here we seek a reverberant character that sounds appropriate for the music and evokes an architecturally grounded image of a realistic space. It is not necessarily Notre Dame or Boston Symphony Hall. It is some kind of church or hall that is convincing and realistic. Maybe it is a little smaller than the real space. Maybe it is a little darker (sonically or visually or both). As creative engineers, we try to stimulate the imagination and sonic memory of the listeners and let them fill in some of the details themselves.

Evoking realistic spatial qualities is not limited to symphony halls and opera houses. On any given track, we might decide we need the sound of a canyon with its long echoes, a gymnasium with its shorter echoes, a shower with its distinct resonance, or other such locations. Many of the same digital reverberators that evoke real halls can also be adjusted into configurations evocative of other real-sounding spaces. In these applications, we have total creative freedom to adjust the reverb parameters until our simulation of the desired space is finally realized.

FABRICATING A SPACE

▶ It gets more fun still. Beyond creating a space that is at least conceptually born in architecture, reverb devices are also used for synthesizing a reverberant character that simply may not exist in nature. In the fictional sound world that is pop music, we may wish to enhance the music with lushness or some other form of beauty, held only to a standard that it "sound good." We dial in settings on a digital reverb that violate the physics of room acoustics. For example, we might grab a pattern of early reflections typical of a small room and combine it with a late decay typical of a large hall. Wait: we're not done. That decay might be brightened (high frequencies emphasized) for a pleasing timbre. Digital reverbs make these parameters adjustable. Such an amalgamation of acoustic elements is fabricated despite the implications that there now exists a small room geometry (early reflections) inside a large hall (long decay) in which air absorption has been miraculously overcome (the bright quality).

Such a reverb can be made to sound glorious coming out of loudspeakers. When reverberation is synthesized in a device, it is able to make the acoustically impossible electroacoustically available. Rooms can't do it, but loudspeakers can, with a little help from us. As the mix engineer, you are free to abandon any connection to room acoustics and stretch the capabilities of a reverb device to whatever limits necessary in pursuit of your sonic goals, supporting the art of music—not the physics of acoustics.

Reverberation used to evoke spatial qualities is part of the signal processing—along with level settings, panning, EQ, and other effects—used to assemble a complete auditory scene. Within the wave field created by stereo or surround loudspeakers, this processing helps control the essential subjective spatial parameters of depth and distance, the immersive attributes of discrete sound sources and groups of sound sources, and the size and quality of the associated performance environment. We invent, implement, and fine-tune all the spatial variables. Space becomes a part of the composition, holding its own with timbre, rhythm, harmony, and melody. And although the composers and performers generally have a handle on rhythm, harmony, and melody, they often rely on you as the mix engineer to support the composition with appropriate spaces and timbres. Never underestimate your obligation to compose and construct a symphony of spatial qualities in your mix. It is a fundamental part of the mix engineer's job.

TIMBRE THROUGH REVERB

In some situations, reverb is used specifically because of the coloration it brings to a sound. Their frequency response, density of reflections, and resonant nature make springs, plates, and chambers useful devices for coloration. Productions that use these unique kinds of reverb (and the digital devices that emulate them) in this way do not seek to mislead anyone into believing that a real space exists around the instruments. Yes, there is a reverberant wash of energy. That characteristic signature of reverb is clearly audible. However, evoking an illusion of a symphony hall or an opera house is beyond the capability (or, indeed, the modern day intent) of these devices. Yet despite the existence of many powerfully effective digital signal processors that can reliably evoke illusions of real spaces, we still use springs and plates in popular music mixes.

State-of-the-art digital signal processors provide the ability to simulate real spaces with presets labeled "hall," "medium room," and "cathedral," for example. It is interesting to also note that they simulate other reverb processors, with patches called "chamber" and "plate," for example. Although springs, plates, and chambers were invented out of necessity to add reverberant character back to dry recordings, pop-music engineers (and less consciously, pop-music listeners) came to like their sound qualities even as more realistic and natural reverberation technologies were developed.

The coloration of a chamber comes in part from its small size. The highly reflective surfaces of the chamber lead to reverberation. However, the modal density in so small a room is not sufficient, especially at the lower end of the frequency range, to prevent obvious resonances. The room is simply so small that the existence of spectral pockets of strong resonance may be audibly obvious. In the design of a performance hall, this is a well-known problem to be avoided. In popular multitrack production, the mix engineer creatively matches this resonant behavior with tracks of music that are flattered by this coloration.

In this way, reverb devices are used as an alternative to equalizers and filters (see Chapter 3), which directly alter the frequency content of the signal by design. The resonance of a chamber, plate, or spring might add some sort of glow to the track that sounds pleasing.

Beyond resonance, reverbs can influence the perceived timbre of a signal through frequency-dependent reverb time differences. A bass ratio target value of 1.1 to 1.5 is known to contribute to the perceived warmth of the instruments that play in the space. Likewise, employing a digital reverb that creates reverberation without modeled air absorption, high-frequency reverb times can be stretched longer than mid-frequency reverb times. This method adds high-frequency energy to the loudspeaker music, leading to a perceived airiness, sparkle, shimmer, or other pleasing high-frequency quality.

🔊 The frequency response of a spring reverb is far from flat. That strong spectral flavor influences the perceived timbre of the audio track being sent to the spring. The mixture of an audio track with reverberation like this colors the sound. Our sense of the timbre of an instrument is based on a perceptual net sum of the track timbre, plus the reverb timbre. If we are doing our job right, the combination is flattering to the track. A steel string acoustic guitar might be made to sound a bit brighter still through judicious use of a bright, metallic-sounding spring reverb.

The plate reverb offers a variation on this theme. Any plate will have its own set of spectral biases, perhaps offering a modified kind of high frequency emphasis with some midrange complexity. Look for the right track in need of a timbral shift in this direction.

In this way, reverbs are used to subtly shape the perceived timbre of the audio tracks being processed. Chambers, springs, plates, and radically manipulated digital reverbs are well suited to this approach.

TEXTURE THROUGH REVERB

The unique sonic character of some reverbs, especially springs and plates again, leads to their use for elements of texture. Reverb is added not for the spatial attributes it evokes, but for the visceral quality of the sound energy it offers. As mix engineers, we are sometimes motivated to add to a pillowlike softness, a sandpaperlike grittiness, a metallic buzziness, a liquid stickiness, and so forth.

Mature musical judgment is essential for this application, and it requires experience and a good understanding of reverb and synthesis because there is no reverb patch labeled "pillowlike softness." There is, however, the capability to make a track take on a soft, pillowlike texture if we make clever use of the reverb devices in the recording studio. Much as the disciplines of architecture and acoustics must collaborate to create a great-sounding space, the fields of sound engineering and music intersect to find musical, aesthetically appropriate reverb applications.

▶   Examples abound. A reverb might be used to slightly obscure a particular part of the multitrack arrangement giving those instruments an ethereal, veiled texture. The metallic sound of a spring reverb might be used to help a track take on a steely, industrial texture. That pillowlike soft texture can be created through use of a long (reverb time exceeding 1.25 seconds) plate reverb with a long (about 120 ms) predelay and a gentle roll-off of the high frequencies (starting at about 3–4 kHz).

When a composition is orchestrated, a horn chart arranged, or a film scored, elements of texture are a part of what motivates the creative thinking. As mix engineers, we must make strategic use of reverb to help create textures and their associated feelings in support of the music.

SCENE CHANGE

Any reverb effect we have running can change during the course of the song. As we move from intro to verse to chorus, we may increase or decrease the amount of reverb, we might introduce and remove new reverbs, and we might tweak any parameter within the reverb. It is not always a static, always-on effect. In fact, one of the most important mix gestures we have available is the ability to evoke a scene change in support of the composition. Perhaps the introduction occurs in a lush, highly reverberant soundscape. When the first verse begins, the reverb vanishes, leaving a more intimate setting.

💡   This approach knows no limits. The chorus might be accompanied by a transition from intimacy to a live concert feel, and so on. Active manipulation of your reverb choices and parameters in close, creative coordination with the structure of the song is a powerful pop-music effect.

EXTREME REVERB

Applying a given type of reverb to any defined, discrete element of the multitrack arrangement enables you to dial up some extreme reverb. Every instrument in the orchestra must be treated to the same reverb because they play together in a single hall. In multitrack production, we aren't so constrained. We are able to add extraordinarily long reverberation in part because that reverb can be applied to selected tracks only, not the entire mix. Generally, a rock-and-roll drum kit would not be very satisfying to listen to with a reverb time in excess of 2.5 seconds. A noisy wash of reverberant energy would trip and tumble through the mix like Godzilla running through the streets of a crumbling and frightened 1950s city. It would obliterate all details in the drum performance. It would drown out the vocal. Listeners would run away from your mix as fast as they could—no doubt while screaming. Such an extreme reverb can't be applied to the whole drum kit.

But the vocal of a ballad—the vocal alone—may soar heavenward when treated to such a long reverb. The drums, meantime, might be sent to a different, much shorter reverb to better reveal their impulsive character.

💡 The multitrack production process makes it possible for us not only to apply a given reverb to any single track, but also to apply reverb to a single phrase or single note of a performance. The reverb effect can be turned on and off, instant by instant, during the song. It's all or nothing for an orchestra in a reverberant hall, but any degree of fine control is allowed in a multitrack mix using reverb devices.

In this way, it is not unusual for a reverb—when applied to isolated multitrack elements—to climb to reverb times in excess of 10 seconds. It is not uncommon for predelay to range from some 60 ms to beyond 100 ms, or even more. Bass ratio moves freely from a value 0.5 to beyond 2.0. High-frequency reverb times are routinely allowed to rival mid-frequency reverb times, even though this could never happen in a symphony hall. We get to reach for whatever reverb settings we want in support of the sonic art we seek to create, no matter how extreme. Radical, physically impossible reverb settings are applied at any time, to any part of any track in search a better sound.

! To be clear, many styles of music aren't offered so liberal an approach to reverb-based treatment. Some pop, most folk and jazz, and nearly all forms of classical symphonic music are presented through realistic recordings. This realistic approach looks for a space—a real, believable space—to be recorded in an actual room and/or synthesized by signal processors to create a single convincing performance location. In the world of popular music, however, there is typically great freedom to apply spatial attributes and special effects to any element of the multitrack production, unburdened by the realities of room acoustics. Limits are pushed to the extreme, often with thrilling artistic results.

FEATURE THE REVERB

! Although some reverb techniques strive to make the audio tracks easier to hear, other reverb signal-processing approaches seek to make the *reverb* easier to hear. The contrasting reverbs and the reverbs with extreme settings of parameters discussed previously can serve to highlight the reverb itself. Pop music in particular is driven, in part, by a celebration of exactly this kind of ear candy. If you like the reverb for reverb's sake, don't be afraid to make the reverb a little more obvious. Care is needed, of course, because too much reverb makes the rest of a mix dull and boring. With finesse, you can increase the reverb time and/or stretch the predelay parameter toward larger and larger values (50 ms and beyond) and make the reverb easier to hear.

GATED REVERB

Signal processing aggressively violates laws of room acoustics in this application. A reverberant wash of energy is sent through a compressor (see Chapter 5), followed by a noise gate (see Chapter 6) that radically alters the decay of the reverb, resulting in an abrupt, wholly unnatural decay.

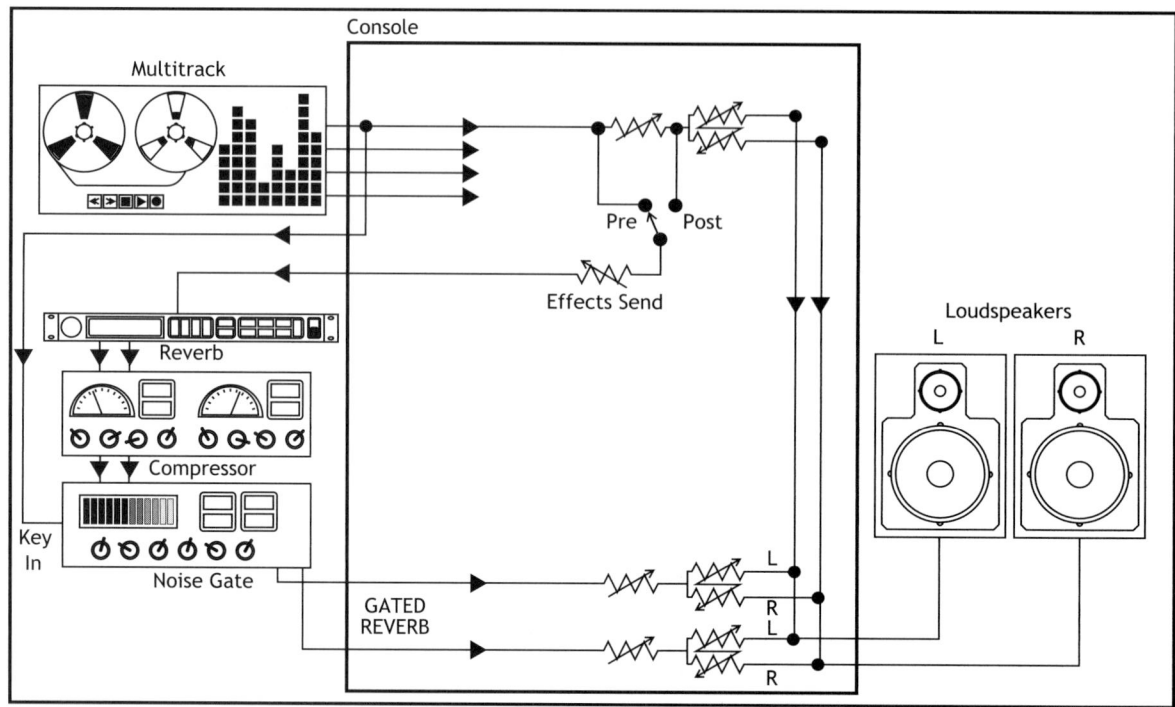

FIGURE 9.4
Signal flow for keyed gating of reverb.

▶ The signal flow for gated reverb system is elaborate, but straightforward (Figure 9.4). The outputs from the reverb device are sent to a stereo (two-channel) compressor, whose outputs in turn are sent to a stereo noise gate, whose outputs are sent to the mix via the mixing console. The gate is keyed open reliably by a close-microphone track for crisp control.

An illustrative example begins with a snare drum recording. Reverb is added to this sound (Figure 9.5a). The resulting reverb is sent to the compressor. The compressor, by attenuating the louder portion of the reverberant wash, flattens out the initial decay of the reverb, giving it a more gradual slope implying a longer reverb time (Figure 9.5b). The gate then cuts off the decay with a slope indicative of a much shorter reverb time (Figure 9.5c).

FIGURE 9.5
Gated reverb.

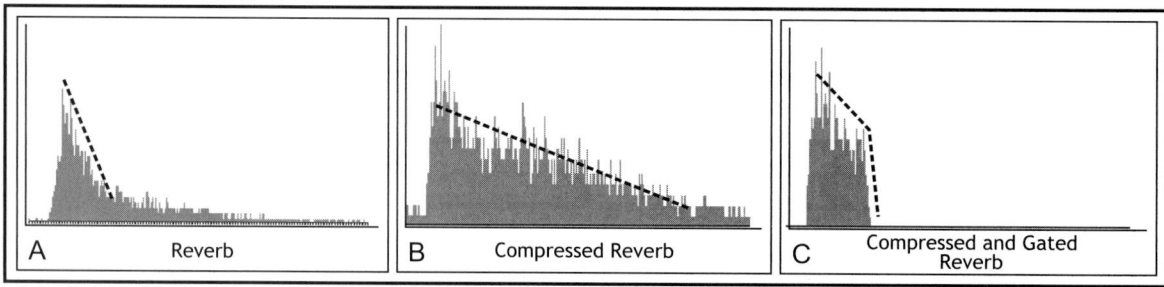

The result is a concentrated burst of uncorrelated energy associated with each strike of the drum, which reshapes the sound of the snare drum into a more intense and more exciting sound. The effective duration of each snare hit is now stretched longer, making it easier to hear.

This effect can be made prevalent in the mix, for all to hear. It's unmistakable. It's unnatural. Recordings from the 1980s elevated this effect to a cliché. If one uses it today, a bit of 1980's nostalgia is attached to the production.

However, gated reverb is also used in more subtle ways. A valid philosophy is to make the gated effect so subtle that it is barely noticeable, if at all, to the untrained listener. The goal is to add a bit of sustain to the sound so that the track becomes easier to hear without having to turn it up. The gated reverb offers an instant of decorrelated energy, widening the apparent size of the event. And the gated reverb offers a burst of new spectral energy, shifting timbre as desired. It's not always over the top. It is always a great way to feature a sound.

REVERSE REVERB

The studio-only creation of reverse reverb requires the temporary reversal of time—no biggie. The goal is to have the reverb occur before the sound that causes it.

Consider a snare drum backbeat within a pop song, falling on beats two and four of a measure (Figure 9.6). The typical addition of reverb to such a track is shown in the lower portion of the same figure. Note that the lower waveform is the reverb from a snare drum, not an impulse response; this is the typical use of reverb for such a track. Reverse reverb takes a different approach.

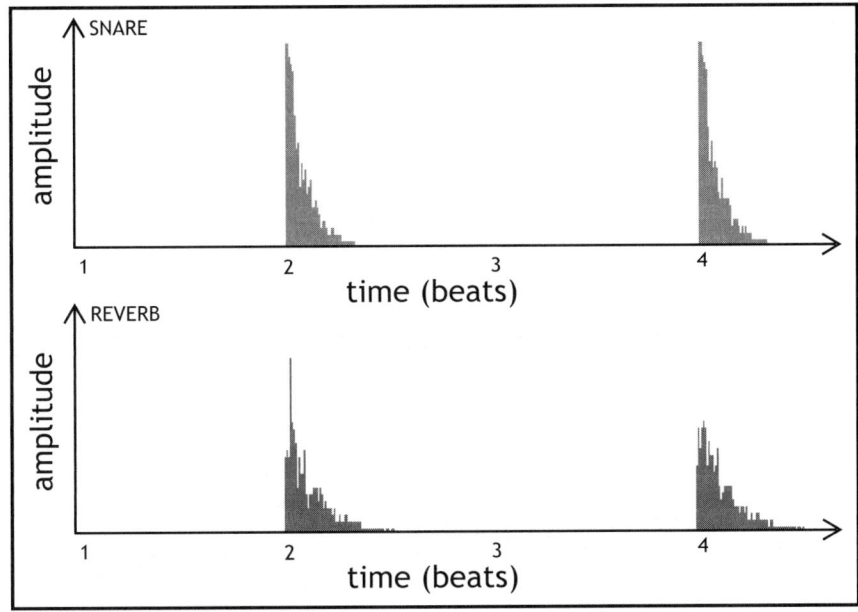

FIGURE 9.6
Adding reverb to a snare backbeat.

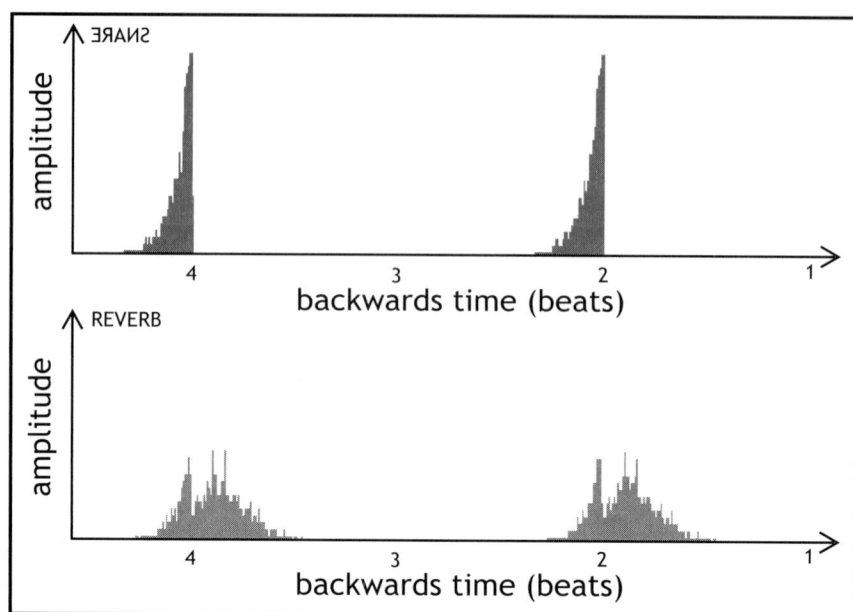

FIGURE 9.7
Adding reverb to a reversed snare backbeat.

First, the snare track itself is played backward in time (Figure 9.7). This step can be done on an open reel analog tape machine by turning the tape over and playing it upside-down. Alternatively, within a digital audio workstation, the audio track is selected and the computer is instructed to calculate the time-reversed waveform. Reverb is then added to this time-reversed snare, creating the signal shown in the lower part of Figure 9.7 and recorded to an available track on the multitrack recorder. Finally, the original snare track and the reverse reverb track are reversed in time (by flipping the tape back or executing another time-reversal command), which restores the snare track to its original place in time. The reverb derived from the time-reversed snare is found to occur before the snare drum sounds (Figure 9.8).

▶ The result is a kind of "preverb." Sound energy swells up and into each snare hit, creating an unnatural but musically effective new snare sound. A short plate reverb was used for these illustrations, but there are no constraints on the type of reverb for this application. Musical anticipation, rhythmic syncopation, small-scale crescendo, and the desire to fabricate an otherworldly sound motivate us when designing this effect.

This overall sound is partially simulated by some reverb devices without resorting to backward playback. Labeled "reverse reverb" or—more accurately—"nonlinear reverb," the amplitude envelope of a reverberant decay is aggressively altered so that the reverb tail seems to go backward in time. That is, the reverb starts off relatively quietly, becoming increasingly louder before abruptly cutting off. Unlike true reverse reverb, nonlinear reverb happens *after* the stimulating sound, not before. The unnatural, antidecaying shape to the reverb tail alludes

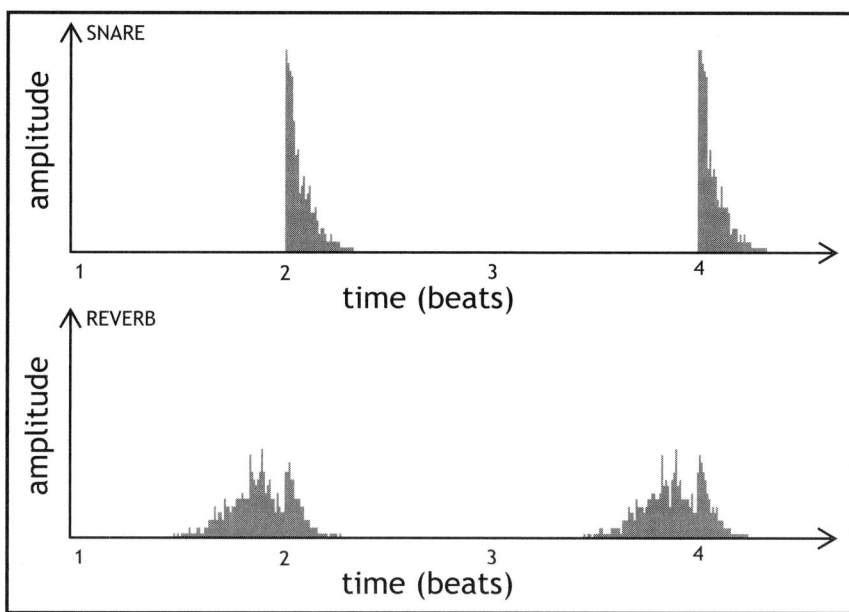

FIGURE 9.8
Creating reverse reverb on a snare backbeat.

to reverse reverb. This trick is a powerful way of lengthening the sustain of short sounds such as percussion, lifting them a bit up out of the mix so that they are easier to hear. It also provides dramatic ear candy without risk of mix-muddying sustained reverberation.

REGENERATIVE REVERB

Reverberation in a hall is initiated by the sounds coming from the musicians within. Reverberation in the recording studio is not so constrained. Figure 9.9 shows a reverb being fed by two sources. It might be any type of reverb: spring, plate, or digital. First, the typical signal routing using an aux send to a reverb is employed. The track for which reverb is desired is sent to the input of the reverb (using effects send 1 in Figure 9.9). The reverb outputs feed other inputs on the mixer, sending the reverberant signal into the stereo (or surround where applicable) mix.

For a regenerative reverb, this standard approach is augmented by a delay. The track for which this kind of reverb is desired is sent not only to the reverb, but also to a delay unit (shown using effects send 2 in Figure 9.9). The delay time is likely set to a musically relevant time interval, such as a quarter note, a dotted eighth note, a quarter-note triplet, and so on. The output of this delay in turn feeds the same reverb used by the original audio track.

The delayed signal is likely lower in amplitude as it reaches the input of the reverb than the original, undelayed signal. The sonic result of this second, delayed feed to the reverb is a subtle extra pulse of reverberant energy in time with music. In addition, regeneration on the delay unit, routing its own output

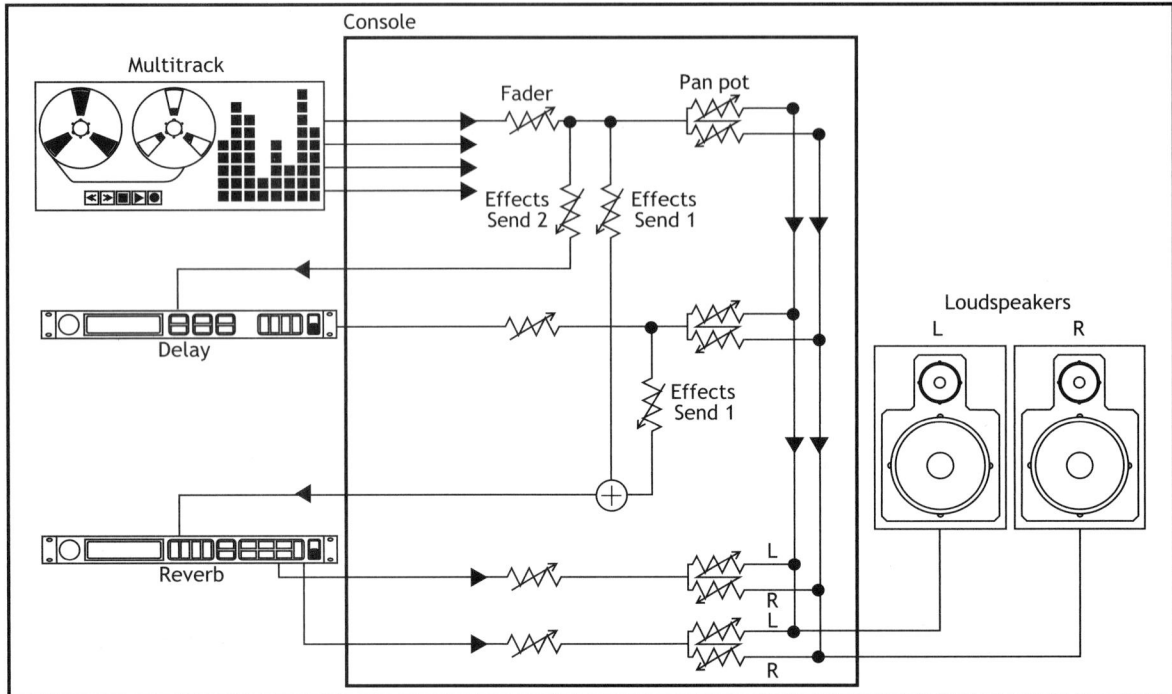

FIGURE 9.9
Signal flow for regenerative reverb.

to its own input, further delays the already delayed signal. This feedback of the delay to itself creates gently decaying, musically timed repetitions of the signal; it is then combined with the direct, dry signal and fed to the same reverb device. The result does not sound like any physical space in existence. This elaborate reverb system creates an ethereal, swirling, enveloping foundation of energy resonating underneath the track.

Although any reverb type is effective for creating regenerative reverb, large hall programs from a digital reverb are most common. Clearly, such a dramatic effect would generally be applied only sparingly to a single element or two of a multitrack arrangement—not the entire mix. It is a matter of taste, but it is often the case that this pulsing, regenerative reverb is placed in the mix at a very low level. Acting almost subliminally, it offers a rich, ear-tingling sound that exists only in the music that comes from loudspeakers, created under the watchful ear of the mix engineer.

DYNAMIC REVERBERANT SYSTEMS

As many of these examples make clear, signals can be processed both before and after the reverberation device to alter and enhance the quality of the sound. Effects devices are placed before the reverb, as was done with the delay unit to create the regenerative reverb shown in Figure 9.9. Effects devices are placed after the reverb, as was done with the compressor and noise gate used to create the

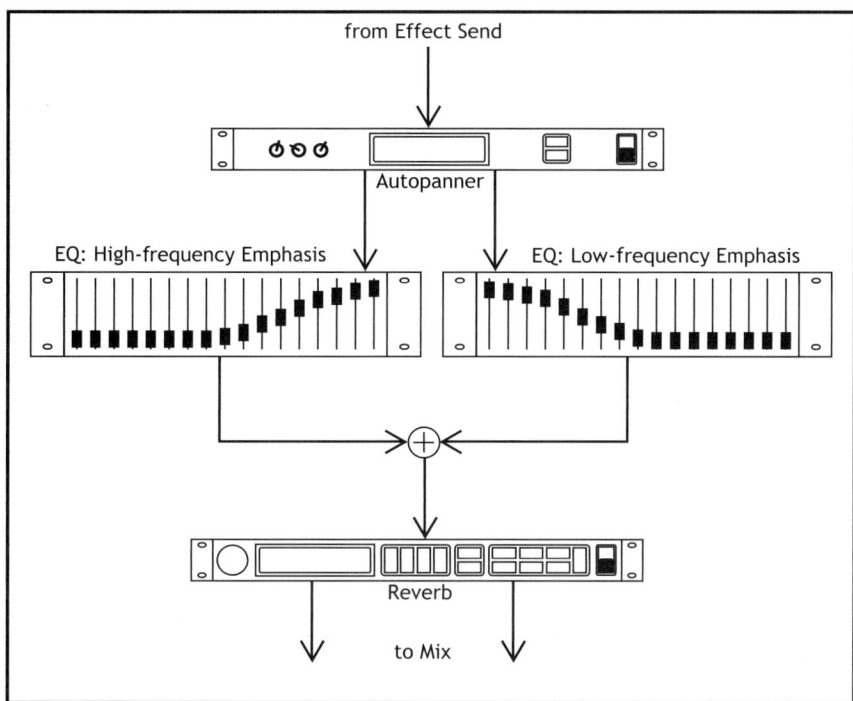

FIGURE 9.10
Signal flow for dynamic reverb.

gated reverb shown in Figure 9.4, which also shows that multiple effects devices may be connected in series to further develop the effect.

This approach can grow still more complicated by introducing additional processors in parallel both before and after the reverb device. A single such system is discussed here as an illustration (Figure 9.10), but an infinite number of options exist for the mix engineer to fabricate a reverb-based sound.

Consider having two parallel feeds into a reverb, each processed differently. For example, each of the two input options might have different EQ curves applied to the signal. Each equalizer is tuned to a different resonant frequency, or perhaps one is bright (high-frequency emphasis) and the other is warm (low-frequency emphasis). Preceding these two filters is an autopanner.

An *autopanner* is a machine-controlled pan pot. It automatically adjusts the amplitude of a signal between two outputs so that as the level is raised on one output, it is lowered by an equivalent amount in the other. It is two synchronized amplitude modulators moving with opposite polarity. Its typical application, used in a mixer, is to create the illusion of an audio track moving left to right and back again as the perceived localization follows the louder signal. In this dynamic reverberation system, the autopanner is applied in a different way.

The effects send from the mixer feeds the autopanner. Each autopanner output feeds a different equalizer. Both equalizer outputs are combined to feed the input to the reverb. The reverb system is shown in Figure 9.10. The resulting reverb is a constantly changing sound, perhaps subtle, perhaps obvious. This dynamic reverberation takes on a life of its own, a soft pad of ever-changing, uncorrelated energy adding life and mystery to an audio track.

9.4 SUMMARY

Reverberation inspires us in many ways, finding a broad array of purposes in our mixes. It is enough of a challenge to add reverb signal processing to evoke a sense of space in our recordings. But reverberant energy coming from room tracks, chambers, springs, plates, or digital devices enables us to manipulate perceived distance, width, duration, ensemble, timbre, texture, and whole classes of ear-pleasing constructions that don't exist outside of the multitrack music paradigm. Reverb is a deep topic that, when mastered, lays out before us an all but unlimited range of mix possibilities—keys to the creative kingdom.

CHAPTER 10
Future-Proof Mix Skills

Caveat crustulum libri: *Beware of cookbooks.*

10.1 STRATEGIES, NOT RECIPES

Seeking out a recipe for mixing in which someone else tells you the best EQ settings for snare drum or the best compressor settings for a lead vocal is futile. Not all snare drums are alike, and no two drummers are the same. They've never heard your vocal, as tracked with your microphone, in your room, for your song. How dare they tell you what to do?

Great mixes come from great strategies. You need a creative vision for the set of tracks provided and the technical expertise to realize that vision. Creative vision plus technical expertise! Each is a profoundly challenging thing to achieve on its own—the result of a lifetime of passionate progress—and you must excel at both.

You need to be such a great musician and such a creative thinker that your vision for the mix is artistically valid and transcends what anyone else would ever dream of. To be a great mixer, you must be as strong and talented an artist as your favorite musicians.

But that's not enough. Your artistry needs vast technical capability. You need to have achieved such total mastery of the studio that you can coax your artistic vision for the music out of the knobs, sliders, buttons, and thingamajigs that fill your studio. This mastery comes from knowing how to change a signal from its current mediocre state to its needed stellar sound. You choose the needed devices and tweak the necessary parameters to accomplish this. You don't plug in someone else's settings. It takes hard work, years of practice, and the sort of passion for audio that got you to Chapter 10 of this book!

There is no single right answer—there are countless possibilities. There is no universality—what works in one mix may not work in another. Your mixes sound good when you solve problems and exploit opportunities unique to the particular project you are mixing and when your musical goals are reached effortlessly, wielding any and all tools necessary to get you there.

Asking some stranger for the best settings for a snare drum reverb on your mix is as useless as asking someone to recommend the best words for a chorus. Who would take the following statement seriously?

> The best lyrics of a chorus in a rock tune are "baby," "love," "yeah, yeah, yeah," and sometimes "sprocket."

Or, for figuring out how many snare hits should be in a rock tune:

> My favorite song has 147 snare hits!

Such an observation is trivially interesting, at best, and offers no relevant information for our next project. Deep knowledge of recorded music as an art form, mastery of tools and techniques and heightened listening skills, plus experience, judgment, convenience, luck, and creative whim are your guides.

It can be tempting, but don't allow yourself to chase the false goals of mix recipes. They are easy to implement, to be sure, but they are unlikely to help your mix. It is a comfort to some:

> My mix is great! I'm using the exact same reverb settings they used on the hit "Imitation, It's Not Just for Apes," by the band Heard of Sheep.

But applying recipes to a mix removes you from the mix—your musicianship, your opinions, your technical savvy, your artistic influences, and your mix character. Mixing is a personal, nonlinear, creative endeavor expressed through high-tech means, not a *fait accompli* in need of accomplishment.

10.2 LIFELONG LEARNING

If we aren't memorizing settings, what are we to do?

Doctors read constantly, challenge and assist colleagues online and in person, attend conventions, experiment, practice, and generally commit to getting better and better—for their entire career. So do mix engineers. Don't plan to become a great mix engineer in, say, five years. Plan to keep getting better and better for your entire career.

If you love music and technology, if you are a geek with a guitar, if your laptop cranks as much music as math, you likely have found that a career that includes work as a mix engineer is just too much fun. We are lucky to have a career we genuinely love. So many people must separate work from pleasure; they work now so that they can enjoy the benefits later. The weekdays must be survived, the weekends enjoyed. Meantime, most of us in the sound recording industries can't separate the work from the pleasure, the weekday from the weekend.

But don't let all that fun fool you. We still have to work at it. The best mix engineers are both contagiously passionate and tirelessly intellectual about their craft and their art. Like a medical doctor, they read constantly, challenge and assist colleagues online and in person, attend conventions, experiment, practice, and generally commit to getting better and better—for their entire career. Join in. Let your tireless passion for audio motivate an organized journey of lifelong learning.

You can become a better mixer over time, but it requires earnest, proactive effort from you. We don't learn *by* doing. We learn *from* doing. After your mix is done, after the client has gone, become your most thorough critic. Think back on the process. Listen back to the results. Tease out as many lessons learned as possible. The person who mixes 100 times might be a better mixer on the 101st mix. But the person who thoughtfully reviews and studies those 100 mixes is definitely a much better mixer on the 101st mix.

10.2.1 Capability versus Ability

We are lucky to be in the audio industry today. The tools we have at our disposal are more powerful, yet more affordable, than at any other time in the history of audio. When audio went digital, our equipment stopped being just esoteric, specialized, audio-only contraptions. Instead, we began to use the same work environment—the computer—as bankers and photographers and students and scientists. Now, all the benefits in digital signal processing and all the innovations in personal computing devices benefit us directly. Studio tools are built like everyone else's tools, and they just keep getting better sounding, more flexible, more powerful, and easier to use. Meantime, esoteric audio-only tools are still developed just for us—professional audio engineers. The best equipment—time-proven since the birth of the recording industry, combined with the state-of-the-art tools we might download today—combine to give us the capability to do things that engineers only dreamed of during the first century of sound recording. That we can walk out of a store with a laptop possessing more audio capability than EMI Studios on Abbey Road owned when the Beatles recorded their last album is truly thrilling. Lucky us.

We must never confuse the acquired *capability* to do things with the genuine *ability* to do things. Buying a word processor doesn't make you a novelist. Owning a plane is no guarantee that you can fly one—far from it. That our audio gear can do so much doesn't mean we'll be able to do it well. Just because our studio can mix 128 tracks doesn't mean we also have the necessary skills to create a great-sounding mix. Just because the tools permit automatic tuning, sound replacing, and tempo manipulation doesn't mean we should do those things—or that it'll sound good if we do. Capabilities present opportunities, but we must have the artistic maturity and technical mastery to properly take advantage of them.

So feel free to acquire the best studio tools you can, but always accompany that infrastructure growth with skills development. The more gear you have available, the more experience, practice, and study you'll need. Ability trumps mere capability. We live in an audio world of excess capability paired with a dearth of ability. To me, that spells opportunity. When your mix artistry aligns with your studio technology, you'll be able to achieve great things musically.

10.2.2 Artistic Maturity

As mix engineers, we must take our musicianship seriously—music studies, a music degree, and proficiency on more than one instrument are common. Being a musician is a great advantage, and getting beyond the self-taught hobbyist to a gigging performer makes you an even better mixer.

You must also make sure your definition of musicianship includes the recording studio as musical instrument. You should seek to advance your knowledge of music recordings as an art form and of studio techniques as musical proficiency.

It is likely that every challenge we encounter in our project has already been addressed by mix engineers who came before us on similar projects. It is possible that every new idea we have about an approach to a mix has already been tried by others. Learn from the mistakes and the successes of others and your mixes will not disappoint the artist or their fans.

Just as any great poet reads the works of other poets, we must be avid listeners to great recordings by others. We need to know our history—enjoy and study the most important recordings in our chosen styles of music—the time-proven icons. We must also stay out on the cutting edge—avidly seek out and analyze the most popular contemporary recordings getting the most critical attention and earning awards and other forms of recognition.

Great mix engineers know the details of all the great mixes; we are well-schooled musicians on the art of recorded music (see Appendix D).

10.2.3 Technical Mastery

In the meantime, we must accompany our artistry with ever-growing geek cred. Knowing how to work every tool in the recording studio is a pleasant—if long-term—challenge. Take it seriously. Spend time in the studio digging deeper into the various devices you own—practice time, study hall. Know the tools backward and forward. Know what they can and can't do well. Master fully the gear you own.

Wait, there is more: we must also study the gear we don't yet own, old and new.

Vintage equipment remains relevant to making music recordings, both today and tomorrow. Your recordings grow richer when you also know how to use recording devices no longer made, because great studios have them and many projects need them. You must know the proper care and feeding of vintage audio tools. Know your tech history so that you can make the most of a studio booking in a world-class, high-end recording studio with all the best tools and the best toys.

New equipment is invented for us every day. Ours is an ever-advancing field. Look around any great studio (hopefully yours qualifies) and you'll see recording equipment that you rely on every day that didn't exist three to five years ago sitting next to essential tools that are more than a decade or two old. The next three to five years promise even more. To remain a great engineer, each of us must be on a mission to master all of the most important tools released daily.

10.3 THE CASE CREDO

Mix smart by checking in on your author's list of fundamental mix principles (described in this section). Mixing is so wonderfully open-ended that we occasionally get stuck, unable to find the next step. When you hit a creative speed bump, lose momentum, and get disoriented, go through these overarching guidelines again—part mantra, part philosophy, part pep talk. There are ten, so it will look good carved into marble tablets (optional).

10.3.1 Honor the Art

Our mix ideas must be inspired by and supportive of the music. Don't force ideas for things you've always wanted to do as an engineer into a song that doesn't need the effect. Listen intensely to the multitrack material for each song, looking for insight into the songwriter, the band, their guest artists, and their followers. They have a passion for something so strong that they've put their vulnerable selves out in the open for all to see and hear in the form of this recording. Know what it's about, know its place in the history of recorded music, honor all their influences, and dare to be different as appropriate. Make the mix arrangement support the musical aspirations. Our use of effects can sweeten harmonies, but we must know when to defer to deliberate dissonances. We can use EQ, distortion, delays, and reverbs to embellish, announce, emphasize, foreshadow, contrast, advance, or in any other way enrich the listening experience, but we are successful only if we have a deep understanding of the musical tactics and the emotional intentions of the artist. You must play the recording studio as a musical instrument, listening to the band and honoring the songwriter every bit as much as the rest of the band members do. Music matters; technical gimmicks do not.

10.3.2 Stay Balanced

Balancing a mix isn't a one-time, set-it-and-leave-it process; it is a constant part of the mix session. From the beginning of the session to the end, we must work hard to ensure that the key tracks—typically the vocal, snare, kick drum, and bass—remain easy to hear and enjoyable without effort from the listener. Any change to level, panning, EQ, compression, reverb, or any other effect on any track should be followed by a quick listen and tweak to any and all faders as necessary to keep the mix balanced. No matter how ornate and complicated your multitrack mix becomes, always be sure that the core elements—that lead vocal and/or soloist, the snare, the kick, and the bass—stay audible and live at relative levels that make musical sense. Never allow these fundamental tracks to get lost under a blanket of guitars, a cloud of distortion, or an ocean of reverb. Add necessary mix complexity while always tending to fundamental balance.

10.3.3 Attention: Front and Center

The vocal, the snare, the kick drum, and the bass are essential elements of most styles of music. They should be dead center—or close to it—most of the time. If you occasionally dare to locate key tracks like the lead vocal, snare, kick drum, or bass elsewhere, you should have a very good reason, and it should probably be temporary. It is a fundamental part of listening that we humans turn our heads and look toward important sounds. Our hearing system is most acute when we face the sound head-on. When we put the important parts of the mix front and center, we honor that basic human urge to face the music.

There are also practical benefits. Data compression encoder/decoder schemes generally dedicate more data to a dead-center element than to a slightly off-center element. Mastering engineers have greater technical control and creative flexibility when the core stays centered. And for the listeners, dead-center phantom images have the convenient feature that there is a healthy level for that track in both the left and the right loudspeakers. Listeners sitting off center will hear—at least at some reasonable level—all of the key tracks. Panning an essential track like the vocal off to the right risks the left-of-center listener to hear a mix with a weak, unintelligible vocal. For the most part, keep the important stuff front and center.

10.3.4 Look Both Ways Before You Mix the Track

Tracks interact! The sound of track 96 is influenced by what is going on in those other 95 tracks. And—back at ya—the sound of track 96 influences qualities of those other 95 tracks. You either build a mix of tracks that form a mutual admiration society, or you stuff them together into a barroom brawl.

As we build a mix, fader by fader, track by track, effect by effect, we look *outward* from each new signal to assess its impact on the other sounds, and we look back *inward* from all the other tracks to assess how they influence the sound of the new mix element.

This mutual influence suggests two rituals, which will keep your head spinning: first, when you push up a new fader, don't just listen to that new sound, tempting as it may be—not yet, anyway. Listen first to the new track's impact on the other tracks already present in your mix. Listen from the new track outward.

For example, when you push up the fader for the snare track, don't just listen to the snare; listen to what the introduction of the snare track does to the overall sound of the kick drum and the rest of the drum kit. With leakage from the rest of the kit into the snare microphone(s), the snare track is sure to interact with the overall drum sound. When you introduce the keyboard part, listen first for any change it invokes on the vocal, the snare, the guitars, and elsewhere. The keyboard track isn't just about the keyboard. It interacts with the other tracks, too.

Second, once you're satisfied that the new track isn't undoing all you've achieved in the rest of the mix, turn your attention to the sound of the new track itself. Process the track as needed to fulfill your vision for the mix, but don't fail to notice how the other tracks influence this one. Sure, you might press the solo button to evaluate this track in isolation, but you make your final mix decisions based on the full context of the mix. You must decipher what the other tracks do to the new track and make any necessary adjustments. Listen from the other tracks inward.

10.3.5 Get Louder, Without the Fader

Place faders at the lowest levels that make technical and musical sense and make your tracks easier to hear through more sophisticated means than simple volume. Turning it up is too easy and mucks up the rest of the mix. One loud track makes it hard to hear the rest of the mix. Alternatives to volume include strategic applications of ambience, reverb, compression, gating, EQ, and more. Tailoring other qualities in the sound to keep them audible is more work and requires mastery of the subtle aspects of mixing, but resisting that urge to simply push up the fader means the other tracks won't have to play catchup.

10.3.6 Contrasts Communicate Clearly

A passion for power and the desire to make a strong impression tempts the mix engineer to hype each and every track—to make the mix loud by making each and every track loud, which is a mistake. Thrill the audience with volume by including moments of quiet. Make things energetic through moments of calm. Make some tracks bright by making others less so. Make some mix elements wide by keeping others point-source-narrow. Pull some tracks forward, close, and in-yer-face by placing others more distant. You'll best communicate the extraordinary by including a tasteful amount of ordinary.

10.3.7 Change Is Emphasis

It is generally easier to hear things that are changing within a mix than elements that are static—level, panning, spectral content, delay, amplitude modulation (tremolo), frequency modulation (vibrato), reverberation, and so on. Dynamic effects are very effective—tremolo, flanging, chorus, autopanning, and similar. These effects have time-varying qualities built in, making them easier for listeners to pick out of a competitive, crowded mix. When two similar-sounding tracks fight to be heard, give one a bit of tremolo, and—courtesy of this attention-getting change—the two tracks become different enough to be heard independently.

Build in song form–inspired changes to your mix as well. A snare drum that gets a dose of extra ambience each chorus, returning to a slightly drier sound each verse, can offer the listener—even subliminally—an extra bit of sonic thrill courtesy of that slight change. A stereo piano that snaps to mono on the solo seems to grow stronger when it most needs it. The best mix isn't a fixed collection of settings. It is an interactive performance—the mix moves with the music.

10.3.8 Perfection Is a Cruel Distraction

The mix engineer is required to keep track of all levels of detail in the mix and to be zoomed into the highest resolution possible. That mastery of the details serves only one purpose: to maximize the musicality of the overall mix.

Visual perfection is all but irrelevant. Staring at a screen full of bright colors and flickering animations, we are tempted to click it until it "looks" right—everything lined up and perfect. But the visual is supposed to be a tool for us in the creation of something sonic. What we *see* matters nothing if it doesn't help us improve what the end user *hears*. The temptation for some sort of visual perfection is strong, but off-mission for the mix engineer. While the meters and flickering lights beckon for our attention, we must defer to our ears to know what matters most. Sometimes it is best to close your eyes and listen.

Sonic perfection can be the enemy of sound art. Detail-obsessed, we are tempted to right every audio wrong we find. We'll want to replace every ugly sound, pitch-correct every sour note, scrub out the noise between every phrase, edit the groove into a mathematically flawless tempo, and keep going. Just because we *can* doesn't mean that we *should*. Performance variations, blue notes, and messy sounds can be what the listener remembers most about a mix. Emotions often live in the near misses and the outright mistakes. Great sounds are, well, great. And bad sounds can be so bad that they are, um, great. It's middle-of-the-road mediocrity—where we fail to inspire the listener—that curses our mixes. Doing what everyone else does, playing it safe, and forcing your mix to imitate the same sounds that you hear all the time in other productions is a sure path to mediocrity. Keep your ear on the big picture, the finished product, the overall mix. Achieve a better artist-to-listener connection by focusing on the human expression of the performers over any perceived technical perfection in the mix.

10.3.9 Limits Are Opportunities

You know it's gonna get good when you are backed into a corner. Great music happens when musicians push the limits and bend the rules. A never-ending desire to be unique and to be better pushes all creators—songwriters, guitarists, painters, architects—to do something no one has ever thought of, to use materials and instruments in ways no one thought possible, to find the limits of their own abilities, and to stretch just a tick beyond. We must mix with the same work ethic and the same creative drive.

When we are asked to choose between seemingly limitless possibilities, we wander, and the mix sags. When we find our options narrowed or when a device can't crank it any further, we shouldn't complain about not having another plug-in or a better box. Instead, we focus, we prioritize, and the music soars. Step away from the long list of presets on the digital reverb and convert the bathroom into a reverb chamber. Lacking convenient presets, chambers make you work—and think—for your reverberant reward. Stop buying more plug-ins and signal processors until you've mastered the ones you already own—until you've pushed them in every conceivable direction for seemingly every mix application. Every great studio has the same basic tools. You need to know how to use them well, in usual ways. But you'll be a better mixer when you find out what each and every signal processor can't quite do and you explore your way around it. Great instruments respond well to unusual performance gestures. Our creativity grows more productive when we push on the edges. Your studio is a musical instrument. Know how to coax it to its expressive limits.

10.3.10 None of the Above

You've followed the rules—now break them. Rules aren't very durable in art. Know what works so that you know the effect of deliberately doing it wrong. This advice does not give you permission to be sloppy! For every mix gesture that is a rule-bender, know the risks associated with the decision. Then be sure that it has an impact on the listener that supports the intent of the artist, or compensate as needed elsewhere in the multitrack arrangement. When the guitars drown out the vocals (failure to follow the "Stay balanced" credo), you sell fewer discs and downloads. When the lead vocal is panned hard right (a violation of the "Attention: front and center" credo), drivers and passengers listening in the car sitting off center have a compromised experience. But if the guitars might be more important than the lyrics for this one tune, for this band, and for their fans, you might get away with it. If the first verse is a flashback or dream sequence, maybe this verse, and this verse alone, should be panned off center for dramatic impact. When you have a reason—and you know the risks—feel free to go against the norm, stick it to the man, defy, protest, surprise, provoke, rebel. Have fun.

No amount of mix inspiration can be realized if you don't hook things up correctly and use available resources wisely. Philosophically, there are two approaches to adding effects to a mix: parallel and serial processing.

A.1 PARALLEL PROCESSING

Consider first the use of reverb on a vocal track (an effect discussed in detail in Chapter 9). The right dose of reverb might embellish and support a vocal track that was recorded in a highly absorptive room with a close microphone. It is not merely a matter of support, however. A touch of just the right kind of reverb (known in the studio world as "magic dust") can enable the vocal to soar into pop-music heaven, creating a convincing emotional presence for a voice fighting its way out of a pair of loudspeakers. Quite a neat trick, really. The distinguishing characteristic of this type of signal processing is that it is *added* to the signal, it does not *replace* the signal. We hear the recorded signal plus the output of the reverb.

This structure is illustrated in Figure A.1. The dry (i.e., without reverb, or more generally, without any kind of effect) signal continues on its merry way through the DAW or the console as if the reverb were never added. The reverb itself is a parallel signal path, beginning with some amount of the dry vocal, going through the reverb processor, and returning elsewhere on the mixer to be combined to taste with the vocal and the rest of the mix.

A.2 SERIAL PROCESSING

Consider, in contrast, the use of equalization (described in Chapter 3). A typical application of equalization is to make a spectrally mediocre or problematic track beautiful. A muddy floor tom, courtesy of a narrow cut around 180 to 240 Hz, is made clearer, more open. An edgy trombone gets a carefully placed, gentle roll-off beginning somewhere high, around 8–12 kHz to become more listenable. This type of signal processing changes an undesirable sound into a new and improved version. In the opinion of the engineer who is turning the knobs, it "fixes" the sound. We don't want to hear the problematic, unprocessed version any more: just the improved, equalized one. We want to hook it up so that the processed sound *replaces* the old sound.

To do this, the signal processing is placed in series with the signal flow, as shown in Figure A.2. Adding clarity to a floor tom is not so useful if the muddy drum sound is still in the mix as well. And the point of equalizing the trombone

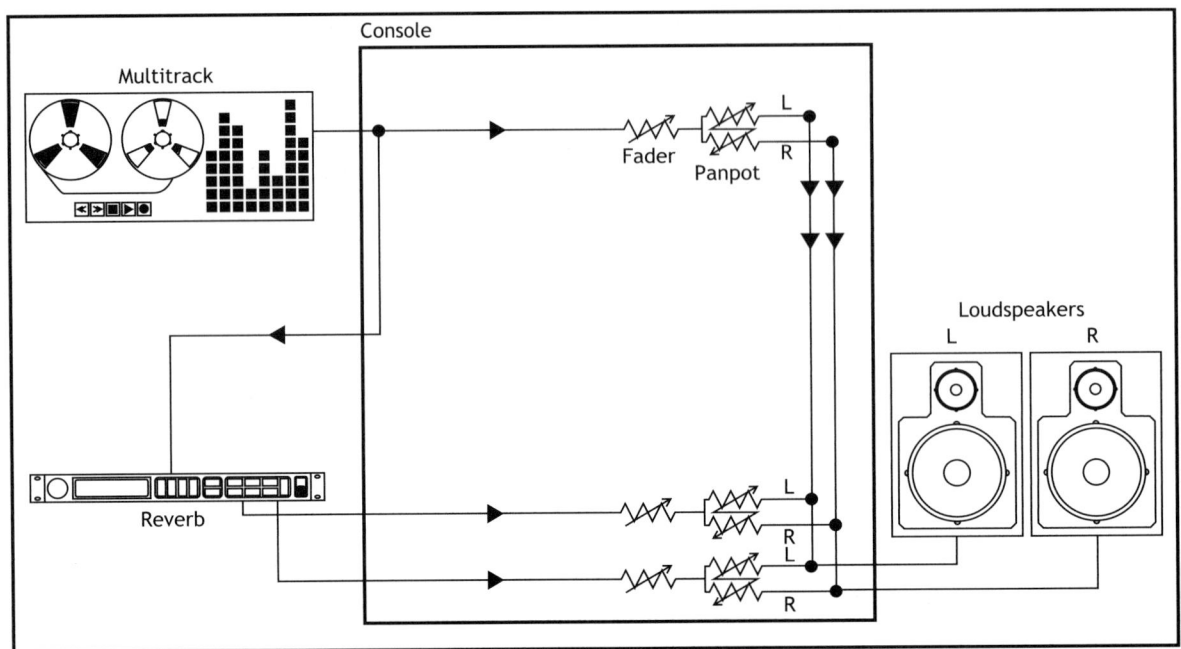

FIGURE A.1
Parallel signal processing.

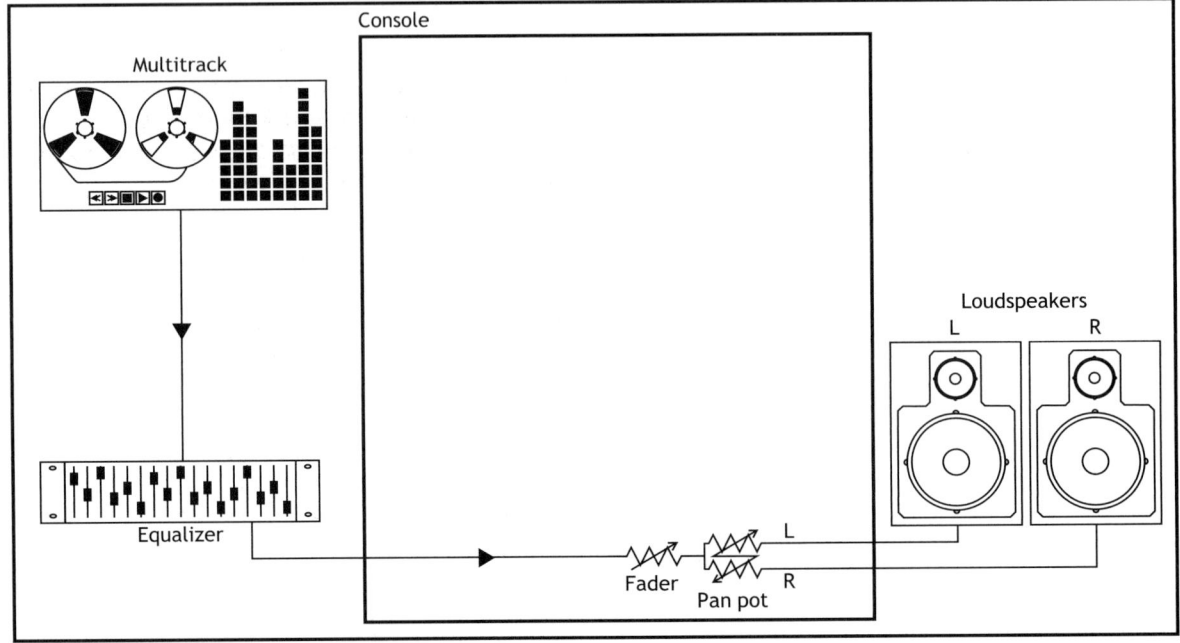

FIGURE A.2
Serial signal processing.

track was to make the painful edginess of the sound go away. The equalizer is dropped in series with the signal flow—between the multitrack machine and the console, for example, so that only the processed sound is heard. For the equalizer, it is muddy or edgy sound in, and gorgeous, high-fidelity sound out. Equalizing, compressing, de-essing, wah-wah, distortion, and such are all typically done serially so that your listeners hear the affected signal and none of the unaffected signal.

Parallel processing, like reverb, adds to the sound. Serial processing, like EQ, replaces the sound.

A.3 EFFECTS SENDS

Not surprisingly, these two flow strategies—parallel and serial—require different signal flow approaches in the studio. For parallel processing, some amount of a given track is sent to an effects unit for processing. Enter the *effects send* (see Figure A.3).

Also known as an *echo send* or *aux send* (short for auxiliary), it is a simple way to tap into a signal within the mix environment, be it DAW or console, and send some amount of that signal to some destination. Probably available on every strip of the console or DAW, the effects send is really just another fader. This particular fader determines the level of the signal being sent to the signal processor. Reverb, delay, and such are typically done as parallel effects and therefore rely on effects sends. Figure A.3 shows the mixer's realization of the parallel structure introduced in Figure A.1.

FIGURE A.3
The effects send creates parallel signal processing.

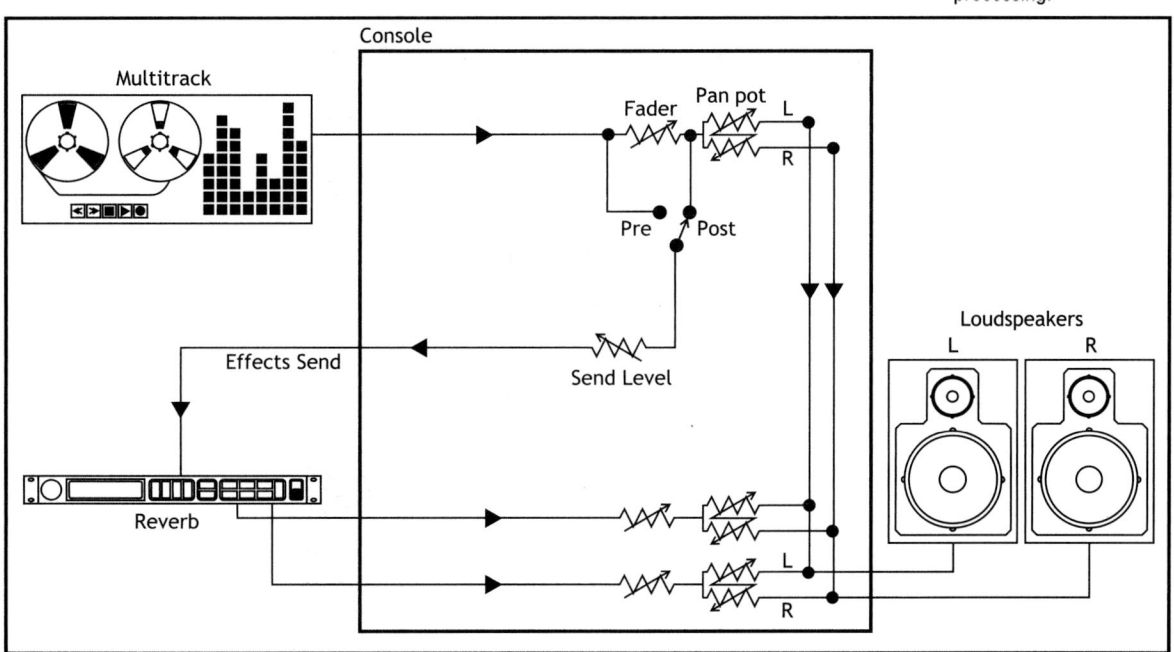

There is more to the echo send than meets the eye, however. It is not just an "effects fader." An important benefit of having an effects send level control on every strip of the DAW or console is that, in theory, you can send some amount of every component of the project to a single effects unit. To give a common example, if your studio invested several thousand dollars on a single, super-high-quality, sounds-great-on-everything sort of reverb, then you are probably going to want to use it on several tracks. Unless your studio has several very high-quality (i.e., very expensive) reverbs, it is not practical to waste it on just the snare, or just the piano, or just the vocal. The effects send makes it easy to share a single effects unit across a number of tracks. Turn up the effects send level on the piano track just a little to add a small amount of this reverb to the piano. Turn up the effects send level on the vocal a lot to add a generous amount of the same reverb to the vocal. In fact, the effects send levels across the entire console can be used to create a separate mix of all the music being sent to an outboard device. It is a special purpose mix that we do not usually listen to directly; it is a mix for the reverb, not for any of us. And the reverb takes that particular mix—that particular combination of piano, vocal, ukulele, and so on—and introduces reverberation accordingly. The hotter the signal going into the reverb unit, the more reverberation generated by the reverb algorithm.

As we are rarely satisfied with just one kind of effect in a multitrack production, you will probably want to employ a number of different signal processors within a single project. Each one of these effects devices might need its own effects send. That is, you might have one device that has a lush and long reverb dialed in; another that is adding a rich, thick chorus effect; and perhaps a third unit that is generating an eighth note echo with two or three fading repetitions. The lead vocal might be sent in varying amounts to the reverb, chorus, and delay, while the piano gets just a touch of reverb and the background vocals get a heavy dose of chorus and echo and a hint of reverb.

The mix environment, hardware or software, must have more than one effects send to do this. In this particular case, three effects sends are required on each and every strip of the mixer. The solution, functionally, is that simple: use equipment with more effects sends. It is an important feature to look for on consoles and digital audio workstations as the number of effects sends available largely determines the number of different parallel effects devices one can use simultaneously during any mix.

A.3.1 Prefader Sends

A useful feature of many aux sends is that they can grab the signal before (i.e., pre) or after (i.e., post) the fader (see Figure A.3). Though rarely a mix issue, the prefader send is desirable for creating headphone mixes. In this way, performers in headphones can play along to their mix without being distracted by fader rides done in the control room. With a prefader send, the headphone mix is independent of the levels in the control room mix.

A.3.2 Postfader Sends

When you are using the aux send for effects, not headphones, you'll typically choose a postfader send. Consider a very simple two-track folk music mix-down: fader one controls the vocal track and fader two controls the guitar track (required by the folk standards bureau to be an acoustic guitar). The well-recorded tracks are made to sound even better by the oh-so-careful addition of some room ambience to support and enhance the vocal while a touch of plate reverb adds fullness and width to the guitar (see Chapter 9 for more on the many uses of reverb).

After a few hours—more likely five minutes—of tweaking the mix, the record label representatives arrive and remind us that, "It's the vocal, stupid." Oops; we are so in love with the rich, open, and sparkly acoustic guitar sound that the vocal track was a little neglected. The label's issue is that listeners will not be able to reliably understand all of the words. It is pretty hard to sell a folk record when the vocals are unintelligible, so it has to be fixed. Not too tricky though. Just turn up the vocal.

Here is the rub: although pushing up the vocal fader will change the relative loudness of the vocal over the guitar and therefore make it easier to follow the lyrics, it also changes the relative balance of the vocal versus its own reverb. Turning up the vocal leaves its prefader reverb behind. The dry track rises out of the reverb, the vocal becomes too dry, the singer is left too exposed, and the larger-than-life magic combination of dry vocal plus heavenly reverb is lost. The quality of the mix is diminished.

The solution is the postfader effects send. If the source of the signal going to the reverb is *after* the fader (see Figure A.3), then your fader rides will also change levels of the aux send to the reverb. Turn up the vocal, and the vocal's send to the reverb rises with it. The all-important relative balance between dry and processed sound will be maintained. Effects are generally sent postfader for this reason.

You are really making two different decisions for the vocal: the level of the vocal appropriate for this mix and—separately—the amount of reverb desired for this vocal. The fader determines the vocal level. The aux send determines the corresponding amount of reverb. Flexibility in solving these two separate issues is maintained through the use of the postfader echo send. The occasional exception will present itself, but generally headphone mixes use prefader sends and effects use postfader sends.

If your mixer is hardware-based, your project may eventually run out of aux sends. Fear not. You've got an opportunity to overcome this apparent hardware shortcoming. The trick is to use the track assignment network to get to additional sends. Sending a signal to track bus 1 is the way to record to track 1, of course. When mixing, not multitracking, you can use that track bus as a way of accessing additional effects units.

Consider the production challenge of adding a delay to the ukulele, using an additional delay unit not previously connected to the mixer. Making matters worse, all aux sends have been used. The ukulele signal is sent to the stereo mix, of course. In order to add a new delay effect to the instrument, send some of it to track bus 1 as well as the mix bus. Patch track bus 1 send into the delay input. Patch the delay output into an available return to the mix bus on the console. In this way, the ukulele goes to both the mix bus and the signal processor, using repurposed track busses as extra aux sends on the console.

A.4 INSERTS

Parallel processing of signals requires the thorough discussion of aux sends as mentioned previously; serial processing has a much more straightforward solution. All that is needed is a way to place the desired signal processor directly into the signal flow itself.

One approach is to crawl around behind the gear, unplugging and replugging as needed to hook it all up. Of course, it is preferable to have a patch bay available. In either case, you just plug in the appropriate gear at the appropriate location. Want an equalizer on the snare recorded on track 3? Then, track 3 out to equalizer in; equalizer out to monitor module 3 in as was shown in Figure A.2. Placing the effects between the multitrack output and the console mix path input is a pretty typical signal flow approach for effects processors.

Adding additional processing requires only that they are daisy-chained together. Want to compress and EQ the snare on track 3? Simple enough: Track 3 out to compressor in; compressor out to equalizer in; equalizer out to monitor module 3 in.

Elaborate signal-processing chains are assembled in exactly this manner. If you are hooking up a compressor and want to use the console's built-in equalizer, it gets a little trickier: track 3 out to compressor in; compressor out to monitor module 3 in. That's all fine and good if you don't mind having the compressor compress before the equalizer equalizes. There is nothing universally wrong with that order for things, but sometimes we want equalization before compression.

Enter the insert, shown in Figure A.4. The insert is a patch access point within the signal path of the console that lets the engineer insert plug-ins or outboard processing right into the mixer. It has an output destined for the input of the effects device, generally labeled *insert send*. The other half of the insert is an input where the processed signal comes back, called *insert return*. Using this pair of access points, signal processing can be placed within the flow of the mixer.

Using the insert or patching the processing in before it reaches the console are common serial-processing techniques.

A.5 HYBRID PARALLEL EFFECTS

Let's summarize the signal flow philosophies so far: effects sends are used for parallel effects, and inserts are used for serial effects. You know we need to bend these rules a bit, don't you?

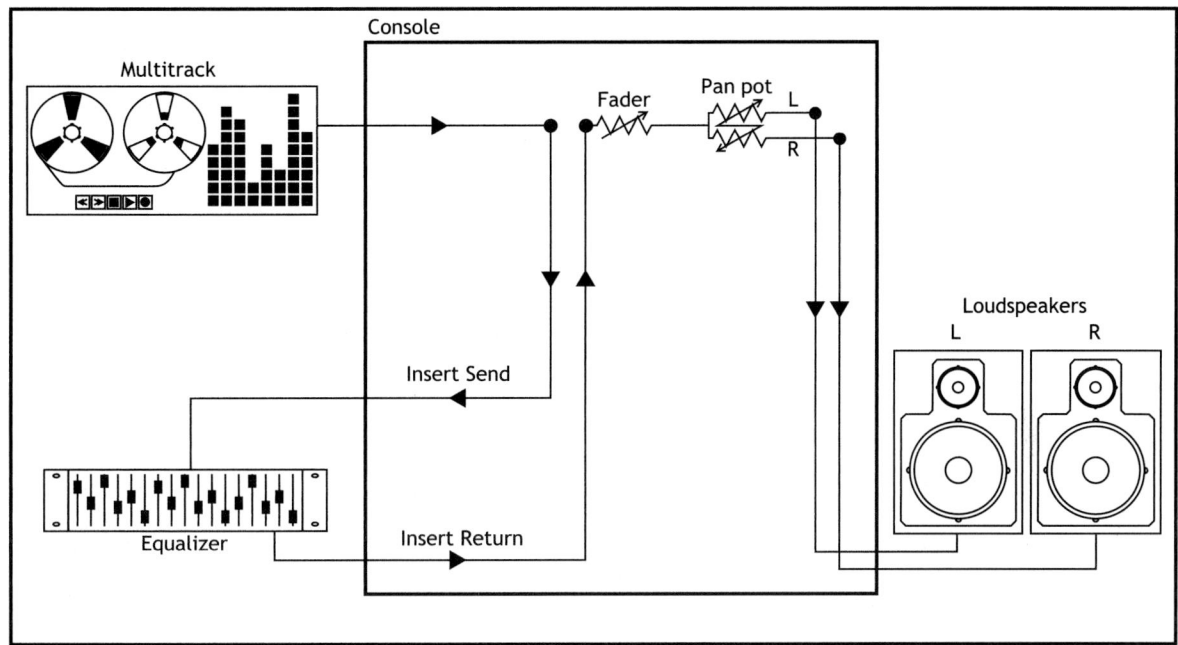

FIGURE A.4
The insert creates serial
signal processing.

Compression and expansion signal processing sometimes (not always!) motivates a kind of hybrid parallel structure. Parallel compression and expansion refers to the signal routing strategy in which you monitor both your unprocessed *and* your processed signal.

You hear the uncompressed snare and the compressed snare. You hear the open room tracks and the gated room tracks. That's a kind of parallel effect.

But because they are dynamics processors in which the action of the signal processing depends very directly on the level of the signal going into the compressor or expander, we are a little nervous to use aux sends. Aux sends have those pesky aux send level controls. If we use an aux send to feed a dynamics device, then raising or lowering the send level is functionally the same as lowering or raising the threshold parameter. And fader rides on a postfader send will also change the way the signal hits the dynamics processor.

Two solutions exist. These count as advanced signal flow, so don't do this until you are quite comfortable with the traditional uses of sends and inserts discussed earlier.

A.5.1 Insert Send Only

On an analog mixing console, the insert send is used to get audio into the compressor or expander. However, that compressor or expander output is not then returned via the insert return, as expected. Instead, the output of the compressor/expander feeds any available return to the mix bus: a spare monitor fader or an effects return. Nothing gets patched back into the insert return. In this way, the insert send acts as a prefader, unity gain source of the signal to be compressed or expanded without interrupting its flow.

The uncompressed or unexpanded track now heads to two places: it continues on its way through the mixer without compression or expansion, and it is split at the insert send and routed to the compressor or expander. You now have the unprocessed signal under your control at the original fader and the compressed or expanded signal under independent control at the new return—parallel compression and expansion.

A.5.2 Prefader Effects Send

Working on a digital audio workstation, any plug-in inserted behaves normally, fully replacing the sound with the output of the effect, which won't lead to any parallel effects. We need a different way to hook up parallel compression and expansion. Instead, an effects send is used to get into the signal processor. Importantly, that send is prefader, and the send level control is generally bumped up to unity gain, never to be touched again. That step feeds your signal to the dynamics processor, and any fader rides you perform do not affect the signal's interaction with threshold in the processor. The output of the compressor or expander now feeds a spare strip in your DAW and hits your mix bus under the control of its own fader, pan pot, mute, and any added effects—parallel compression and expansion.

What's good for the DAW is often good for the console. The unity gain, prefader send strategy is also perfectly valid on an analog mix surface. When an insert send isn't available, this will do the trick. Just be sure to make it a prefader send, and—unlike the other effects sends you'll have going—don't touch the send level to the compressor or expander or you'll find you radically change the effect. Set the send level (typically to unity gain) and leave it there!

It's confusing at first. When we want more or less reverb, the adjustment is to raise or lower the associated aux send. But when we want more or less parallel compression or expansion, don't touch that aux send! Simply raise or lower the fader where the effect is returned. Dynamics processors, reacting to level, get special treatment when we need them in parallel with their source tracks.

A.6 LATENCY

Latency describes the signal processing delay associated with creating any effect in the digital domain. The more sophisticated the signal processor, likely the longer the latency. Be aware of latency. Although delay-based effects are supposed to have delay, we might be surprised at first to find that digital equalizers, compressors, distortion processors, pitch modifiers, and so on all introduce at least a little bit of delay. That delay may have unintended production consequences, leading to image shifts, phasing, and comb filtering.

Parallel effects always require careful attention to latency. In the all-analog domain, latency is rarely an issue, but when the work environment is digital or a mix of analog and digital, problems might arise. In parallel compression and expansion, for example, you are monitoring the same track in two different places, each with differing amounts of signal processing. That track hits your mix without the effect on one fader and with the effect on the other. Digital signal processors take some amount of time to implement their algorithm. Digital-to-analog converters used to get the signal out of the DAW and into an analog signal processor followed by the analog-to-digital converters needed to get that analog processor output back into the DAW introduce additional delay. It might be just a few samples of delay, but this processing latency that accumulates on the processed track means the parallel tracks are no longer time-aligned. Panned to different locations, the image pulls toward the earlier, less delayed track. Panned to similar locations, the combination may cause a phasey, comb-filtered sound to occur where one was not deliberately intended.

For parallel effects to be of use, you must either like the accidental spectral impact of the resulting comb filter—a happy accident unlikely to be the solution every time—or you must have a way to correct for the latency. Most high-end DAWs offer the ability to automatically correct for latency by sliding all the other tracks a little later in time or by advancing the individual processed track earlier by reading ahead on the hard disk. Either way, it must do so by knowing, measuring, or calculating precisely the correct amount of time shift needed to offset the latency.

In a pinch, if you have the time and patience, this step can also be done manually by nudging each processed track earlier until you hear it perfectly aligned. To do this, you can align it by flipping the polarity of one track, summing the two tracks to mono, and nudging the track in time to the point of maximum cancellation. Unflip the polarity back, and you can confidently mix the tracks knowing that any time offset between the processed and the unprocessed track has been minimized.

A.7 COMPUTER RESOURCES

Every signal processor we use in a DAW requires computer assets in the form of processing cycles and memory. And the very act of accessing the chips associated with this processing and storage means the computer has to move data—lots of data—around. It ain't easy. It ain't free. Mixing on a DAW requires good stewardship of computing resources.

The previous discussion on effects sends pointed out that a single outboard processor might be used across many tracks—the single extraordinary reverb used to decorate vocal, snare, *and* ukulele. The original motivation, in a world of outboard hardware signal processing, is to make the most of a single signal processor. A similar logic still applies to plug-ins.

It is true that owning a single plug-in reverb means you can use it several times at once; just click to instantiate another. In this way, an engineer might insert a different reverb on each and every track of their multitrack mix. Just because the DAW allows you to run reverbs in this way doesn't mean that you should. Every one of those reverbs requires its own necessary computing power, dwindling your DAWs finite stock of signal-processing capability.

Using effects sends to send multiple tracks to a shared reverb is vastly more efficient computationally. So although you should use as many different reverbs (or any other effect, for that matter) as your creative needs require to make the mix you envision, you should also look for ways to use aux sends and share effects among tracks where appropriate. Avoid using inserts to grab effects and dedicate them to that track and that track only. In this way, you preserve computing horsepower for running the rest of the effects and the rest of the session.

A.8 OFFLINE PROCESSING

Another strategy for easing the burden on the DAW's computing resources is to perform some signal processing offline. That is, you tweak the processor until you know it is correct, then record the processed audio onto another track. Active effects demand more computer resources than simply playing back a processed track.

If you run out of computing resources before you run out of mix ideas, print the ones you are most confident in so that you can deactivate the computationally expensive effects and just play back the affected results.

The downside to this approach is that you can no longer interact with the effects. Printed, they sit static in your mix. To adjust a parameter, you have to go back to the original track, with the original active effect, readjust it as needed, and then reprint the processed audio to a new track. It is a slow way to work, and such extra steps rob us of the momentum we need to stay creative while mixing.

Think of offline processing as a way to deal with those processes that don't require as much fine-tuning in relation to the rest of the mix and a way to clumsily coax a little extra signal processing output from your computer when resources are maxed out. It might enable you get an essential extra bit of inspiration into your mix, helping it compete with the mixes coming out of that bigger studio down the street.

We might mix for years, successfully creating recorded works of art, without fully appreciating what an audio signal actually is. But your technical mastery of all things audio will empower your creative power when you wish to experiment and bend rules. When you want to know more than enough about audio signals, read this appendix. It's optional bonus material for overachievers.

When a guitar is strummed, a drum struck, or a trombone blown, we know sound will follow. The motion of the soundboard of the guitar, the vibration of the heads of the drum, and the resonance of the air within the plumbing of the trombone ultimately drive the air nearest our eardrums into action. We hear the air vibrate near us due to a chain of events that started perhaps some distance away at any such musical instrument or sound source. It is a separate matter, but we likely hope the sound made is music.

B.1 MEDIUM

The air between a musical instrument and a listener is a springy gas. When squeezed together, it pushes back apart. If pulled apart, it snaps back together. Picture air as a three-dimensional network of interconnected springs, as in Figure B.1. Any push or tug at just one point causes the whole system to jiggle in reaction. A continuous vibration of any one particle leads to a corresponding continuous vibration of the whole system. Motion of one element causes it to compress and stretch neighboring springs, which in turn push and pull against other springs further down the line.

Sound in air is a pressure wave with compressions (increases in air pressure) and rarefactions (reductions in air pressure) analogous to the squeezing together and stretching apart of elements of this vibrating spring system. Particles of air push and pull on one another very much as if connected by springs. Displace a bit of the air in one location, such as onstage, and it causes a chain reaction throughout the space to the audience. As long as the source and receiver are near enough to each other, air motion at the instrument's location will eventually cause, however faintly, a bit of air motion at the listener's location.

Slight increases in pressure occur when air particles are squeezed closer together. A loudspeaker cone, kick drum head, or piano soundboard moving toward the listener will do this. Decreases in pressure occur when air particles are pulled apart—the loudspeaker cone, kick drum head, or piano soundboard moving away from the listener.

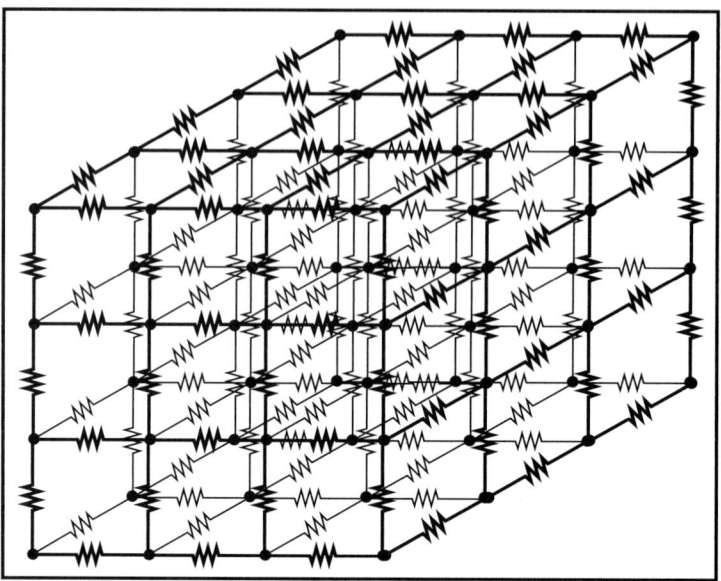

FIGURE B.1
Air is a springy medium, like a three-dimensional network of springs.

B.2 AMPLITUDE VERSUS TIME

The physiology and neurology associated with the human hearing system search constantly for changes in air pressure. Passing through the ear canal, changes in sound pressure push and pull on the eardrum, triggering a chain reaction that ultimately leads to the perception of sound. The pressure of a sound wave is the same type of air pressure associated with pumping air into a tire: PSI (pounds per square inch) in the some parts of the world or kPa (kilopascals) elsewhere. Micropascals (µPa) is the preferred order-of-magnitude expression of air pressure for sound that humans can healthily hear.

FIGURE B.2
A pure tone—a sine wave.

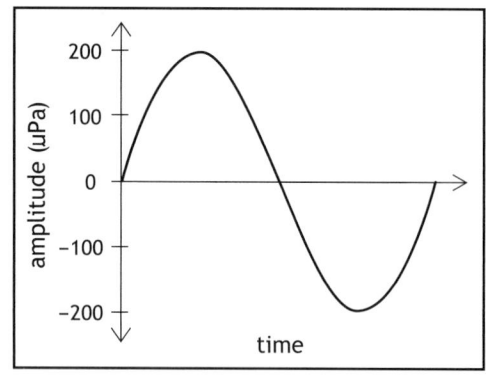

A common way to graph sound plots air pressure as the vertical axis and time as the horizontal axis. Such a graph describes sound at a single, fixed location in space. As sound occurs, the air pressure at that point increases and decreases several times per second. The familiar plots of sound in textbooks and comic books accurately portray this concept.

The screen of your DAW shows audio as a squiggly line across the computer screen. It might be a sine wave as shown in Figure B.2, or the more general waveform of Figure B.3. A line is drawn to zig and zag, up and down, describing air pressure as it is occurs over time. The higher parts of the curve represent instances of increased pressure (compression), and the lower parts represent decreased pressure (rarefaction). A straight, horizontal line describes no change in pressure (i.e., no sound).

A lack of sound does not mean there is no air pressure, rather only that there is no *change* in air pressure. When the air pressure is unchanging, our eardrums aren't moving. In other words, we have nothing to hear.

The horizontal axis in these figures meets the vertical axis, not at a pressure of zero micropascals, but at the ambient atmospheric pressure around you today. The negative amplitudes in these curves represent a reduction in air pressure below ambient, not negative pressure. The air pressure is always positive; sound represents varying degrees of positive. The precise value of pressure where the x-axis meets the y-axis is not an audio concern; it is a matter for those who track weather. The ambient pressure published in weather reports is the centerline for the pressure oscillations of our music.

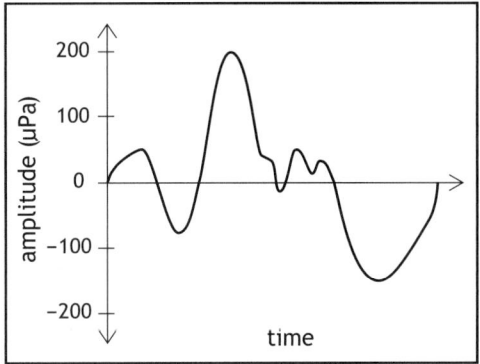

FIGURE B.3
A general waveform.

B.2.1 Amplitude Confusions

Discussing the amplitude of a signal, it is natural to want to assign it a numerical value. Reducing the amplitude of a signal to a single number gets a little tricky. What is the amplitude of the signal shown in Figure B.2? The plot shows that at it highest point, it reaches an amplitude of +200 µPa. Similarly, the lowest pressure shown is –200 µPa. One might correctly describe the signal as having a *peak amplitude* of 200 µPa or a *peak-to-peak amplitude* of 400 µPa. Because this signal is perfectly sinusoidal, the peak amplitude or the peak-to-peak amplitude fully describes the general amplitude of the signal, even though it is constantly changing. The amplitude of the waveform in between the peaks follows the known pattern of a pure tone.

The slightly more complicated waveform of Figure B.3 unravels this amplitude notation methodology. Its positive peak is still 200 µPa, but its negative peak is –150 µPa, with several intermediate positive and negative peaks in between. If this signal is a musical waveform, it will surely keep changing shape, with local maxima and minima that change as the song plays. There is no single consistent positive or negative peak. As most audio signals lack the perfect symmetry of a sine wave, a better way to express the amplitude of an audio waveform is required.

Perhaps the *average* amplitude would be helpful. This approach is frustrated by the fact that audio spends about as much time above zero as below. In the case of the sine wave (see Figure B.2), the average amplitude is exactly zero. No matter what the peak amplitude is (it may be raised or lowered by any amount), the average amplitude remains zero.

In search of a number that describes the amplitude and does not average zero, it might be tempting just to ignore the negative half of the wave. Averaging only the positive portion, a nonzero figure can at last be calculated. This issue remains problematic. The negative portion of the cycle also contributes to the perception of amplitude. Turning up the volume while music is being played back causes the negative portion of the waveform to become more negative still. More extreme amplitudes, positive or negative, may be interpreted as louder. The more extreme air pressure changes lead to more extreme motion of the eardrum. Humans are impressed by amplitude whether a pressure reduction pulls the eardrum outward or a pressure increase pushes the eardrum inward. It's amplitude either way. So the negative swings in air pressure must contribute to any numerical expression of amplitude as much as the positive ones and therefore should not be ignored.

Musical signals, though lacking the perfect symmetry of a sine wave, share this tendency to average zero. The springy air, in reaction to the driving action of a loudspeaker, compresses and stretches. Each pressure increase is followed by a pressure decrease. At the end of the song, the air returns to ambient pressure, the loudspeaker cone returns to its original position, and the eardrum returns to its starting point.

One way to allow the negative portion of the oscillating wave to contribute to the amplitude calculation is to average the absolute value of the amplitude. Make all negative amplitudes positive, keep all positive values positive, and find the running average. The resulting expression for amplitude can track the perception of amplitude reasonably well. *VU meters* do exactly this, averaging the absolute value of the amplitude observed over the preceding 300 milliseconds.

There is further room for improvement: *root mean square* (RMS). Measuring RMS amplitude properly allows both negative and positive parts of the wave to influence the resulting number for amplitude. RMS might best be understood by working through the acronym in reverse. *Square* the amplitudes to be measured, so that a positive value always results. Take the *mean* (a.k.a. average) value of the amplitudes observed. Finally, take the square *root* of the result to undo the fact that the contributing amplitudes were all squared before being averaged.

RMS amplitude is more convenient for scientists and equipment designers, as it is this type of average amplitude that must be used in calculations of energy, power, heat, and so on. Audio engineering rarely needs such precision. The simpler absolute value average of the VU meter is almost always a sufficient indicator of amplitude.

B.2.2 Time Implications

The amplitude versus time plot reveals fundamental information about audio waveforms. A pure-tone sine wave (see Figure B.2) consists of a simple, never-changing pattern of oscillation. Measure the length of time associated with each cycle to determine the waveform's *period*. Count the number of times it cycles each second for a determination of its *frequency*. Period is the time it takes for exactly one cycle to occur, with units of seconds per dimensionless cycle, or simply seconds. Frequency describes the number of cycles that occur in exactly one second, with units of dimensionless cycles per second. Therefore, units for frequency live entirely in the denominator (per second, or /s) and have been given the alternative unit of hertz (Hz).

Note that counting the number of cycles per second (frequency) is the opposite of counting the number of seconds per cycle (period). Mathematically, they are reciprocals:

$$f = 1/T \qquad\qquad (B.1)$$

and

$$T = 1/f \qquad\qquad (B.2)$$

where f = frequency, and T = period.

B.3 AMPLITUDE VERSUS DISTANCE

The springiness of air ensures that any localized changes in air pressure near a sound source will cause a chain reaction of air pressure changes, above and below the current air pressure, all around that source. Even a slight disturbance of air pressure will ripple outward. In order to describe the state of air pressure along some distance, a different pair of axes is needed: air pressure versus location or air pressure versus distance.

At a fixed instant in time, a plot is made of the air pressure as a function of its location in space. Figure B.4 shows such a snapshot. It looks familiar visually, though it is in fact quite different in meaning. The horizontal axis in Figure B.4 is distance in space. It is not what is shown on the screen of your DAW. It is a snapshot of your room when sound is playing within.

The amplitude versus distance expression of sound leads to another fundamental property of waveforms: *wavelength*, which is the distance traveled during exactly one cycle. Drive 55 miles per hour for one hour, and the distance covered is exactly 55 miles. Distance traveled can be calculated through the multiplication of average speed by time. The speed of sound in air (under normal temperature and pressure) is 344 m/s. The always-friendly metric system does fail us a bit here, as the speed of sound in feet per second is 1,130 ft/s. For rock and roll, it is often acceptable to round this down to an even 1,000 ft/s.

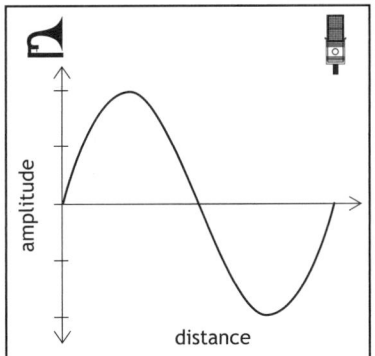

To calculate the wavelength, multiply this speed-of-sound figure by the appropriate amount of time. Recalling that the time it takes a wave to complete exactly one cycle is, by definition, its period:

$$\lambda = cT \qquad\qquad (B.3)$$

where λ = wavelength, c = speed of sound, and T = period.

FIGURE B.4
A snapshot in time shows amplitude over a distance from sound source to receiver.

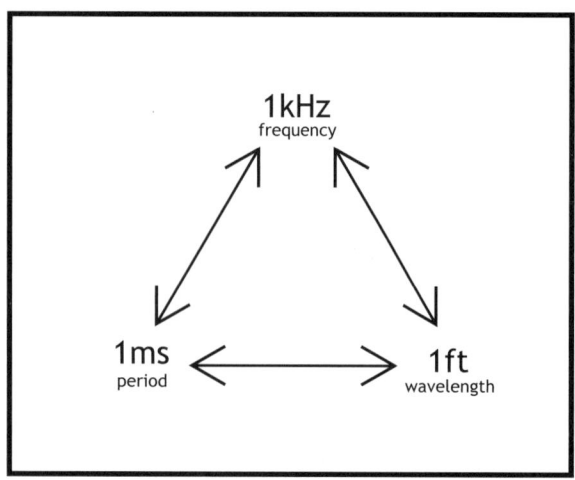

FIGURE B.5
The alignment of 1s.

Expressing wavelength as a function of frequency (f) requires substitution of frequency for period. Using Equation B.2:

$$\lambda = c/f \tag{B.4}$$

Precise calculations are straightforward, but it is worth noting that wavelengths can be juggled in one's head in the heat of a recording session without resorting to pencil, paper, or calculator, provided that the speed of sound sticks to the fair approximation of 1,000 feet per second.

A representative middle frequency is a 1-kHz sine wave. Using Equation B.4,

$$\lambda_{1,000} = (1,000 \, \text{ft/s}) / (1,000 \, \text{Hz}) \tag{B.5}$$

and recalling that the units underlying Hz are cycles per second (/s),

$$\lambda_{1,000} = (1,000 \, \text{ft/s})/(1,000/\text{s}) \tag{B.6}$$

This result leads to the final result for the convenient, approximate wavelength for a 1,000-Hz sine wave:

$$\lambda_{1,000} = 1 \, \text{ft} \tag{B.7}$$

This middle frequency, 1,000 Hz, which has a period of 1 ms, conveniently has a wavelength of approximately 1 foot. This alignment of "ones"—1 kHz, 1 ms, 1 ft—is a useful point of reference that an engineer can bring to every recording session (Figure B.5).

B.4 AMPLITUDE VERSUS FREQUENCY

Plots of amplitude versus time and amplitude versus distance are helpful and are used throughout this text. An important third way of describing signals must also be understood. When music is enjoyed, listeners are certainly aware that there are changes in amplitude over time at whatever location they currently occupy. Without a computer screen in front of them offering the information visually, listeners don't consciously pay attention to the fine details shown in the amplitude versus time plots.

The physiology of human hearing in fact analyzes sound as a function of frequency. The mammalian hearing system breaks sound up along the frequency axis. A pure tone is perceived as spectrally narrow and activates only a localized portion of the hearing anatomy. More complex sounds containing a range of frequencies, such as music, stimulate a broader portion of the hearing. Separating sound into different frequency ranges allows for the evaluation and enjoyment of sound across a spectral range. Humans simultaneously process the low-frequency sounds of a bass guitar in parallel with the higher-frequency sounds of a cymbal, all the while sorting out the complex detail of a vocal occupying a range of frequencies in between.

Amplitude versus frequency (Figure B.6) is therefore an important graphical representation of sound. This plot must make assumptions about space and time. Generally, location is fixed, creating a plot that represents the sound at one place only—perhaps the comfortable couch the listener uses when listening to his or her favorite music. In addition, time must be constrained to some finite duration.

The right-hand side of Figure B.6 shows the amount of amplitude in a signal as a function of frequency. Is it for the last second of the signal? The preceding minute? The entire song? These time increments are all perfectly valid. An engineer might want to know the spectral content of the signal for any window of time.

In fact, such a display can be updated continuously, as the audio occurs. *Real-time analyzers* (RTAs) do exactly this. When they are set to a "fast" setting, they describe the signal that just occurred over the last 100 milliseconds or so in the intuitive terms of amplitude versus frequency. When they are set to "slow," that window in time expands to about 1,000 milliseconds.

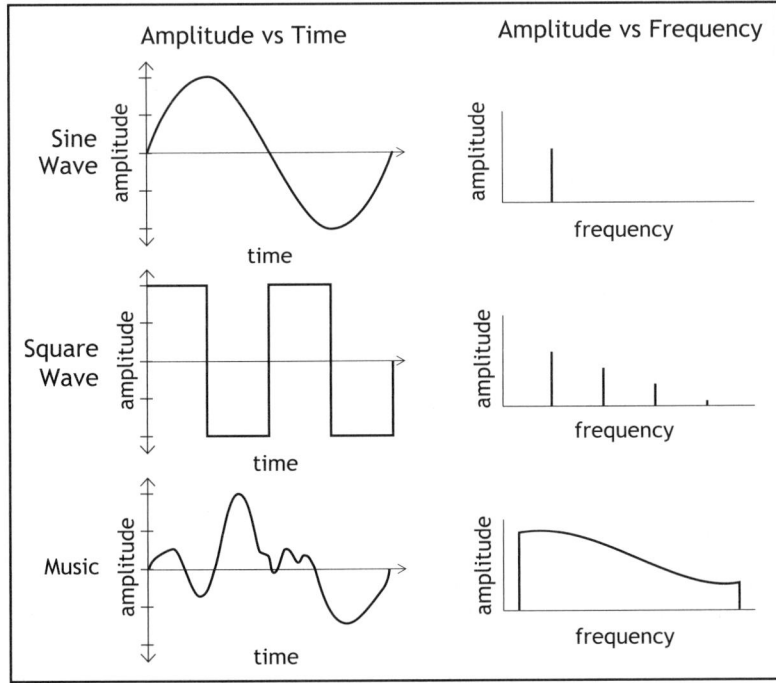

FIGURE B.6
The frequency content of a signal.

The amplitude versus frequency plot therefore represents the signal at a fixed location and for a specific duration, identifying the distribution across frequency during that part of the signal.

B.5 COMPLEX WAVES

Although vocals, guitars, and ukulele waveforms are of more musical interest, simple building-block waveforms are worthy of study (Figure B.7). The purest tone is a sine wave. This specific pattern of amplitude, repeating without fail, contains just a single frequency. The sine wave is plotted (Figure B.7a) using:

$$Y(t) = A_{peak} \sin(2\pi ft) \tag{B.8}$$

where $Y(t)$ is the amplitude (y-axis) as a function of time t (in seconds), A_{peak} is the peak amplitude (likely in volts or units of pressure), and f is the frequency of the sine wave (in hertz). For this and all waveform equations, readers more comfortable with degrees of phase instead of radians should simply

FIGURE B.7

Four 100-Hz waveforms: (a) sine wave, (b) square wave, (c) sawtooth wave, and (d) triangle wave.

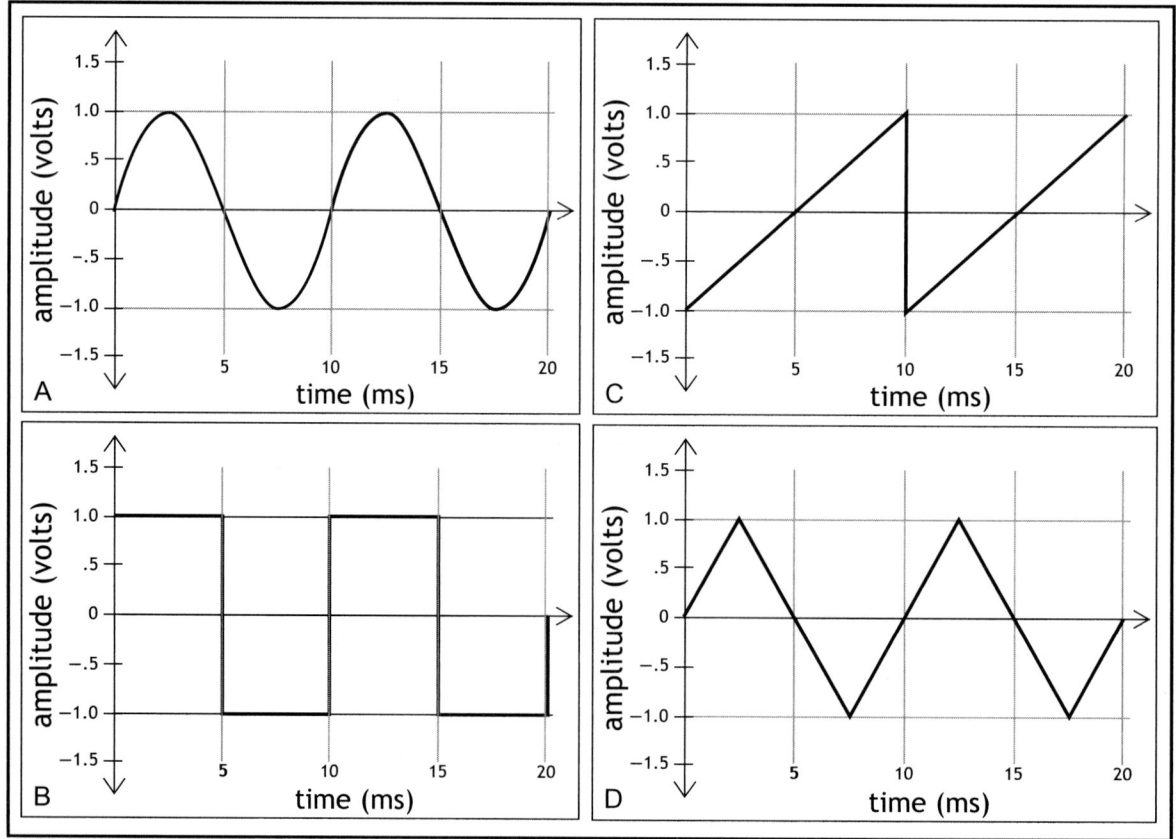

replace 2π radians with 360 degrees within the argument of the sin function. Therefore, Equation B.8 would become:

$$Y(t) = A_{peak} \sin(360ft) \qquad (B.9)$$

Sine waves may be combined (Figure B.8) to create more complex waves. Beginning with a fundamental frequency of 100 Hz, a second and third harmonic are added, each with unique amplitude. The resulting waveform is simply the continuous algebraic sum of the amplitude of each of these two sine waves as time goes by:

$$Y(t) = A_1 \sin(2\pi ft) + A_2 \sin(2\pi 2 ft) + A_3 \sin(2\pi 3 ft) \qquad (B.10)$$

where A_n is the amplitude of the nth harmonic. In Figure B.8,

$$Y(t) = \sin(2\pi ft) + 0.5\sin(2\pi 2 ft) + 0.25\sin(2\pi 3 ft) \qquad (B.11)$$

and f = 100 Hz. Any number of sine wave components may be part of this mathematical exercise. Specific recipes of sine waves should be noted, such as those of the square wave, the triangle wave, and the sawtooth wave.

B.5.1 Square Waves

A square wave is described by the following equation:

$$Y(t) = 4A_{peak}/\pi \sum_{n=1}^{\infty} (1/(2n-1)) \sin(2\pi(2n-1)ft) \qquad (B.12)$$

It is an infinite sum of precisely these harmonic components. The harmonics must fall at exactly these frequencies—these specific multiples of the fundamental. If any frequency is shifted, even a little, the wave becomes nonsquare.

As important as the frequency is, each harmonic must also have the specified amplitude. In this case, the amplitude of each harmonic, numbered n, decreases

FIGURE B.8
A 100-Hz complex wave with three harmonics.

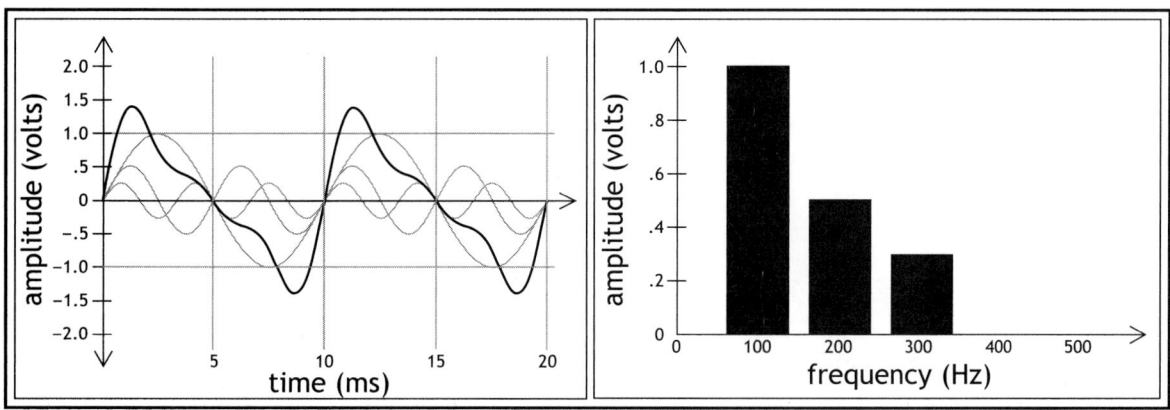

inversely with n. That is, the nth harmonic is scaled by the factor $1/n$. The term out front, $4/\pi$, serves to give the full bandwidth square wave a convenient peak amplitude of unity. If you increase or decrease the amplitude of any or several of the contributing harmonics, the waveform becomes less square.

It is important to note that the square wave contains only odd harmonics. Harmonics that are even multiples of the fundamental frequency are simply not part of the recipe. The presence of any amount of any even multiple of the fundamental frequency would make the resulting wave less square. Equation B.12 has become a little clumsy in an effort to force the harmonics to always be *odd* multiples of the fundamental frequency. The term $(2n - 1)$, which appears twice in the equation, enables the series to step through values of n and create only odd multiples of f. If the series is restated using only odd numbers, m, it might be easier to follow:

$$Y(t) = 4A_{peak}/\pi \sum_{m=1,3,5,...}^{\infty} (1/m)\ \sin(2\pi mft) \tag{B.13}$$

A square wave with a fundamental frequency of 100 Hz and a peak amplitude of 1 volt (Figure B.7b) uses Equation B.12 or B.13 to create:

$$Y_{100}(t) = 4/\pi\ [\sin(2\pi 100t) + (1/3)\ \sin(2\pi 300t) + (1/5)\ \sin(2\pi 500t) +$$
$$(1/7)\ \sin(2\pi 700t) + ...] \tag{B.14}$$

The significance of the harmonics is shown in Figure B.9. The contribution of ever-more upper harmonics, in strict adherence to the amplitudes and frequencies specified, is clear through visual inspection. The waveform becomes increasingly more square as the bandwidth reaches upward and the number of harmonics included in the summation grows.

This makes clear the need for wide-bandwidth audio systems when square waves (think MIDI, SMPTE, and digital audio) are to be recorded and transmitted. A cable that rolls off the high frequencies of the signal within will attenuate the necessary harmonics that make up a square wave, in effect making a square wave less square. A perfectly square wave is achieved only through the rather impractical inclusion of an infinite number of the prescribed harmonics.

B.5.2 Sawtooth Waves

The sawtooth wave might be considered a variation on the square wave theme. Retain both odd and even harmonics, continue to diminish the amplitude of each nth harmonic by $1/n$, rescale the overall amplitude to preserve unity peak amplitude, and a sawtooth wave with positive slope results (Figure B.7c):

$$Y(t) = -2/\pi \sum_{n=1}^{\infty} (1/n)\ \sin(2\pi nft) \tag{B.15}$$

A 100-Hz sawtooth wave is built up through increasing bandwidth in Figure B.10. Careful calculation through up to $n = 50$ is shown, but the proper sawtooth does not occur until $n = \infty$.

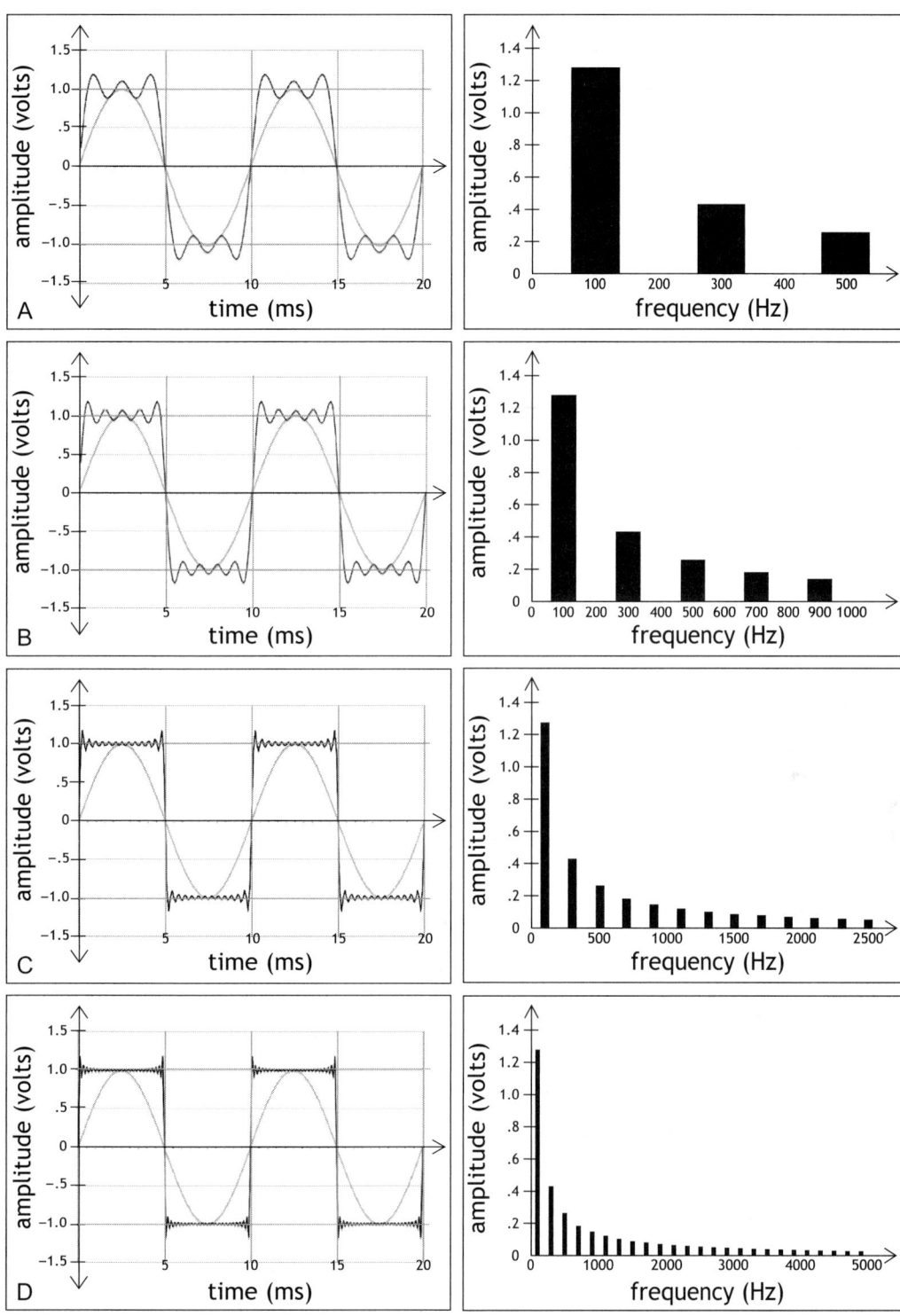

FIGURE B.9

A 100-Hz square wave through the addition of harmonics up to: (a) 500 Hz (3 harmonics), (b) 1,000 Hz (5 harmonics), (c) 2,500 Hz (13 harmonics), and (d) 5,000 Hz (25 harmonics).

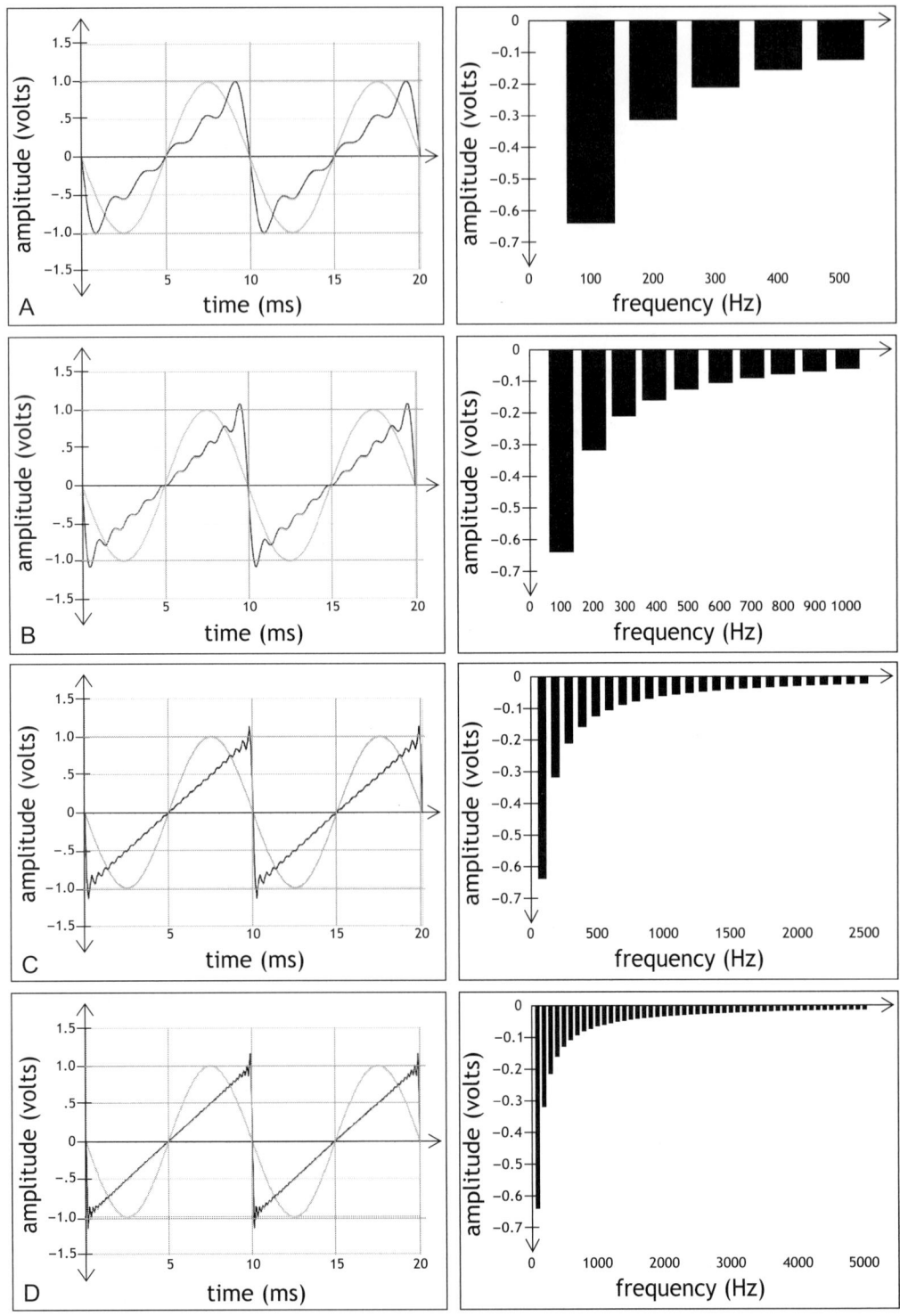

FIGURE B.10
A 100-Hz sawtooth wave through the addition of harmonics up to: (a) 500 Hz (5 harmonics), (b) 1,000 Hz (10 harmonics), (c) 2,500 Hz (25 harmonics), and (d) 5,000 Hz (50 harmonics).

The traditional beginning of a sine wave is the instant at which the amplitude is crossing up through zero toward positive amplitude. The minus sign in Equation B.15 dictates that all sine wave components of a sawtooth initially head in the negative direction instead. For this reason, comparison is made in Figure B.10 to a negative sine wave—that is, a sine wave multiplied by –1.

The sum of all sine waves in the equation (an infinite number of precisely scaled multiples of fundamental frequency f) causes the net wave to leap to –1 before steadily rising toward +1. The instant when the sawtooth wave reaches +1 is also the fortuitous instant when each and every component harmonic happens to be beginning a negative cycle anew. This symphony of sine waves crossing upward through 0 but multiplied by –1 causes the net amplitude to snap to –1 again. The pattern repeats.

B.5.3 Triangle Waves

The triangle wave (Figure B.7d) comes from a different set of carefully scaled odd harmonics:

$$Y(t) = 8A_{peak}/\pi^2 \sum_{n=1}^{\infty} \left(-1^{(n-1)}/(2n-1)^2\right) \sin(2\pi(2n-1)ft) \qquad (B.16)$$

In addition to the requisite scaling to achieve unity peak amplitude, and the use of the term $(2n - 1)$ to generate odd harmonics, notice the additional need for an alternating polarity among the harmonic components. The term $-1^{(n-1)}$ causes the harmonics to switch sign with each increment of n. The polarity of every other harmonic is positive; the polarity of each harmonic in between is negative. The summation of these particular components—some adding to the total while others subtract—leads to a triangle wave.

Figure B.11 demonstrates the significance of adding additional harmonics to the fundamental sine wave. As the amplitude of successive harmonics falls proportional to $1/n^2$, this complex wave is more dependent on lower harmonics than the square and sawtooth waves, which is evident in two ways. Note the towering significance of the lower harmonics on the right-hand side of Figure B.11. Note also how the wave obtains its characteristic sharpness and comes quite close to resembling the full bandwidth shape with just 13 harmonics.

Very much as multitrack music is built from a mix of component production elements such as drums, bass, keys, and vocals, individual pitched waveforms that make up each multitrack element are themselves made up of a specific mix of sinusoidal components. It is our job to make art from these humble ingredients.

B.6 DECIBEL

It is difficult to do anything in audio and not encounter the decibel. As discussed shortly, the decibel offers a precise calculation that quantifies properties of an audio signal in a very useful form. One may never trouble to dig out these equations and perform a decibel calculation during the course of a recording session. But the hardware designers and software code jockeys who create the effects

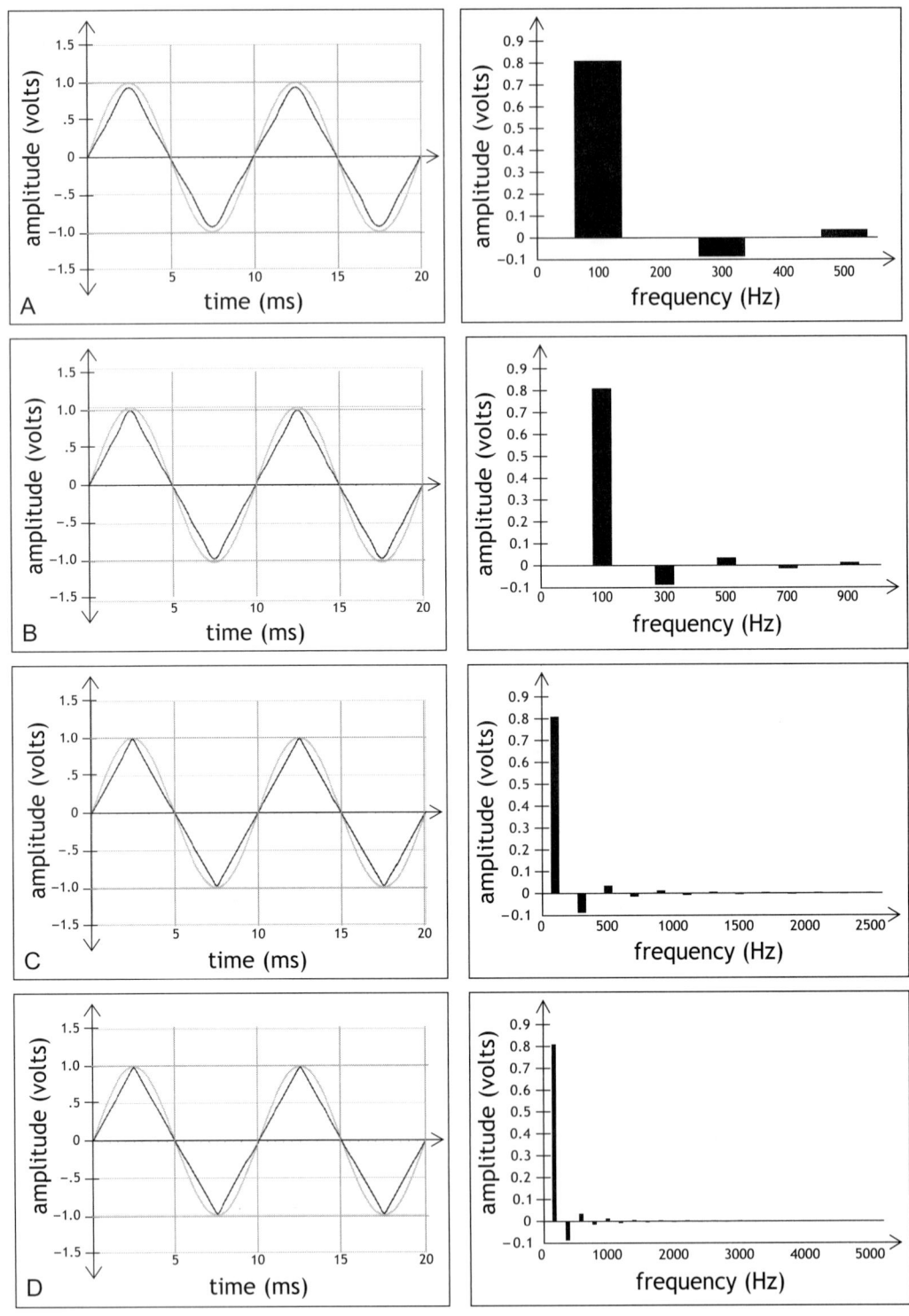

FIGURE B.11
A 100-Hz triangle wave through the addition of harmonics up to: (a) 500 Hz (3 harmonics), (b) 1,000 Hz (5 harmonics), (c) 2,500 Hz (13 harmonics), and (d) 5,000 Hz (25 harmonics).

processors and recording devices that fill the studio certainly do. If an audio engineer is to speak comfortably and accurately about decibels, it helps to know a little of the math that makes it possible. Those who are bored or frustrated by the math should at least know that someone went to a lot of trouble to find a way to express the level of the signal in a way analogous to the expression of pitch. The decibel offers a perceptually meaningful description of amplitude— one that the ears and brain can make sense of.

The decibel appears in some form on almost every faceplate and every user inter- face of every signal processor in the recording studio. Understanding the mean- ing of quantities in decibels is essential to understanding sound effects. There's an equation that absolutely defines the decibel (dB):

$$dB = 10 \times \log_{10}(power_A / power_B) \qquad (B.17)$$

The English translation of that equation goes something like, "The decibel is ten times the logarithm of the ratio of two powers." This straightforward statement is rich with meaning.

The equation for the decibel has two features built in. First is the *logarithm*. The mathematical properties of this function are considered in detail shortly, but it is important to understand the motivation for digging up the *logarithmic*, or *log*, function in the first place. The log is part of the decibel equation to make the math more convenient. It makes the vast range of amplitudes typical of audio much easier to deal with.

The second key element of the decibel equation is the *ratio* of powers within parentheses. The decibel equation uses a ratio so as to be consistent with the human perception of power and related quantities. The equation attempts to create a number that describes the amplitude of a musical waveform. For the decibel to be useful, the resulting number needs to have some connection to the human perception of this property of sound.

These two features—log and ratio—make the decibel a versatile and useful way to express the amplitude of our musical waveforms.

B.6.1 Logarithm

The log represents nothing more than a reshuffling of how the numbers are expressed. The two following equations are both true and say very nearly the same thing:

$$10^y = x \qquad (B.18)$$

$$\log_{10}x = y \qquad (B.19)$$

Equation B.18 is relatively straightforward: 10 raised to the power y gives the result x. For example:

$$10^3 = 10 \times 10 \times 10 = 1,000 \qquad (B.20)$$

The logarithm enables us to undo the calculation mathematically. Starting with the answer from Equation B.20, 1,000, the log function leads back to 3:

$$\log_{10}(1,000) = 3 \tag{B.21}$$

Said another way, Equation B.19 answers the question, "What power of 10 will give us this number, x?" To take the log of 1,000 as in Equation B.21 is to ask, "What power of 10 gives us 1,000?" The answer is 3: $10^3 = 1,000$, so $\log_{10}(1,000) = 3$.

What power of 10 will give us 1,000,000? With an eye for powers of 10, or perhaps with the help of a calculator, it is easily confirmed that the $\log_{10}(1,000,000)$ = 6. Ten raised to the sixth power gives us one million, as Equation B.9 would describe it. Now calculate the power of 10 that gives 100 trillion: $\log_{10}(100,000,000,000,000) = 14$.

Herein lies the motivation for logarithms in audio. They make big numbers—potentially very big numbers—much smaller: 100,000,000,000,000 becomes 14. It converts actor's salaries into football scores.

The log function is an acquired taste. Those with little or no exposure to them will likely find the logarithm awkward at first.

There's an interesting twist, so follow along in Figure B.12. The log function calculates the power of 10 needed to create a number—any number. So the $\log_{10}(100) = 2$ and the $\log_{10}(1,000) = 3$, and the log function can also find values in between. For example, the $\log_{10}(631)$ is about 2.8. In other words,

$$10^{2.8} = 631 \tag{B.22}$$

To know this, a calculator, a computer, a slide rule, a class geek, or some tables full of logarithm answers are required. These aren't calculations easily done in the head, with the help of counting on fingers and toes.

FIGURE B.12
The logarithm mathematically connects a potentially very large number to a much smaller number.

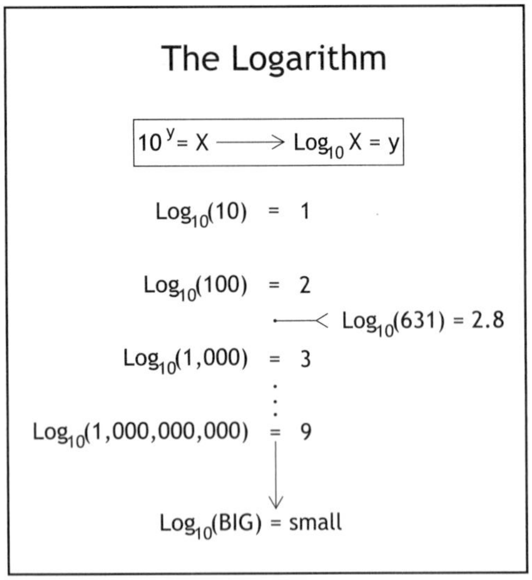

The Logarithm

$$10^y = X \longrightarrow \log_{10} X = y$$

$\log_{10}(10) = 1$

$\log_{10}(100) = 2$

$\log_{10}(631) = 2.8$

$\log_{10}(1,000) = 3$

$\log_{10}(1,000,000,000) = 9$

$\log_{10}(\text{BIG}) = \text{small}$

The logarithm mathematically connects a potentially very large number—and it can be any number greater than zero—to a much smaller number, which is useful because the range of amplitudes humans can hear is truly vast. The smallest sound pressure that average healthy humans can hear, rounded off for convenience, is about 20 micropascals. Compare that amount to the amplitude of air pressure associated with the onset of physical pain in our hearing system. (Please note: the risk of hearing damage starts well before the pain begins, so listen safely and wisely. Please don't risk hearing damage.) Pain starts happening at about 63,000,000 µPa (micropascals). The difference between detection and pain in the human experience of air pressure is many millions of micropascals. Listening to a conversation at normal levels might occur with amplitudes of around

20,000 µPa. It is reasonable to monitor a pop mix at about 630,000 µPa. One might occasionally crank it to more than 6,000,000 µPa. Even at this level, the neighbors aren't complaining, the drummer wants it louder, and, if it doesn't go on for too long, there isn't much risk of hearing damage yet. Jet engines and power plants are much, much louder still, on the order of hundreds of millions of micropascals.

The problem with these numbers is clear: They are too big to be useful in the studio. "Yeah, let me push the snare up about 84, pull the strings down about 6117, and see if the mix sits right at 1,792,000." This is too awkward; the decibel comes to the rescue. The mathematical log function exhibits the following helpful property:

$$\log_{10}(\text{BIG}) = \text{small} \tag{B.23}$$

The log of a big number results in a much smaller number. Human hearing is capable of interpreting a vast range of amplitudes. The numbers used to describe amplitude become unwieldy if not routinely subjected to the logarithm, so it's a fundamental part of the decibel equation.

B.6.2 Ratios

Also built into the decibel equation is a ratio (consult again Equation B.17). Mathematically, a ratio strategically reduces two numbers to a single, informative number.

Consider pitch. Musical harmony is built on ratios. The octave, for example, represents a doubling of pitch: a ratio of 2:1. The orchestra tunes (generally) to A440. Also identified as A4, this is the first A above middle C. A440 is a musical note whose fundamental frequency is exactly 440 Hz. Double that frequency to 880 Hz to create a note exactly one octave higher. To lower the pitch by exactly one octave, reverse the ratio (e.g., 1:2, or mathematically, $\frac{1}{2} = 0.5$). That is, cut the frequency in half. An octave below A440 has a fundamental frequency of 220 Hz.

All the musical intervals represent ratios. The numerical value of the ratio represents a scaling factor that when multiplied by the starting frequency, finds the new frequency needed to reach the desired musical interval. As the intervals deviate from the octave, the ratios are no longer built on simple whole numbers, and the exact ratios depend on the type of tuning used. For example, a perfect fifth is a ratio of 3:2 ($\frac{3}{2} = 1.5$) in just intonation. In equal-tempered tuning, a perfect fifth is achieved by multiplying by about 1.498. A perfect fourth represents the ratio of 4:3 ($\frac{4}{3} = 1.3333$) in just intonation, but the slightly different 1.3348 for equal temperament. No matter which form of tuning is selected, it is always a ratio that rules. In order to create a note that is a specified musical interval away, multiply the starting pitch by the value of the appropriate ratio.

It doesn't make musical sense to think in terms of the actual number of hertz between two notes. A big band arranger won't ask the trumpet player to play

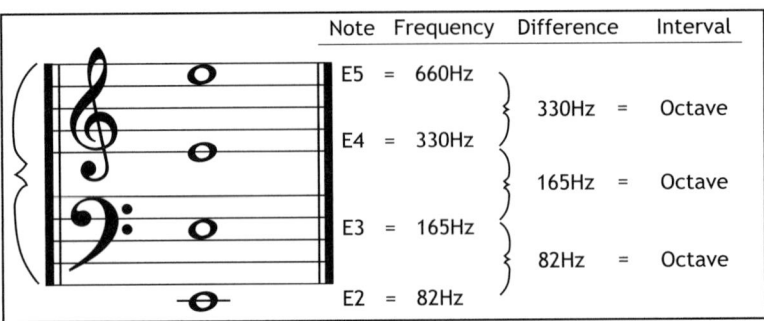

FIGURE B.13
Frequency changes of the octave.

217 Hz above the tenor sax. Instead, ratios are used. Score the horn a minor third above, or an octave above, and the trumpeter can oblige.

Looking at Figure B.13, consider each note's fundamental pitches mathematically, not musically. Start two Es below middle C, labeled E2. It has a fundamental frequency of about 82½ Hz—a meaningless observation for the performing musician, perhaps, but an important one for the recording musician. Up one octave, the pitch is exactly twice the starting pitch. That's the very definition of an octave. One octave up leads to E3, with a fundamental pitch of about 165 Hz (2 × 82.5). One more octave up is E4, the first E above middle C. E4 has a fundamental pitch of about 330 Hz. E5 is yet another octave above and has a pitch of about 660 Hz. Four pitches, one musical value. They are all labeled "E," and they all sound very similar, musically speaking. In harmony, E at any octave performs very nearly identical functions.

What's the difference between one E natural and the next E up? An octave. But there's a subtle illusion going on here. Using Figure B.13, watch as the absolute numbers fail and the ratio takes over. The "distance" (as measured in hertz) from E2 to E3 is 82 Hz. This 82-Hz difference has meaning to the human hearing system; it's an octave. Starting at E3 and going up to E4 traverses an octave again. However, measured in hertz, this octave represents a difference of 165 Hz. E4 to E5 is an octave, worth 330 Hz.

An octave equals 82 Hz in one instance, 165 Hz in another situation, and then 330 Hz in the third case. The octave cannot be expressed in hertz unless the starting pitch is known. It can always be expressed as a ratio: 2:1.

The musical significance of the octave is well known, offering the most consonant (i.e., least dissonant) pairing of two notes of different pitches. Experienced musicians also attach specific sensory meaning to many (probably all) of the other intervals: the buzz of the perfect fifth, the warmth of the major third, and the bittersweet mood of the minor seventh. Such complex, advanced human feelings about the pitch differences between two notes reduce almost insultingly to some pretty straightforward math. To go up an octave, multiply by 2. Done. It doesn't matter what the starting pitch is.

Minor headache: beyond the octave, the numbers aren't so neat. To go up a perfect fifth in the most common form of tuning in pop music, equal temperament, simply multiply by about 1.49830708. To go up a major third (in equal-tempered tuning) multiply by the unwieldy (and rounded-off) 1.25992105. The numbers are rather unappealing. But the fundamental principle is comfortingly straightforward. Don't add a certain number of hertz to go up by a certain musical amount. Instead, multiply by the numerical value of the appropriate ratio.

The idea of the ratio is built into our musical pitch-labeling scheme. Notes are described on the familiar musical staff and labeled with the familiar short, repeating alphabet from A to G. Peek at the numbers and something peculiar is revealed. If we plot the musical staff using linear mathematics, in which all the lines and spaces of our traditional notation system are spaced an equal number of hertz apart rather than simply an equal distance apart, we get the rather strange-looking staff shown in Figure B.14. The traditional notation scheme masks the actual quantities involved—for good reason. The *musical* relevance of the notes is captured in the notation system. The relationship between C and G is always the same: it's a perfect fifth at any octave at any location on the staff. Therefore, it is shown that way on paper. It isn't musically important how many hertz apart two notes are, but it is certainly important how many lines and spaces apart they are, as arrangers well know.

FIGURE B.14
The linear staff.

The keyboard of the piano presents the same illusion, physically. Figure B.15 shows a piano in which the number of hertz between the notes determines the physical size and location of the keys, which are unplayable and unmusical. The layout of a proper keyboard repeats a pattern based on the musical meaning of the notes, not the linear value of the frequencies of the notes.

Ratios are a part of music. On sheet music and on the keyboard, the ratio is a proven, convenient way to take physical properties and rearrange them in a way that is consistent with their musical meaning. As with pitch (frequency), so it is with amplitude (voltage or pressure). That is, human perception "consumes" musical pitch in a relative way. It is the ratio relationship between notes that creates musical harmony, not their value in hertz. The human perception of amplitude behaves similarly, so a way to quantify the amplitude of audio signals that has musical meaning is needed. The decibel, built in part on the ratio, accomplishes this.

Research has shown that in order to double the apparent loudness of a signal, the power must increase approximately tenfold (Figure B.16) Starting with a power of 1 watt, doubling the apparent loudness leads to 10 watts. Doubling

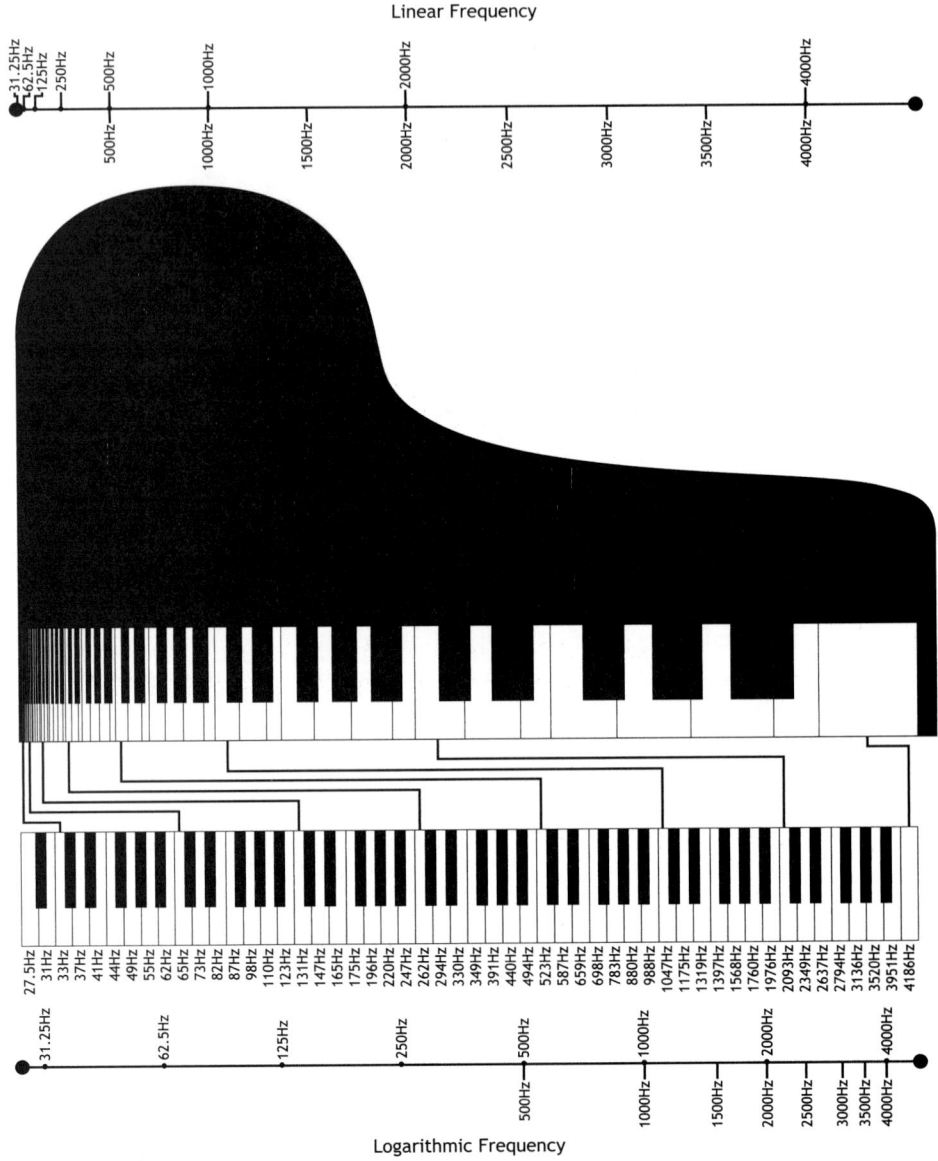

FIGURE B.15
The linear piano.

the loudness from 1 watt required an increase of 9 watts. Repeat this exercise starting at a different power. Beginning with 10 watts, doubling the loudness requires that the power be scaled up ten times to a new value of about 100 watts. This doubling in loudness requires an increase of 90 watts. The next doubling, to 1,000 watts, comes courtesy of a 900-watt addition of power. In all cases, the perceptual impact was the same: the signal became roughly twice as loud. This is the power amplitude analogy of the octave. Here, we are talking about loudness, not pitch. But just as the perception of pitch is driven by a ratio (multiply by 2 to go up an octave), so is the perception of power (multiply by 10 to double the apparent loudness).

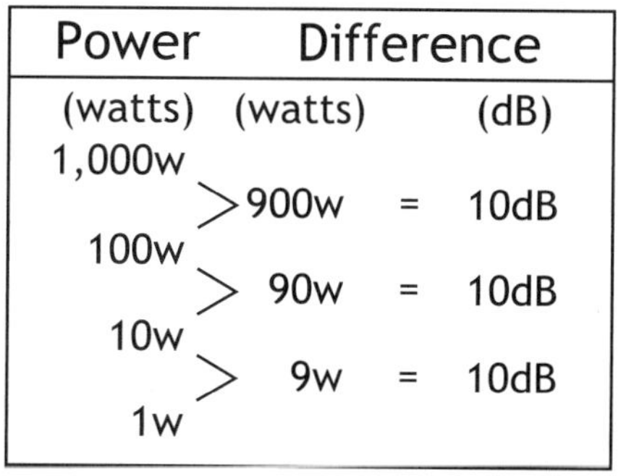

Power	Difference	
(watts)	(watts)	(dB)
1,000w		
	>900w	= 10dB
100w		
	> 90w	= 10dB
10w		
	> 9w	= 10dB
1w		

FIGURE B.16
The power changes of perceived doublings.

The equation for the decibel, therefore, has a ratio built in. As Figure B.16 reveals, the decibel difference between each of the power settings is always the same: 10 decibels. For each equivalent perceptual change, the actual number of watts needed is different, depending on the initial power setting. The decibel equation brings consistency in numbers to a consistent sensory event. The amount of change required to double the loudness—to have the same perceptual impact on our listening systems—is always 10 dB.

Decibels provide engineers the amplitude equivalent of the musical pitch labeling scheme. They convert the physical quantity into a numerical expression that is highly consistent with the perception of that quantity. Through experience, audio engineers develop a very specific idea in their mind about what a 3 dB or 6 dB increase in level sounds like. Experienced musicians are able to start at one pitch and find by ear any other pitch, be it up an octave, or down a fifth, or any other interval away. Music schools offer ear training to teach this ability for pitch. Audio schools offer audio ear training to accomplish the same thing, in the amplitude domain.

B.6.3 References

A close look at the decibel equation reveals that it is a single-number expression for two numbers. That is, 30 dB represents a comparison of one number to another. It doesn't make sense to say that a power of, say, 1,000 watts equals 30 dB. To use 1,000 watts in the decibel equation (Equation B.17), a second wattage must be put into the ratio. Starting with a reference of 1 watt, it is correct to calculate that 1,000 watts is 30 decibels higher than 1 watt:

$$10 * \log_{10}(1,000/1) = 30 \qquad (B.24)$$

The decibel is meaningless without mentioning two numbers. It is always necessary to compare two numbers with the decibel. Often, an engineer wishes to make statements that compare amplitude to the current value, as in, "Turn the snare up 3 dB." That is shorthand for, "Make the amplitude of the snare 3 decibels louder than the current amplitude." If one were to resort to the equation, the current amplitude is used in the bottom of the ratio (the denominator), and the top of the ratio (the numerator) gets the amplitude of the new, louder snare that is desired. Of course, this equation is never used during a session. The faders on a computer screen or on a mixer are labeled in decibels already. Someone else already did the calculations for us.

If a signal isn't being compared to its current value, then it is compared to some reference value. A good starting point might be a reference of 1 watt; 10 watts is

10 dB above this reference, and 100 watts is 20 dB above the same reference. So the correct way to express decibels here is something like, "100 watts is 20 decibels above the reference of 1 watt."

It gets tiring, always expressing a value in decibels above or below some reference value. Here's the time saver. If the reference is 1 watt, express it as dBW (pronounced, "dee bee double you"). The "W" tacked on to the end identifies the reference as exactly 1 watt. This trick shortens the statement to "100 watts is 20 dBW." Done. The reference, which is required for the decibel statement to be meaningful, is attached to the dB abbreviation.

Note that the statement is not "100 watts is 20 dB." That is incorrect. A single number is not expressible as a decibel. There must be some value stated as a point of comparison. Using "dBW" instead of just "dB" is the subtle addition that gives these statements meaning.

So although Equation B.17 is the general equation for the decibel, a more specific equation using a reference of 1 watt is helpful:

$$\text{dBW} = 10 \times \log_{10}(\text{power}/1\text{ watt}) \qquad (B.25)$$

Other subequations exist, with different reference values. For example, sometimes 1 watt is too big to be a useful reference power. Use the much smaller milliwatt (0.001 watt) instead. If the power reference is one milliwatt, the suffix attached to dB is a lowercase "m," for milli:

$$\text{dBm} = 10 \times \log_{10}(\text{power}/0.001\text{ watt}) \qquad (B.26)$$

A little physics lets us leave the power domain and create expressions based on the quantities we see more often in the studio: sound pressure and voltage. Sound pressure decibel expressions use the threshold of hearing (20 micropascals) as the reference pressure, and we tack on the suffix "SPL" to express *sound pressure level* in terms that have perceptual meaning:

$$\text{dBSPL} = 20 \times \log_{10}(\text{pressure}/20\ \mu\text{Pa}) \qquad (B.27)$$

Note that the equation changes a little. Instead of multiplying the logarithm by 10, sound pressure statements require multiplication by 20, as a result of the physical relationship between acoustic power and sound pressure. Power is proportional to pressure squared. The sound power terms within Equation B.17 become sound pressure squared instead. It is a property of logarithms that this power of 2 within the logarithm may be converted to multiplication by 2 outside the logarithm. Hence the 10 becomes 20 for decibel calculations related to pressure. Likewise, we can use decibels to describe the voltage in our gear. Electrical power is proportional to voltage squared, so again we use a 20 instead of a 10 in the decibel calculations for voltage.

In the voltage domain, a few references must be dealt with:

$$\text{dBu} = 20 \times \log_{10}(\text{voltage}/0.775\text{ volt}) \qquad (B.28)$$

$$dBV = 20 \times \log_{10}(\text{voltage}/1 \text{ volt}) \hspace{3em} (B.29)$$

It is a quirk of history that the unwieldy reference of 0.775 volt was chosen. The interested reader can use Ohm's Law to apply a standard 1 milliwatt of power across a load of 600 ohms (which was a standard in the early telecommunications industry, not audio). A voltage of 0.775 V results. Even as the idea of a 600-ohm load lost its significance in the modern professional audio industry, the quirky standard voltage remains. Someone, tired of the clumsiness of that number, chose an easier to remember reference: 1 volt. Good idea. Confusing result. Too many standards. It sort of misses the point of a "standard," doesn't it? The output voltages specified in the back of the manual for any signal-processing device might be expressed in dBu—for example, +4 dBu. Or it might show up in dBV, like –10 dBV. In both cases, the manual is just indicating the nominal output level relative to a chosen industry reference voltage.

B.6.4 Zero Decibels

The meaning of zero decibels must be considered. When a meter of a signal processor reports a value of 0 dB, it does not mean that no signal is present. It in fact means that the amplitude is identical to the reference.

Working in dBu, consider an input that is 0.775 volt. Expressing this input in dBu using Equation B.28, we get:

$$dBu_{0.775} = 20 \times \log_{10}(0.775/0.775) \hspace{3em} (B.30)$$

$$dBu_{0.775} = 20 \times \log_{10}(1) \hspace{3em} (B.31)$$

Importantly,

$$\log_{10}(1) = 0 \hspace{3em} (B.32)$$

Therefore, the special case of having an input identical to the reference voltage leads to:

$$dBu_{0.775} = 20 \times 0 \hspace{3em} (B.33)$$

$$dBu_{0.775} = 0 \, dBu \hspace{3em} (B.34)$$

When the signal hits zero—be it 0 dBu, 0 dBV, or any other decibel reference—the amplitude equals the reference.

B.6.5 Negative Decibels

The log of one is zero (see Equation B.32). The log of a value greater than unity is a positive number; it's greater than zero. The log of a value less than unity is a negative number; it is less than zero.

So when a signal is greater than the reference value being used (0.775 volt, using dBu for example), the decibel calculation produces a positive result. Any positive expression of decibels indicates a signal is higher in amplitude than the reference.

When a signal is less than the reference, the decibel calculation gives a negative result: +3 dBu indicates a signal that is 3 decibels greater than 0.775 volt, and −3 dBu indicates a signal that is 3 decibels less than 0.775 volt.

Discussing amplitude with units of micropascals or volts, though technically correct and quantitatively useful, is not productive in the recording studio. A different expression of amplitude is preferred. The decibel gives audio engineers a way to communicate matters related to amplitude that is perceptually meaningful and musically useful. So don't turn it up 2 volts, turn it up 6 dB!

B.7 DYNAMIC RANGE

Musical dynamics are so important to a composition and performance that they are notated on every score and governed closely by every band leader, orchestra conductor, and music director. Making clever use of loud parts and soft parts is a fundamental part of performance, composition, and arranging, in all styles of music from classical to jazz and folk to rock. In the studio, we must concern ourselves with a different, but related sort of dynamics (Figure B.17): *audio dynamic range*.

Exploring the upper limit of audio dynamic range comes naturally to most musicians and engineers. The music is often turned up until it distorts. It's as if the instruction manual required it. The volume setting on everything from guitar amps to home stereos is often found to push the limits, flirting with distortion. It is a basic property of all audio equipment: turn it up too loud and distortion results.

At the other amplitude extreme lives a different audio challenge. If the musical signal is too quiet, the inherent noise of the audio equipment itself becomes

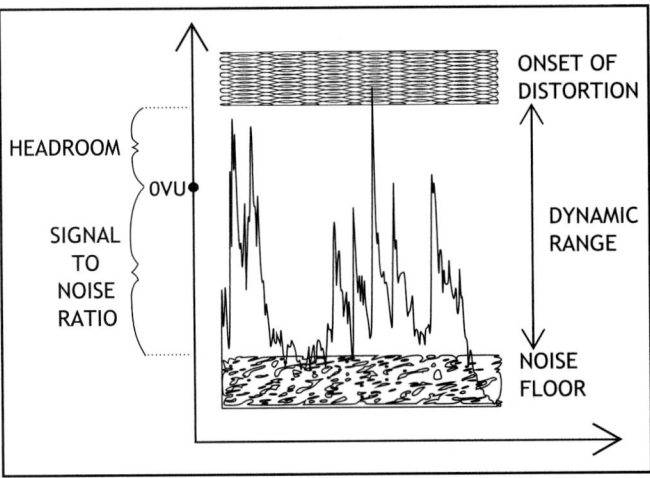

FIGURE B.17
Audio dynamic range.

audible and possibly distracting. Cassette tapes with their characteristic hiss and LPs with their crackles and rumble demonstrate the challenge that a noise floor presents. In fact, all audio equipment has a noise floor—equalizers, compressors, microphones, and even patch cables. Yes, even a cable made of pure gold, manufactured in zero gravity during the winter solstice of a leap year and costing half a year's salary, will still have a noise floor, however faint.

A constant part of the recording craft then is using all these pieces of audio equipment in that safe amplitude zone, above the noise floor but below the point at which distortion begins. That safe zone is, in fact, the audio dynamic range. It defines and—using decibels—quantifies the range of usable amplitude between the noise a piece of gear makes and the level at which the piece of equipment starts to distort.

The target in between these two extremes is typically labeled 0 VU. Zero VU is a nominal level for a signal through a piece of equipment. At 0 VU, the music gets through well above the self-noise of the equipment, but safely under the point where it starts to distort.

If we recorded pure sine waves for a living, we'd raise the signal amplitude up just to the point of distortion, back off a smidge in level, and hit record. Thankfully, musical waveforms are nothing like sine waves. The amplitude of a musical waveform races wildly up and down due to both the character of the particular musical instrument and the way it is being played. Musical instruments lack the amplitude predictability of a sinusoid.

Some signals are more predictable than others. Electric guitar amps cranked to the limit have very little dynamic range. Many guitar sessions find the engineer recording the way Nigel does, with the amp set to 11. Many (but not all) guitar amps sound best when they are cranked to within inches of their lives, which leaves no room for audio peaks to get through at a higher level. The meters on the mixing console and the multitrack recorder simply zip up to 0 VU at the downbeat and then barely move until the end of the song. Chugga chugga crunch ch-chugga. Chugga chugga crunch ch-chugga. The meters don't budge until the guitarist stops playing. Crunchy rhythm rock-and-roll guitars are a case study in limited dynamic range.

Percussion, on the other hand, can be a complicated pattern of hard hits and delicate taps. Such an instrument is a challenge to record well, presenting extremely wide and difficult-to-predict dynamic range.

Every instrument offers its own complicated dynamics. The musical dynamic range of the instrument must somehow be made to fit within the audio dynamic range of the studio's equipment. Otherwise, the listeners are going to hear distortion, noise, or both.

Accommodating the unpredictability of all musical events, we record at a level well below the point where distortion begins. The amplitude "distance" (expressed in decibels) between the target operating level and the onset of distortion is called *headroom*. This headroom provides the engineer with a safety cushion, absorbing the musical dynamics of the instrument recorded without exceeding the audio dynamic range of the gear used to capture the recording.

The relative level of the noise floor compared to 0 VU, again expressed in decibels, is the signal-to-noise ratio. It quantifies the level of the noise relative to the nominal signal level. The trick, of course, is to send the audio signal through at a level well above the noise floor so that listeners won't even hear that hiss, hum, grit, and gunk that might be lurking low in the piece of equipment.

Making effective use of dynamic range influences how audio engineers record to any format, from analog tape to digital hard disk. It also governs the levels used when sending audio through a compressor, delay, reverb, or any other type of audio equipment. It is a constant trade-off of noise at low amplitudes versus distortion at high amplitudes.

This appendix is a tutorial on mix automation that is appropriate for any automation system, analog or digital, on a console or in a digital audio workstation.

C.1 UNAUTOMATED MIXING

So what's so automatic about automation? An automation system plays back an engineer's mix moves—however elaborate—automatically. That is all automation can do. It repeats the mix done by someone else. It's nothing without the engineer. In order to see what can be done with automation, it makes sense to take a look at what can be done without automation. Mixing without automation is called *manual mixing*. The following is pretty typical:

- *Intro:* All vocals cut, extra reverb on the strings, fade organ in. Bass enters at bar four.
- *Verse 1:* Guitar, drums, and lead vocal in, keep background vocals out, less reverb on the strings, pan organ left, and make room for horns on the right at bar 12.
- *Chorus 1:* Lead vocal double comes in, six background vocal tracks up and perfectly blended (three-part harmony, doubled), gated room sound added to snare, strings out, acoustic guitar in.

That's a lot for one engineer to do all at once, so the assistant engineer, the studio manager, and the Chinese food delivery person help out. There's a lot to remember, so notes are scribbled on the track sheet, the console, and the Chinese food menu. It's hard to listen critically while doing so much, so in truth the engineer doesn't hear the mix objectively until the master is played back later.

Oops, the horns were too loud on verse 3? Try again. Do everything the same way as the last mix, but get the horns right on the third verse. This sort of thing isn't easy. It takes several more passes to get close to that last good version, and the whole time, the mix team is trying to remember the horns and not forget anything else.

When everyone has run out of the ability to remember another thing, it is time to start mixing in pieces: "Well, we finally got the intro right. Now, let's move on to the first verse." The song is mixed section by section and later edited together into what sounds like a single pass.

That's manual mixing, which can be summarized as:

- *All hands on deck.* Only an octopus with golden ears could make all these moves at once, so the production pulls in the help of others in the control room.

- *Extensive documentation.* Make creative but informative notes about everything that changes in the course of the mix, putting tick marks next to the faders for their key levels at each part of the song, sticking red tape on the pan pot that gets twisted left in every verse, slapping a sticky note on the reverb that gets cut in the bridge, and so on.
- *Trial and error.* Print several passes to the master deck. Then have a listening session with the band and choose the best one.
- *Cut and paste.* Be prepared to edit the good pieces from several good mixes into a single best mix.

Manual mixes often become an intense process of choreographing the contortions of the various helpers grabbing knobs and faders on cue while speed-reading the notes, scratches, and scribbles all over the studio. This process can be an adrenaline-filled experience, pulling the engineer into the musical performance on the multitrack; the engineer starts to feel like part of the band. It is important that the excitement not cloud the engineer's judgment. Sometimes these engineering thrill rides aren't fun for anyone except the engineer. Sometimes the music suffers. The thrill of mastering the complicated logistics in a manual mix can mask any opinion the engineer has about the music. One must remember that those who listen to the mix later will not, for the most part, have any idea what occurred in the studio. Listeners will react to the sound of the art, not the complexity of the craft.

Automation to the rescue. Because it can control any number of faders, automation can be several sets of hands doing several mix moves at once. Because it "remembers" different settings and fader positions by storing them, automation can make all the crazy documentation unnecessary. Automation obsoletes trial and error and cut and paste, turning each mix pass into a controlled, repeatable fine-tuning of the mix.

C.2 AUTOMATED MIXING

There are degrees of automation capability. In the world of digital workplaces (digital audio workstations, digital hard disk recorders with built-in mixing capability, or digital consoles), it is often possible to automate nearly every knob, switch, slider, or parameter in the mixer. This is more difficult to do with an analog work surface on which it is quite likely that only the faders and cut buttons are automated. Somewhere in between is MIDI automation, where simple note on/note off and other performance gestures are used to drive a mixer instead of a synthesizer or sequencer.

C.2.1 Faders and Cuts Automation

Some consoles automate only the faders and the cut buttons. It is possible to spend more than $200,000 and only get faders and cuts automation. That might seem disappointing at first. There is a lot more to a mix than faders moving and mute buttons cutting tracks in and out. What about pan positions, EQ, reverb time, and other settings? Of course these other settings are important to a mix. But a key to successful automated mixing is being honest with oneself in assessing whether these settings really need complex changes throughout a mix. Keeping the mix moves simple helps keep the mind free enough and calm enough to think creatively and listen carefully. It's possible to have too much of a good thing, even in rock and roll. By the way, it is safe to say that at least 80 percent of all the hit pop and rock records made in the 1980s and 1990s were mixed with fader and cut automation only. Wonderfully elaborate and complicated mixes can be built with this relatively limited amount of automation capability. Here's how.

ALTERNATIVE SIGNAL PATH

The horn arranger has the creative mandate to build colors and feelings through the controlled use of different kinds of horns. Which horn plays which part of the chord? How will the chords be connected to each other? The arranger answers these sorts of questions creatively to produce a horn chart. Multitrack mixes are also arranged; they are arranged by the mix engineer. The engineer decides the sound and texture of each track, using signal processing to tweak or mangle the sound as desired. The mix engineer decides which part plays when, using the cut buttons.

The mixer/arranger often wishes to push the sonic development of the song further by changing the signal-processing structure of a given track for a special part of the song. For example, it might be desirable for the vocal to take on a less aggressive persona during the bridge. This can be accomplished through the use of a different EQ contour, less compression, more reverb, and a touch of chorus on the reverb tail—signal processing details that weren't a part of the vocal sound during the rest of the tune.

Not surprisingly, automation is the mixer's/arranger's tool. What perhaps is surprising, however, is the discovery that such elaborate changes to different elements of a mix can be achieved through simple faders and cuts automation. Often it isn't necessary to automate the equalizer as it transitions from its primary sound to the less radical tone desired in the bridge. Nor is it necessary to automate the rest of the signal-processing components as the compression decreases, the reverb increases, and the chorused reverb appears. All that is needed is a parallel vocal channel, based on the same original vocal track, but sending it through the signal-processing chain required for the bridge. During the bridge, automated cuts simply mute the aggressively treated vocal sound that is open for the rest of the tune and turn on the parallel, sweet, and gentle one. The action of two mute switches, turning one vocal patch off while the other is turned on, affects this significant change to the mix.

The sonic result feels like an elaborate mix move and, hopefully, a compelling musical statement. But on the console, it is created through the use of one additional channel using different effects and a couple of mute and unmute commands of the automation system.

AUTOMATED SEND

A variation on the previous theme is the automated send. It may not be necessary to create an entirely different effects structure to accomplish a creative twist in the mix. For example, it might be desirable to have extra reverb on the acoustic guitar during the intro but back off once the band kicks in. In this case, the acoustic guitar track is routed to an additional fader that sends the guitar to the reverb only, not the mix bus. Most consoles can do this. The input into the additional channel is a *mult* (short for "multiple," the mult is a copy of the signal created simply by splitting the signal at the patch bay) of the guitar sound. The aux send patched to the reverb is turned up, but this channel's output is specifically not assigned to the mix bus. The cut button on this extra guitar channel amounts then to a reverb on/off button; the fader is a reverb send level. Pull the fader down for less reverb; push it up for more. This automated send offers the engineer a way to layer in areas of more or less effects, again using only straightforward faders and cuts automation.

The opportunities for automated sends are limited only by the engineer's imagination—and good taste. Add a triplet echo to key words through some automated cuts on a send to a delay. Consider the delay accent on the promising line, "My baby's got a new pair of ear plugs … ear plugs … ear plugs … ear plugs." The automated send comes from a module with a mult of the vocal, feeding a delay. It is muted the entire tune until it unmutes for the words "ear plugs," sending it to the delay, only to mute again for perhaps the rest of the tune. Alternatively, the engineer can ride a fader-sending signal only to reverb and gently add width and depth to some background vocals during the chorus. In all of these cases, the mix engineer is creating very sophisticated layers to their mix using faders and cuts automation only.

MIDI EQUIVALENTS

Depending on the extent of the desired sonic change, it may be possible to use MIDI messages sent to the effects units to achieve some mix goals. The engineer creates and saves two different reverbs on the same effects device and uses program change commands to switch to the "verse reverb" and back to the "main reverb." These reverbs might be the same exact patch with a single parameter, like reverb time, changed.

Some of the more clever effects units allow not just program changes but also the changing of many parameters while the effect is running. That is, using a MIDI controller, the reverb time might be shortened slightly without any audio artifacts; the reverb smoothly transitions into the smaller sound. Once again, detail and complexity are added to the character of the mix through the use of very simple commands: program changes or mod wheel motions.

AUTOMATED COMPS

Ever done a session on analog tape in which the lead vocal ends up scattered across several tracks? Actually, the real question is whether anyone has ever done a session when it wasn't. Take 1: sounds really good, especially the last verse and chorus—the singer really dug in there. Save that track and do another take.

Take 2: everything is sounding good and the missing parts of track 1 are all now covered, except a couple of words drift flat. Now record those problematic words, but use a third track as punching in risks erasing a portion of the keeper part. Naturally these tracks are to be "comped" into a single track: bounce the best part from each of these three source tracks onto a fourth track, feeling no pressure when punching in and out at each transition. If a punch is missed, the engineer just tries again. The original magic moments are safe on other tracks. As shown in Figure C.1, the process has created a best, composite take from the many different options tracked.

Automate the cuts associated with creating that composite track for a more relaxed experience. Better yet, don't bounce the tracks to a new one. Using automated cuts, create a virtual comp that plays back the best take at all times, even those single words over on track 17 that were resung to correct pitch. Used wisely (i.e., not constantly), this approach can bring out a single, convincing, consistent, and powerful performance. Faders and cuts automation are all that is needed.

FIGURE C.1
Creating a vocal composite from three takes.

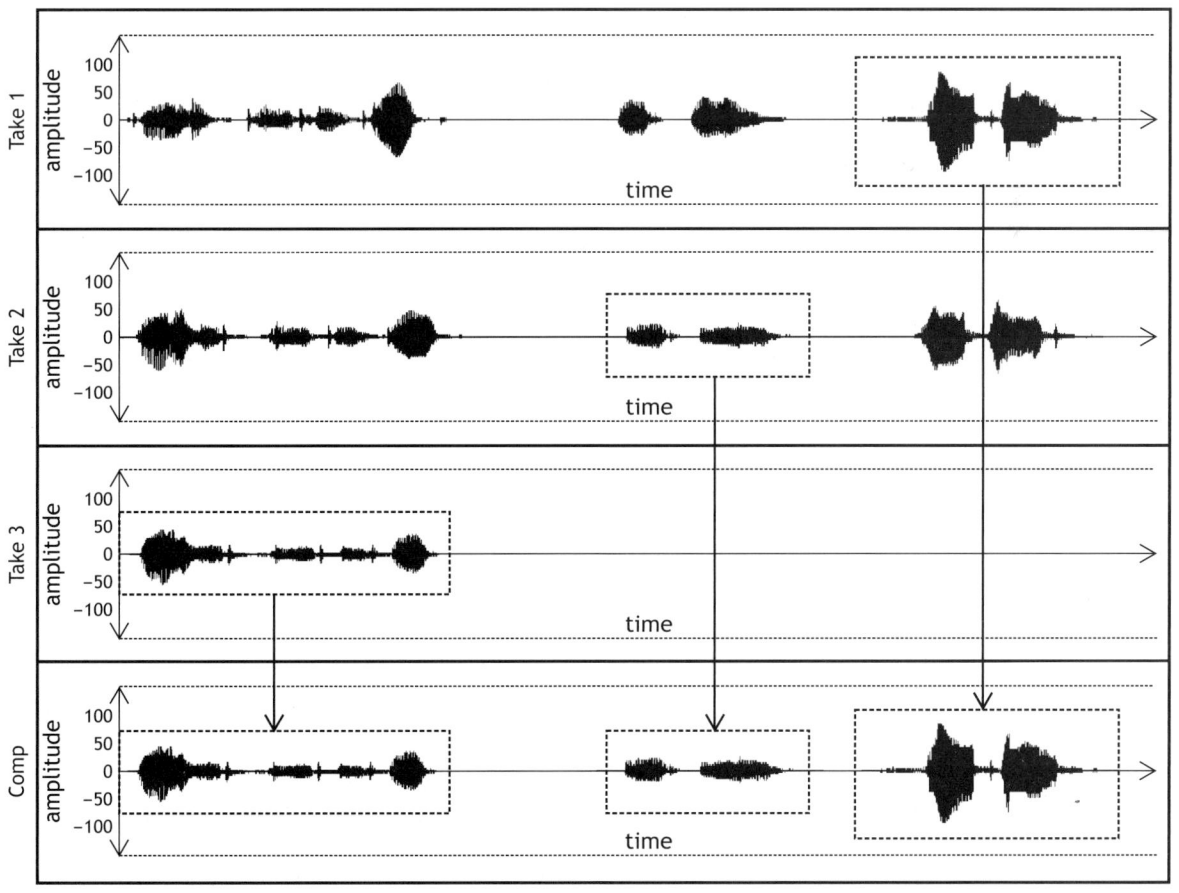

PRINTED MIX MOVES

Sometimes—and this might be once every dozen mixes or so—the engineer gets really inspired and throws together a mix move too complicated for the automation system—and too magic to let go. The solution, tracks permitting, is to record the effect. In the interest of creating a distinct sound for the vocal on the last chorus, one might feel inspired to run it through an old guitar amp with tremolo, using a wah-wah pedal, sticking the amp in the shower (the water is off!). Wait: there's more. It's too tempting to use that strange-looking ribbon microphone that the tech says, "Doesn't work," but that the assistant engineer says, "Sounds freakin' wicked." Without really knowing what "freakin' wicked" means, one tries it and it sounds good. While the singer plays the wah-wah pedal and the engineer pans it in time with the music, this crazy effect is printed to multitrack. Not only will it be impossible to recreate the amp tone, microphone placement, and other bits of this sound tomorrow, there is the possibility that the microphone will quit working altogether. This mix move isn't saved in automation; it's saved to multitrack. Simple automated mute commands will bring it into the mix anytime.

C.2.2 Everything Automation

Of course, when the audio or the mixer live in the digital world, automation as a technology becomes a lot like a word processor or video game. If software controls the music, there is almost no limit to what the automation system can do. DAWs and digital mixers offer this sort of opportunity. The alternative signal path explored in faders and cuts automation is still useful, but it lives in software, not hardware. To have an alternative vocal sound, just program the changes in: automate the equalizer, compressor, reverb, and chorus. There's no need for an entirely different signal path for the vocal to have this altered state—just a different kind of signal processing. In addition, the automated sends discussed earlier become trivial to set up if the echo send level control itself can be automated. That is, one needn't dedicate an entire channel to automating an effect. For more ambience on the intro, ride the automated echo send level up on the intro. Nice and easy. Everything automation is a real ally to the engineer's creativity. The mix engineer can do more sonically because the automation system can do more logistically. Automate the EQ or the compressor for de-essing. Automate a mid-frequency sweep for wah-wah. The imagination is the limit.

The previous discussion of automation approaches for faders and cuts automation reveals that it sometimes takes an elaborate signal flow structure just to create a small effect. The engineer often has trouble feeling inspired if every sonic idea ("Wouldn't it sound great if ...") is followed by an analysis of the automation manual ("How the heck is that automated?"). With everything automation, sound engineers don't have to restrain their creative impulses to enhance or manipulate a sound. The mix moves can be programmed almost as soon as the engineer dreams up the type of sound effect wanted.

The possibilities are endless, but the risk is that it's too much of a good thing. Too many options can paralyze the mixing engineer: "Let's see. I'll push the lead vocal fader up on the word 'baby,' roll off a bit of high end on the hi-hat, cut the snare track between each hit, pan the piano and guitar back and forth in triplet time, add a nasally EQ to the bass, and shorten the delay on the ukulele for the first two bars of the first verse. Then, on the third bar, I'll …"

Clearly, the music could suffer. There is a real temptation to explore mix moves because the gear can do it, not because the music needs it (see "Case Credo," Chapter 10). With everything automated, the recording studio has lost a sort of "reality check" that the more limited automation system imposed. It takes restraint and maturity. Most engineers are at least occasionally seduced by the equipment in this way, no matter how committed they are to the music. The equipment will take over or just plain interfere with the process if left unchecked. The trick is to know when to explore the automation possibilities and when to take a step back and let the music be.

C.2.3 Snapshot Automation

Everything automation is made less intimidating through what is known as *snapshot automation*. Although it is often musically desirable to have the bridge of the tune live in another sonic world (e.g., different fader, pan, EQ, and effects settings), often it is not necessary to continuously modify all the controls with subtle shifts from one setting to the other. Snapshot automation can store the configuration of every parameter of the console at one instant (snapshot A) and then store all the settings wanted for the bridge of the tune (snapshot B). The snapshot automation system enables the mixer to toggle from one setting to the other, smoothly, all at once. The mix engineer doesn't have to program the changes in fader by fader, pan pot by pan pot, effects parameter by effects parameter, and so on. Just store the best-sounding chorus setup and the best-sounding bridge setup and let the snapshot automation system connect the dots.

The clever snapshot system gives the user some control over how the change from one snapshot to another is made. Adjustable cross-fade or morph times, perhaps with adjustable slopes, can make the snapping process itself more musical. The result is often a very complicated set of mix moves accomplished with very little tedious automation programming. It often more than adequately serves the musical needs of the mix.

C.3 MIX MODES

Although there are as many different automation systems as there are brands of consoles and workstations, they have enough in common that the reader can be prepared to use any automation system through the following orientation.

C.3.1 Write or Read

Either the engineer is performing the mix moves, or the automation system is. Terms like "write" or "read" are common ways of making this rather important distinction. Think of it as *automation record* and *automation play*. When a mix move has been designed and rehearsed, the engineer enters automation write mode and records the move in. Automation read mode makes the automation system reproduce the mix move.

Write and read can usually—and sometimes dangerously—be done globally across the entire console or workstation. Alternatively, write and read may be applied in a more focused way: a few channels at a time, fader by fader, cut button by cut button.

Like any piece of software, there are some quirks that seem strange at first but become more natural with experience. For example, some automation systems need you to write at least a rough version of the whole mix for all faders and mutes on the entire console before one is allowed to go back and do those smaller tweaks on a single fader within a small part of the tune. Sometimes something as trivial as the start time of the mix is an unmovable anchor for the automation system. Getting expressive ideas into a computer is never easy. The artist's approach to creating computer graphics is at least a little different from how they draw on paper. Similarly, entering a mix into the computer requires some navigation through menus and mastering of peculiar syntax that slightly alters the routine from just manual mixing. Once these quirks are mastered, one writes and reads mix automation moves as naturally as one records a track, rewinds and does another pass, plays it back, moves onto another track, and so on. Expect at least a short learning curve.

C.3.2 Write: Absolute or Relative

When recording mix moves, there are usually two broad approaches, often described by words like *absolute mode* and *relative mode*. *Absolute mode* tells the system that the engineer wants the automation to store the exact mix parameters currently being adjusted. Any previously written automation moves at this point of the song are completely forgotten, erased, and replaced by the new mix moves.

Relative (a.k.a. *update* or *trim*) mode lets the engineer revise an existing mix through nudging and tweaking. If the mix engineer likes where the vocal is sitting in the mix but wishes to push a couple of key words up a little louder, a relative mode might be a good approach. Entering write mode in this way doesn't make the automation system disregard the mix information already recorded; rather, it updates it based on the new relative moves. Push the fader up and it adds that increase to the move already there. These relative mode trims can be accumulated through additional automation passes to create a very complicated set of fader rides in a simple and intuitive way.

Using Figure C.2, consider this fairly typical scenario. The first pass sets the general level of the vocal in the mix. Absolute mode writes this into the computer as a starting point. The second pass might address some overall arrangement issues: turn the vocal up from this basic level at every chorus and have it be a little lower at the bridge. Relative mode would be a good way to do this. With all other faders

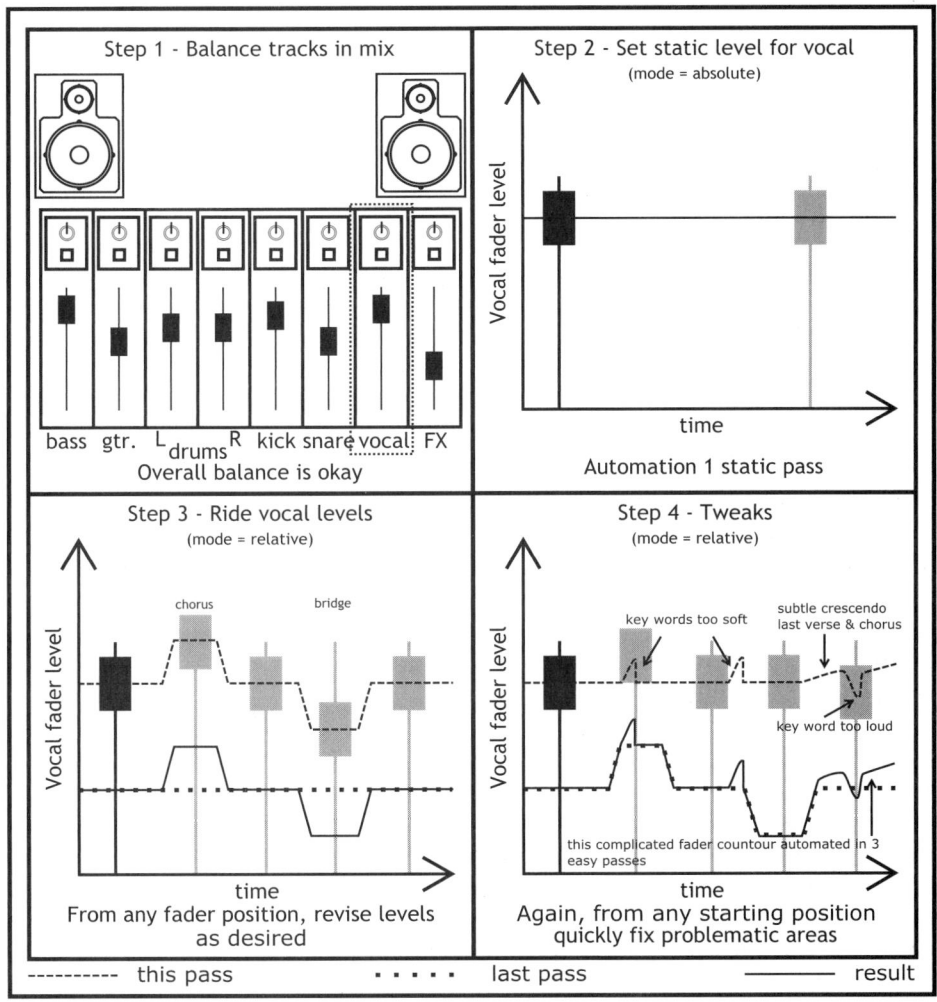

FIGURE C.2
Fader automation in four easy steps.

safely in read mode, enter write mode on the vocal with the fader in any position, push it up in the choruses, pull it back down to the original starting point in the verses, and lower it in the bridge. Because the mix is being trimmed in relative mode, the automation system looks only at the *changes* in the fader's position, not the actual, absolute position of the fader itself. On the third pass, the engineer might then modify those already changing fader settings on a word-by-word basis, reducing the words that poke out of the mix and raising the words that are getting lost. Relative mode would be the appropriate choice again. As the mix plays, there's no need to find and match fader levels already written; just write automation moves in relative mode and make these quick fader rides to revise and refine the evolving fader moves already stored in automation.

C.4 AUTOMATION STRATEGIES

It is essential to bring some amount of order to the mix approach. The ability to do everything makes it difficult to do anything. Here's a common way to break it down into bite-size increments that grow into an elaborate, musical mix.

C.4.1 Balance

Consider the various stages of building a mix. First, one lays out the rough balances that start to make musical sense. Listening to the entire song, one finds fader positions and pan pot settings that enable the song to stand on its own. The pop music vocal and the snare sit pretty loud in the mix, dead center. Kick and bass are also in the center, likely at a lower level. The other pieces of the arrangement fill in underneath and around these critical tracks. Guitars and keys must not mask the vocals or harm the intelligibility of the lyrics. The musical role of each and every track must be understood musically and the mix balanced accordingly, so that elements playing together blend or achieve counterpoint as needed. Work hard to find a balance that is fun to listen to, supporting the music while revealing the complexity and subtlety of the song. Most important, the engineer does not yet automate the mix.

C.4.2 Effects

With the fine-tuning that comes from the addition of various forms of signal processing, the engineer builds up a mix that tastefully highlights every element of the pop arrangement that needs it, while achieving subtle blending and layering for the behind that scenes, supporting tracks. Various forms of distortion, EQ, compression, limiting, expansion, gating, tremolo, flanging, chorus, echo, pitch shift, reverb, and any number of other effects are introduced, tested, rejected, adjusted, and refined. The sonic traits that will ultimately define this mix are developed now. Every effect introduction requires the engineer to check the balance again. Effects on the snare likely affect its apparent loudness. Effects on the snare are likely to influence its relative balance versus the guitars and the vocal. The engineer constantly and iteratively works to keep the multitrack arrangement balanced as effects are added and deleted.

Please keep in mind that so far, the whole mix is static. That is, the mix engineer has a pretty decent sounding tune coming out of the loudspeakers without doing any clever fader moves and effects twiddles. These two steps "Balance" and "Effects" actually do the most to determine the overall sound of the mix. The engineer works hardest here and still hasn't entered automation yet. As they are so important, these first two phases should take the most time and consume the most creative energy. In fact, because automation is so darn fun, this is rarely the case.

C.4.3 Cuts

With the mix very near where it needs to be sonically (i.e., the producer, engineer, vocalist, and the rest of the band pretty much approve of the sound), the engineer will at last begin to automate the mix. The first step is to apply the appropriate cuts to channels that either aren't being used or that were decided against for performance or arrangement reasons. This step amounts to making the mix arrangement official.

If the producer doesn't want horns in the first chorus, the engineer automates the mutes on all of the horn tracks accordingly. If, in the engineer's judgment, the bridge sounds better using doubled harmony vocals, the appropriate mutes and unmutes are automated. If the singer likes the second chorus from the third take of the lead vocal, the engineer "comps" it in using cuts automation.

Now with these cuts happening automatically on cue, the producer and engineer can listen carefully to how the song feels to make sure those are the right decisions. It comes as a surprise at first, but the manual mixing process of diving for the cut buttons to mute and unmute tracks at the appropriate times is enough to interfere with one's hearing, psychologically. That is, it is hard for the engineer to form a certain opinion of the mix idea while also remembering all these manual moves. It is much more effective to plug the cuts into automation, sit back, and listen to the song unburdened by any other activities or distractions. Hands folded, eyes closed, the engineer can join the producer and the artist as they decide how much they like the unfolding production. Does the arrangement make musical sense? Does it grow musically? Does it sag and feel empty in the bridge? Is the detail of the snare lost when the doubled electric guitars enter each chorus? Is the piano in tune with the bass? All these important questions can best be answered when one is just listening and not pressing buttons.

C.4.4 Rides

To complete the multitrack arrangement of the song, the cuts described previously are followed by some general fader rides in which the engineer does things like push the vocal up in the choruses, pull the piano down during the guitar solo, and such. Generally, these are pretty subtle rides. These fader moves are aimed at the musical interpretation of the mix, trying to make the song feel right, whatever that means. A little ride here and another one there helps shape the energy level and mood of the mix.

C.4.5 Tweaks and Special Effects

Only after the musically compelling and well-organized automated mix built carefully through the four steps described above is complete should the mix engineer attempt the more elaborate, the silly, and the downright crazy moves that are now so tempting. At this point, the engineer can add quarter-note delays to keywords of the lead vocal. Now is the time for the mixer to bump up the lost notes of the solo and ride the faders word by word, syllable by syllable, as needed. At last, the engineer can experiment freely with some more elaborate effects to dress up the bridge. With the fundamental elements of the mix being faithfully replayed by the automation, the mix engineer's mind is free to explore the complicated stuff: "In the last chorus, lets run the returns of the long piano reverb through a distortion pedal. Then run that through a noise gate being keyed open by a swing eighth-note delayed snare hit. Then, if we have any more patch cables, it would sound cool if we… ." Anything goes at this phase of the automated mix.

C.5 PLAYING THE MIXER

It isn't enough, after all this, just to know the theory of how to operate the automation system. The mix engineer has to develop some performance ability on the automated mixing console.

C.5.1 Practice

Truly musical mixes come only after the engineer is comfortable with the automated mixer, mentally and physically. A goal of automation is to make the console more like a musical instrument—a device on which all good engineers can perform. It follows then that engineers need to *practice*. Like practicing scales, the mix engineer needs to have a set of typical moves down pat and be able to do them quickly, under pressure, without thinking. The "scales" of mixing to be practiced are things like the quick cuts associated with comping, appropriate level rides on a vocal track or horn solo through a performance, musical fades for the end of a tune, quick fader moves to attenuate an unwanted squeak or highlight a subtle phrase, setting up an automated send, nudging a mute or unmute a fraction of a second earlier or later, and so forth.

Building on this set of often used basic moves, the engineer also develops the techniques needed to attempt more unusual moves. Automation—just like piano—requires practice, practice, practice. It may sound a little silly, but it is also recommended that engineers find some calm time to simply noodle around with the automation system. Think of practicing a musical instrument. Scales, arpeggios, études, and prepared pieces are part of that discipline. But all musicians stop the drills during practice sessions, and just jam. Engineers should do the equivalent on the mixer. Experiment with tracks when the client isn't there. Design and explore elaborate automation moves just to see if it can be pulled off. Practicing with the mixing console in this way does more than make the engineer a better mixer. The engineer also develops the ability to work quickly and to improvise on the mixing console. Musicians and producers notice this ability in an engineer.

C.5.2 User Interface

The techniques developed for mixing are intimately related to the type of mixer used, be it an analog work surface, a digital console, a digital audio workstation, or some combination thereof. Some interfaces welcome a pretty natural approach; others require special techniques. To get an understanding of how to automate all controls (faders, pan pots, aux sends, equalizers, compressors, and so on), it is helpful to first look closely at faders alone. They serve as an excellent example that provides insight into automating the rest of the console.

The faders on an automated console take instructions from two possible sources: the engineer or the automation system. Know when you are in control, and when you've relinquished it to the machine.

MOVING FADERS

The performance techniques—the physical, dexterity part—are slightly different based on this sole criterion: do the faders move? Consider moving fader automation first, as it is a little more intuitive. When the console has the ability to move the faders, there is less chance for confusion. Moving faders offer the engineer that much-desired feature of WYSIWYG—what you see is what you get. That is, one can see the mix moves during playback. Guitars too loud in the bridge? One glance at the guitar faders reminds the engineer: oops, I forgot to pull them down.

Not only is there excellent visual feedback about the mix, but there is also a nice physical feature. Most moving fader systems will automatically leave read mode and enter write mode whenever a fader is touched. Moreover, they can return to read mode as soon as the engineer lets go of the fader, which gives the engineer a real opportunity to perform. Couldn't hear that word? Wind the mix back to a point a couple of bars before the word. Play the mix and simply nudge the fader up on that word. When the engineer touches the fader and pushes it up, the automation system instantly starts recording the mix move. When the engineer lets go, the fader returns naturally to the old level and the automated mix playback resumes. It couldn't be more intuitive. Automation systems without moving faders can achieve the same mix move, but without motorized faders, it is less intuitive how the change gets programmed in.

The ease of use as well as the immediate visual, aural, and physical gratification that moving fader systems offer makes them the most desirable way to work. Most world-class studios have moving faders. All digital audio workstations offer a click-and-drag equivalent. And thankfully, many less-expensive automation systems have moving faders too. But moving fader systems aren't without drawbacks. Perhaps most obvious is cost. Consoles are already expensive devices full of seemingly countless components. Putting a reliable, accurate, consistent, small, and quiet motor on every fader is not cheap. If they are cheap, then one has to worry about their reliability, accuracy, consistency, size, and noise. Those world-class studios build this into their big-ticket studio fees. And they probably have spare motors in the tech room. The solution is to have high revenue that justifies this—or to shop very carefully. Such a financial commitment to automation isn't always the best choice when it might be more productive to buy another compressor rather than maintain a set of servomotors.

There is an additional potential drawback to moving faders, and this one's a little scary. If the automation system can move the faders when the engineer tells it to, then it might also be possible for the automation to move the faders when the engineer didn't (mean to) tell it to.

Trade shows and studio tours often show off demos of the moving fader systems on large-format consoles. In the demo, the faders move together, not for a musical mix, but for some sort of visual effect—sort of a Radio City Rockettes sort of entertainment. The faders arrange themselves into a sine wave that moves left to right and then right to left across the console. First slow, then fast. Those are nifty demonstrations. But imagine spending a not unreasonable four or more hours finding the balance for a tune and, just before entering automation, someone accidentally hits the "demo" button. Buzz, click, hum. The faders start doing the wave like they're at the Super Bowl. A careful balance is destroyed by the preprogrammed dance of the faders. It sounds silly, but it really happens. Worse, it's not just the demo that wipes away hours of hard work.

Imagine the following situation. The engineer has found a decent static balance for the tune when, accidentally, the multitrack winds back too far, into the previous song. If the engineer isn't careful, locating to the previous song can cause the faders to race to the appropriate levels from *that* mix. The automation has seen that time code address before; it knows what to do. The balance for this tune is gone, baby, gone. The solution is to know how to turn the motors off. If fader motors are disabled until after the careful balance is safely stored in the automation memory, then those important fader levels won't be lost. Even after doing this, your author has to admit that he still feels a little rush of terror every time he turns the motors on.

VCA FADERS

If the faders can't be moved by the automation system, how can they be automated? Good question. The solution relies on a voltage-controlled amplifier (VCA). VCAs are faders whose amplifiers boost or attenuate a signal based not on the position of a fader control, but on the value of a control voltage. The result is that the engineer and the automation system share control of the fader level. The engineer uses the slider on the console to adjust the control voltage; the automation computer uses software. Either control voltage then determines the fader level, the engineer's or the automation system's. But it's tricky mixing when the faders aren't moving. What one sees definitely is not what one gets.

No problem. With some adjustment to technique and some practice, most engineers find VCA automation perfectly easy to use. As with true audio faders, the first automated pass is done in an absolute mode. All subsequent passes, however, are generally done in a relative mode. A typical series of mix automation passes is shown in Figure C.2. Once the static level is written in absolute mode, additional adjustments are made with the VCA in relative mode.

Broadly, two approaches to VCA automation have evolved. The first approach is to return all faders to their zero position after the first pass. This way, the engineer can cut or boost the level, and the scale next to the fader quantifies just how far it's been moved. If the fader is at –12 dB, the engineer has reduced the level 12 decibels from the last stored fader level. There is little confusion as to how extreme the fader move on any automation pass has become, and one can always return to the level of the last pass by returning to the 0 dB marking. This is a good way to keep track of all changes within a pass on a nonmoving fader system.

On the downside, however, it does remove entirely the visual representation of all the fader levels relative to each other. That is, if all faders are set to zero after the first automation pass, then the careful, overall balance built before automation—in which the relative levels of each and every track in the multitrack production were thoughtfully and iteratively coaxed into position—lives in the computer's memory only, and the engineer can't see it. Some engineers find this feature discomforting. Another option is to do even the later relative mode passes with the faders left where they originally were in the rough balance. That rough balance is usually the level for most of the faders most of the time. Leave the faders at that position for the start of any relative trims and then the actual, physical position of the fader will stay at or very near the fader level that is programmed into the automation. It's a little more confusing to enter write mode at –8 dB, nudge it up briefly to –5 dB, and then remember to return to –8 dB before finishing the revision. With practice, it becomes more natural.

SOFTWARE FADERS

When the faders live on a computer screen, they can move without motors. Much of the benefit of moving fader automation is preserved in digital audio workstations without the expense and maintenance disadvantages. Perhaps the only frustration is that without an external hardware controller possessing motorized faders, one has to click the fader, not touch the fader. For old-school engineers who were weaned on large-format analog consoles, clicking on faders is an acquired taste. For anyone who is reasonably computer-savvy, this isn't a problem. Power users who are quick with the word processor, efficient at surfing the Internet, and agile while playing computer games probably have enough mouse dexterity to find mixing by clicking on a screen perfectly comfortable.

In the software domain, all the other controls on the computer screen can usually be automated as well. The approach is the same, so an engineer's knowledge of write/read and absolute/relative modes is easily applied to pan pots, echo sends, reverb parameters, and other controls.

Yet another automation feature appears when one can interact with it on screen: graphic editing of mix parameters. There are times when one might not need to perform an additional mix pass just to fix a mix problem. Wish the trombone solo were just a little louder? Click on the graphical representation of the level and drag it up a bit. Listen to it to confirm that it's right. Accidentally uncut the track containing the organ solo a hair late, clipping off a bit of the first note? No need to repeat passes and hope through trial and error to eventually time the uncut move at the right instant. Just type in a slightly earlier unmute time in the computer to trim it until it sounds right. These screens, menus, lists, and pictures are often a powerful way to program a mix.

C.6 ADVICE AND WARNINGS

Mix automation gives engineers the power to achieve more. Sloppily handled, that power turns on the engineer, leading almost certainly to lower-quality mixes.

C.6.1 Master the Gear

Most engineers do this sort of work because they love music. The focus during mixdown is to make the most of the music already recorded, using as much gear as it takes and employing every trick ever seen or heard. It is not easy work. The proverbial magic dust that gets sprinkled on a single vocal track might realistically include two stages of compression; a "secret recipe" of EQ; the subtle addition of rhythmic delays to create a pulsing, highly customized reverb; and fader rides that don't just change from verse to chorus, but word to word, and sometimes syllable by syllable. It is easy to spend several hours of mix time on the vocal alone. The quantity and quality of the signal processors used—on just the vocal—might take a year's salary to acquire. It is not easy to dive that deep into the detail of the vocal sound and still keep track of the musicality of the mix. Inexperienced engineers find it frustrating when the bass player walks in while they're working on the vocal. Likely, the bass player is focused on the bass. The engineer has been focusing on the vocal. The unfortunate engineer "glances" at the bass sound to find a twangy bit of sonic meatloaf, overpowered by every other track in the mix, including the shaker.

Automation pulls engineers into a microscopic level of resolution. The music asks them to zoom back out and listen to the whole. Experienced mixers can handle this sort of conflict because they've exercised themselves in this discipline and because they know the gear backwards and forwards. New and intermediate engineers should endeavor to do the same. Recognize that the automation system itself is a very seductive amount of signal-processing capability, and it is often so elaborate as to be at least a little intimidating. Overcome this problem through overpreparation. Great engineers become so comfortable with the syntax, techniques, tricks, and limitations of the automation system that they can freely move between big-picture musical issues and highly focused automation moves. They can listen to the bass, and the snare, and the kazoo, even as they refine the vocal.

C.6.2 Not Too Soon

Avoid the common trap of entering automation too soon. Doing so is tricky, and it takes most engineers a number of painful experiences to get over this temptation. Automation employed carefully (and sometimes that means sparingly) is a lifesaver. When misdirected, automation will fall somewhere on a spectrum from distracting to crippling. That is, at best one is faced with the need to constantly undo or redo moves as the mix unfolds. At worst, the automation system so takes over and interferes with the engineer's ability to make changes that musically, the mix unravels. The solution is to activate the automation system as late as possible in the course of the mixdown session. It is easier to explore mix ideas when the automation doesn't have to be rewritten. For example, the producer might like to hear the background vocals panned left, without the doubled tracks, but with the addition of some heavily flanged reverb panned right. Trying this out is no problem preautomation. The engineer just turns the knobs and pushes the faders until it sounds right. If the mix is already automated, the engineer will have to revise the automation data governing the fader levels, pan positions, and cut buttons on every track involved. That will slow the session down. It may take so long to get the idea sounding good that everyone gives up on it. A good mix idea is suppressed by the interference of the automation system.

Only the most complicated mix ideas need automation. Explore those late in the game after the basic mix arrangement is automated. The creative process associated with most mix ideas will be much more successful if they are explored without automation.

C.6.3 Save Often

The performance gestures that make a mix special live in software. They must be treated with the same level of paranoia that other software documents inspire: save often, make safety backups, and be prepared for frustration and irretrievable loss when working during a thunderstorm with its associated power spikes and flickers.

C.7 SUMMARY

Through organization, study, and practice, automation can become an easy-to-use asset in the studio. It doesn't make manual mixing easier or faster. On the contrary, automation enables mixes to become much more elaborate. Mix sessions can sometimes be much, much slower.

Automation doesn't make everyone a better mix engineer. If the manual mixes were disorganized and disjointed, the automated mixes will be, too. It's only in the hands of a talented engineer that automation makes mixes more interesting and sophisticated. Automation makes the rather uninteresting device known as a mixing console into more of a musical instrument. It stores the engineer's performance. It frees them to improvise. Many engineers don't think of automation as a helpful feature; they think of it as a required tool that empowers them to orchestrate all of the effects associated with making recorded music.

D.1 REFERENCE RECORDINGS

It is wise, in the course of a mix session, to take a break and listen to a reference recording. Is the vocal in today's mix too loud? Is the bass too soft? When everything is a variable in our multitrack mix, it is good to check in on a known quantity. Referencing another recording resets our thresholds, reminds us of what good levels sound like, restores our confidence in the foundational elements of our mix, and enables us to return our focus to higher-level issues.

Maintain a listening list and carry some of those recordings to every session you are a part of. These recordings should be tracks you know and love, recordings whose basic levels and balance in fact set a great example of a correct answer, and recordings that possess well-orchestrated mix beauty.

What are the best reference recordings? The answer is unique to each of us. Ask the great mixers of the world—the award-winning engineers who are mixing the hit tunes today—and they are very likely to recite a list of their best records, *their own* best records. That is, their reference recordings are projects they themselves worked on. Far from being a vain gesture, when mix engineers references their own recordings—their own accomplished, critically acclaimed, time-proven recordings—they are teaching us a great strategy. You know your work better than anyone else does. You know your work better than you'll ever know anyone else's. So use your best mixes as references to help out your next mix.

If you recorded Elvis, if you heard the vocal live in the room with the artist, you heard the track straight off the microphone, and you heard the mix the day it was printed, you'd have a strong sense of the sound of his vocal. Play that Elvis record on today's session, and you are recalibrated, resetting your ability to grok all of the timbral complexity that lives in a vocal. If you didn't track Elvis decades ago, it is less informative to give it a listen now.

Early in your career, before you have found the level of success you know you'll achieve and before you are quite satisfied with the artistic merit and technical validity of all of your mix decisions, you listen to works by the greats. For the music you love, study it. Listen on as many systems in as many different locations as you can—control rooms, living rooms, cars, trains, and anywhere else you can think of. Without the benefit of having been a part of those sessions, you need to learn intimately how they sound. They'll then serve as a useful reference to help you when you are mixing a new tune in an unfamiliar studio.

In this way, you build your own personal listening list. It should include mixes by others that you, colleagues you respect, and critics you trust all find to be among the highest achieving works of recorded art you know. Over time, add to that list some of your own productions: your best mixes, where you bring the insider's advantage of knowing every nook and cranny of the project.

D.2 MIX ICONS

Music is highly derivative. Guitarists today will allude—through technique, tone, and melody—to a guitar god who preceded them. A composer will borrow the form and ideas of a standard tune that came before theirs, and invent his or her own variation on the theme. The same tradition of offering a respectful nod to those who came before us exists for mix engineers. We aren't stealing their ideas; we are following their inspiration. It is an expected form of behavior in the music-making community. Know your history, refine it, add to it, and contribute for future generations.

Toward this goal, we maintain another listening list—one we don't schlep to every mix session. We must know our audio history, and we must stay current on contemporary recordings. We keep an ear on our colleagues, notice their successes, celebrate their creations, and learn from them. Identify recordings that offer even an *isolated* bit of achievement, that serve as a case study for a particular mix gesture, and that dial up amazing effects while still supporting the overall goal for the song. Keep a list of mix icons.

We start that list in this appendix. You should add to it steadily, over time. Study these tracks closely so that in the right production situation, you can do similarly, or stretch it into a new variation.

You may notice that many of these recordings are from some time ago. Though your author makes no excuses for listening to these recordings to this day, there is a lesson to be learned. Contemporary recordings may be built of 100 or more tracks, all mixed with a complexity achievable only in a digital audio workstation. When we reference recordings of the 1990s and earlier, we are accessing recorded works of art that had some constraints on them: constraints useful for our purposes here.

Before DAWs, track count pretty much leveled out at 24 tracks. The greediest productions would synchronize two 24-track machines together, a process that itself needed two tracks from each machine, netting the production 44 tracks for music. There was a single digital tape format from Sony that offered 48 tracks. So track count was all 48 and below—mostly well below—until the DAW finally became a professional format.

When track count is lower, it can be easier for us to hear what they were up to in the mix session. It is well known that the Beatles recorded *Sgt. Pepper's Lonely Hearts Club Band* on a four-track multitrack recorder in 1967. That is a pretty good landmark. Before that, there were fewer tracks in the studio. Often the mix we hear when we listen to these recordings is the mix that was achieved live, while the band was playing. From the late 1960s to the mid 1990s, tape-based track count made its way up to 24 and—barely—48 tracks. After the mid 1990s, track count exploded to its current state of being all but limitless. You'll find the earlier recordings the easiest to study.

Also before DAWs, automation (see Appendix C) was generally limited to faders and cuts only. Equalizers and other effects didn't receive the measure-by-measure adjustment that is possible today. This fact makes the mix moves of yore less of a moving target for us to find and study.

So we often study mixes from simpler times because they are typically more revealing of the actions of the studio. It doesn't mean you shouldn't include tracks from today and tomorrow on your list of mix icons, but your career today will benefit from thoughtful study of mixes from the early days of sound recording.

Listen, learn, and seek out more such recordings.

D.3 SELECTED DISCOGRAPHY

Artist: Roxy Music
Song: "Avalon"
Album: *Avalon*
Label: Warner Bros. Records
Year: 1982
Notes: This mix makes significant use of Chamber One at Avatar in New York City, a multistory stairwell. Note in particular the lush vocals.

Artist: Bryan Adams
Song: "Summer of '69"
Album: *Reckless*
Label: A&M Records
Year: 1984
Notes: More ear candy from Chamber One at Avatar. The snare hit at the top of the tune offers an exaggerated example, but the rhythm guitar has plenty too.

Artist: Michael Penn
Song: "Figment"
Album: *Resigned*
Label: Epic Records
Year: 1997
Notes: Dramatic change of scene driven by reverb shift between the A section ("Leave it for a while …") and the B section ("Before the day is done …").

Artist: Michael Penn
Song: "Out of My Hands"

Album: *Resigned*
Label: Epic Records
Year: 1997
Notes: Multiple simultaneous reverb qualities: electric bass is close-microphone track and dry. Drums enter with liveness and early reflections of a medium room. Acoustic guitar is also a close-microphone track and dry. Lead vocal is extremely wet, with reverb resembling a very large, dark hall program.

Artist: Paul Simon
Song: "Spirit Voices"
Album: *Rhythm of the Saints*
Label: Warner Bros. Records
Year: 1990
Notes: Reverb on a single note: the conga accent at 0:47 in the intro gets emphasis through a nonlinear reverb.

Artist: Cher
Song: "Believe"
Album: *Believe*
Label: Warner Bros. Records
Year: 1998
Notes: The reference for just how far blatant pitch shifting can be pushed.

Artist: Counting Crows
Song: "Time and Time Again"
Album: *August and Everything After*
Label: Geffen Records
Year: 1993
Notes: This album abounds in good Leslie examples, but the intro of this song isolates the Leslie on the B3 on the right channel. It enters with fast rotation and slows before the lead vocal enters.

Artist: Paul Simon
Song: "The Afterlife"
Label: Hear Music, Concord Music Group
Year: 2011
Notes: Tremolo on 12-string acoustic guitar! A studio-only concoction motivated by a need for a texture, and a groove.

Artist: Peter Gabriel
Song: "Sledgehammer"
Album: *So*
Label: Geffen Records
Year: 1986
Notes: Pitch shift that bass one octave down. Don't hide it. Mix it in with the original bass. Larger than life.

Artist: Garbage
Song: "I Think I'm Paranoid"
Album: *Version 2.0*
Label: Almo Sounds
Year: 1998
Notes: The pitch dives as if analog tape were slowly stopped between the bridge and the third chorus, though this is most likely a digital effect simulating the tape stop.

Artist: The Police
Song: "Synchronicity II"
Album: *Synchronicity*
Label: A&M Records
Year: 1983
Notes: Intro reveals start tape effect. Electric guitar is made to feedback first. Then the tape machine is punched into record from a standstill. It ramps up to speed while recording. Played back at speed, the formerly slow bits are now pitched up. The sound of the machine achieving full speed in record becomes a pitch dive when played at uniformly correct speed.

Artist: Prince
Song: "Bob George"
Album: *The Black Album*
Label: Warner Bros. Records
Year: 1983
Notes: Tracked at a fast speed, the protagonist vocal is low and ominous when played back at regular speed. This all-analog pitch-shifting effect supports a convincing performance by Prince.

Artist: The Beatles
Song: "You Never Give Me Your Money"
Album: *Abbey Road*
Label: Parlophone/EMI
Year: 1969
Notes: From single-track vocals to double- and triple-tracked vocals and harmonies. A good case study. The layering begins after the words "funny paper."

Artist: The Cars
Song: "Good Times Roll"
Album: *The Cars*
Label: Elektra Records
Year: 1978
Notes: This is double tracking taken as far as it can reasonably go. Listen to the vocals when they sing the title "Good Times Roll." One imagines that they filled every available track on the multitrack at this point.

Artist: The Eagles
Song: "Life in the Fast Lane"
Album: *Hotel California*
Label: Asylum Records
Year: 1976
Notes: Over-the-top flanging of the entire stereo mix in the chorus repeats into the coda at 3:38.

Artist: Les Paul and Mary Ford
Song: "How High the Moon"
Label: Capitol Records
Year: 1951
Notes: Sets the pace for doubling and tripling multitrack performances.

Artist: Michael Penn
Song: "Cover Up"
Album: *Resigned*
Label: Epic Records
Year: 1997
Notes: One word, and one word only, gets flanged: "guests" at the end of the second verse.

Artist: The Police
Song: "Synchronicity II"
Album: *Synchronicity*
Label: A&M Records
Year: 1983
Notes: Ghost of an echo on most of the lead vocal.

Artist: U2
Song: "Bad"
Album: *The Unforgettable Fire*
Label: Island Records
Year: 1984
Notes: Not the only example, by far. The Edge uses delay as part of the guitar riff.

Artist: The Wallflowers
Song: "One Headlight"
Album: *Bringing Down the Horse*
Label: Interscope Records
Year: 1996
Notes: The lead vocal is wholly without delay for the entire song, until the third verse. The words "turn" and to a lesser extent "burn," get an almost subliminal dotted quarter-note echo. The slide guitar solo gets a thoughtful quarter-note delay treatment that comes and goes, mid-solo. Solo is panned left, first echo is panned dead center, and the second repetition is panned hard right.

Artist: The Who
Song: "Won't Get Fooled Again"
Album: *Who's Next*
Label: Polydor
Year: 1971
Notes: It's not just the "Yeaaaahhhhh!" It's the layers of echoes underneath.

Artist: XTC
Song: "Then She Appeared"
Album: *Nonsuch*
Label: Caroline Records
Year: 1992
Notes: A modern update of a psychedelic 1960s flanger effect. The title is the hook and gets the effect.

Artist: Arctic Monkeys
Song: "Riot Van"
Album: *Whatever People Say I Am, That's What I'm Not*
Label: Domino
Year: 2006
Notes: The vocal is sonically defiant—in support of the lyrics—yet remains intimate and confidential, with the sort of distortion that comes from running the vocal through a guitar amp.

Artist: Counting Crows
Song: "Mr. Jones"
Album: *August and Everything After*
Label: Geffen Records
Year: 1993
Notes: Classic tremolo guitar in the second rhythm guitar, left side, entering in the middle of verse one. A single-note countermelody pulses in the background of the arrangement, thoughtfully entering just before the word "guitar."

Artist: Sheryl Crow
Song: "Redemption Days"
Album: *Sheryl Crow*
Label: A&M Records
Year: 1996
Notes: Touchstone tremolo, electric guitar on the left channel.

Artist: Peter Gabriel
Song: "Big Time"
Album: *So*
Label: Geffen Records
Year: 1986

Notes: The snare has gated reverb, which is unmistakable, but not so over-done that it hasn't aged well. The clap sound is more extreme. It's a song about hype, after all.

Artist: Green Day
Song: "Boulevard of Broken Dreams"
Album: *American Idiot*
Label: Reprise Records
Year: 2004
Notes: That distorted rhythm guitar probably was not tracked with tremolo. It's almost certainly manufactured at mixdown through use of a delay, with feedback, input to the side-chain.

Artist: Imogen Heap
Song: "Hide and Seek"
Album: *Speak for Yourself*
Label: Megaphonic Records
Year: 2005
Notes: Multiband envelope following, sometimes called *vocoding*, is a key part of the sound of this song. The vocal drives the envelope of the synths underneath. For this process, the vocal is filtered into different bands and the amplitude of the vocal in each band drives the amplitude of the synths in the same band. The keyboard part is made to sing along in tight, machine-like symphony.

Artist: The Smiths
Song: "How Soon Is Now?"
Album: *Hatful of Hollow*
Label: Rough Trade Records
Year: 1984
Notes: Rhythm guitar intro offers tremolo in sixteenth-note time, and a sound that the Smiths' fans spot from five miles away—an iconic sound.

Index

Note: Page numbers followed by *f* indicate figures and *t* indicate tables.